The Early Bronze Age I Tombs and Burials of Bâb edh-Dhrâ', Jordan

The Early Bronze Age I Tombs and Burials of Bâb edh-Dhrâ', Jordan

*Reports of the Expedition to the
Dead Sea Plain, Jordan
Volume 3*

*Expedition sponsored by the
American Schools of Oriental Research*

DONALD J. ORTNER AND BRUNO FROHLICH

*With contributions by
Gillian R. Bentley, Alain Froment, Evan M. Garofalo, Liese Meier,
Victoria J. Perry, and R. Thomas Schaub*

*In association with the
National Museum of Natural History
Smithsonian Institution*

A division of
ROWMAN & LITTLEFIELD PUBLISHERS, INC.
Lanham • New York • Toronto • Plymouth, UK

ALTAMIRA PRESS
A division of Rowman & Littlefield Publishers, Inc.
A wholly owned subsidiary of The Rowman & Littlefield Publishing Group, Inc.
4501 Forbes Boulevard, Suite 200, Lanham, MD 20706
www.altamirapress.com

Estover Road, Plymouth PL6 7PY, United Kingdom

Copyright © 2008 by AltaMira Press

All rights reserved. No part of this publication may be reproduced, stored in a retrieval system, or transmitted in any form or by any means, electronic, mechanical, photocopying, recording, or otherwise, without the prior permission of the publisher.

British Library Cataloguing in Publication Information Available

Library of Congress Cataloging-in-Publication Data

Ortner, Donald J.
  The early Bronze Age I tombs and burials of Bâb edh-Dhrâ', Jordan / Donald J. Ortner and Bruno Frohlich.
     p. cm.
  Includes bibliographical references and index.
  ISBN-13: 978-0-7591-1075-5 (cloth : alk. paper)
  ISBN-10: 0-7591-1075-1 (cloth : alk. paper)
 1. Bab edh-Dhra' Site (Jordan) 2. Tombs—Jordan—Bab edh-Dhra' Site. 3. Human remains (Archaeology)—Jordan—Bab edh-Dhra' Site. 4. Bronze age—Jordan—Bab edh-Dhra' Site. 5. Excavations (Archaeology)—Jordan. I. Frohlich, Bruno, 1945– II. Title.

DS154.9.B32O78 2008
939'.46—dc22
                            2007033132

Printed in the United States of America

∞™ The paper used in this publication meets the minimum requirements of American National Standard for Information Sciences—Permanence of Paper for Printed Library Materials, ANSI/NISO Z39.48-1992.

Contents

Acknowledgments		ix
1.	Introduction	1
	A Brief History of Excavations at Bâb edh-Dhrâ', Jordan	1
	The Excavations in 1977, 1979, and 1981	2
	Scientific Value of the Bâb edh-Dhrâ' Site	6
2.	The Ecological and Social Context of the EB I People	7
	The Physical Environment of the Southern Levant	7
	Human Societies of the Chalcolithic Period	8
	Funerary Traditions in the Late Chalcolithic and Early Bronze Age	9
	Egyptian Societies of the Late Predynastic and Early Dynastic Periods	9
	Mesopotamian Societies of the Proto-urban and Early Urban Periods	10
3.	Methods in the Recovery of and Research on the EB I People	13
	Excavation Methods	13
	Field Conservation Methods	18
	Laboratory Conservation Methods	20
	Analytical Laboratory Methods	20
4.	Cultural Artifacts of the EB I Tombs	25
	R. Thomas Schaub	
	Terminology	25
	Tomb Locations and Types	26
	Stratigraphical Relationship of Tombs	26
	Stratigraphy in Contemporary Occupational Areas	27
	Cultural Artifacts from the EB I Tombs	27
	Distribution of Artifacts Related to the Individuals Represented by the Skeletal Material	42
5.	The Funerary Tradition of the EB I People	45
	Shaft Tomb Architecture	46
	Tomb Chamber Contents and Arrangements	47
	Burial Types	48
	Charnel House Development in EB IB	49
	Burial Traditions of the EB I Southern Levant	49
6.	The Tombs and Burials of the EB I People: Shaft Tombs Excavated in 1977	51
	Donald J. Ortner, Evan M. Garofalo, and Bruno Frohlich	
	Conditions of Burial	51
	Tomb A78	52
	Tomb A79	59
	Tomb A80	64
	Tomb A86	73
	Tomb A87	78
	Tomb A88	79

	Tomb A89	83
	Tomb A91	87
	Tomb A91(92)	88
	Tomb A100	89
	Tomb A101	107
	Tomb A102	110
	Tomb A120 and A121	110
	Tomb C9	113
	Tomb C10	114
	Tomb G2	116
	Tomb G4	118
7.	The EB IB Charnel House (G1): Excavated in 1977	123
	Donald J. Ortner, Liese Meier, and Bruno Frohlich	
	General Osteology	129
	Distribution of Skeletal Elements	136
	Evidence of Disease in the G1 Remains	140
	Summary and Conclusions	142
8.	The Tombs and Burials of the EB I People: Tombs Excavated in 1979	147
	Donald J. Ortner, Evan M. Garofalo, and Bruno Frohlich	
	Tomb A102	147
	Tomb A103	155
	Tomb A105	155
	Tomb A106	161
	Tomb A107	165
	Tomb A108	173
	Tomb A109	178
9.	The Tombs and Burials of the EB I People: Tombs Excavated in 1981	183
	Bruno Frohlich, Evan M. Garofalo, and Donald J. Ortner	
	Tomb A110	184
	Tomb A111	199
	Tomb A112	214
	Tomb A113	216
	Tomb A114	216
	Tomb C11	224
10.	The Osteology of the EB IA People	229
	Bruno Frohlich, Donald J. Ortner, and Alain Froment	
	General Description	229
	EB IA Burials	230
	Metric and Nonmetric Data Collection and Analysis	236
	Group Comparisons	245
11.	The Paleodemography of the EB IA People	251
	Bruno Frohlich and Donald J. Ortner	
	Data Preparation	252
	Probability of Death and Life Expectancy	252
	Estimation of Population Size	260
	Social Composition	261

12. The Paleopathology of the EB IA and EB IB People ... 263
    Donald J. Ortner, Evan M. Garofalo, and Bruno Frohlich
    Trauma ... 265
    Infection ... 269
    Arthritis, Including Osteoarthritis and the Erosive Arthropathies ... 273
    Circulatory Diseases ... 273
    Reticuloendothelial and Hematopoietic Disorders ... 273
    Metabolic Disorders ... 275
    Endocrine Disorders ... 277
    Congenital and Neuromechanical Abnormalities ... 277
    Skeletal Dysplasias ... 278
    Tumors ... 278
    Pathological Conditions of the Teeth and Jaws ... 279
    Nonspecific Abnormalities of the Skeleton ... 279

13. Dental Analyses of the Bâb edh-Dhrâ' Human Remains ... 281
    Gillian R. Bentley and Victoria J. Perry
    Methods and Materials for Analyzing the Dentitions ... 281
    Results of the Statistical Analyses ... 285
    Assessing Kinship Relations at Bâb edh-Dhrâ': Evaluation of the Statistical Results ... 289
    Assessing the Health of the Bâb edh-Dhrâ' Population: Evidence from Dental Pathologies ... 291
    Conclusions ... 295

14. Summary of Findings and Conclusions Regarding the EB I People of Bâb edh-Dhrâ', Jordan ... 297
    The EB I Society of Bâb edh-Dhrâ' ... 297
    Funerary Traditions at Bâb edh-Dhrâ' during EB I ... 299
    Physical Characteristics of the People ... 301
    General Health of the EB I People ... 302
    Conclusions ... 305
    Future Research ... 306

References ... 309

Index ... 319

About the Authors ... 325

Acknowledgments

BOTH PEOPLE AND organizations provided important support for the cemetery excavations conducted in 1977, 1979, and 1981. The Expedition to the Dead Sea Plain, Jordan, was sponsored by the American Schools of Oriental Research (ASOR). The American Center of Oriental Research (ACOR) in Amman, Jordan, is one of ASOR's research centers. At the time of the three field seasons, Dr. James Sauer was director of ACOR, and it is hard to imagine anyone being more supportive of the expedition endeavors and the excavations in the cemetery. The Jordanian Department of Antiquities was directed by Dr. Adnan Hadidi during the excavations. He and his staff provided valuable logistical support and arranged for the employment of the technical men, who were skilled and experienced excavators. Sami Rabadi was the department's representative in Kerak. His daily presence at the site during excavations provided additional expertise about the excavation but was also helpful in maintaining constructive relationships with the people of el-Mazra, where most of our workmen came from.

Funding for expedition expenses came from multiple sources that are recognized in Rast and Schaub (2003) and will not be repeated here, with the exception of substantial grant support to the first author specifically for the cemetery excavations from the Smithsonian Scholarly Studies Program. Subsequent to the excavations, we received crucial financial support for laboratory analysis and curation from the Institute for Bioarchaeology, San Francisco.

1977 EXCAVATIONS

In 1977 the cemetery field crew included several volunteers. Joyce E. Ortner assisted with excavation, conservation, and registration of human remains excavated. Don Ortner Jr., Allison Ortner, and Karen Ortner assisted at the site. Michael Finnegan and Jack Husted had primary responsibility for the excavation of the charnel house burials. Kjell Sandved, the photographer with the National Museum of Natural History, was responsible for photographic documentation of the excavations and for obtaining representative film footage of the excavations. John Meoska and Rebecca Unland were field supervisors for the shaft tomb excavations. Unland also did most of the in-field, top plan drawings of the tomb chambers. Other volunteers who provided valuable assistance included Carl Helms, Brian Prill and C. Larson, John Opie, Michael Several, and Dorthey Hill. Christine Helms assisted in photography and Jeff Raday helped to survey the cemetery. Jennifer Loynd produced the final version of the tomb drawings. Marilyn Schaub was the registrar and was assisted by Linda McCreery and Jeanine Dragan. Sami Rabadi, Jordanian Department of Antiquities representative and archaeologist, was extraordinarily helpful both in providing advice about excavation of tombs and in acting as a liaison with the men from el-Mazra, who did much of the digging. Husni Yusuf Abu-Shwaimeh was another representative of the Jordanian Department of Antiquities who provided advice and assistance during excavation. Technical men from the Jordanian Department of Antiquities who did much of the tomb excavations were Ali Khalaf and Ayish Muhammad Issa.

1979 EXCAVATIONS

The cemetery field crew in 1979 included many of the outstanding volunteer staff from 1977, including John Meoska and Rebecca Unland-Vaage, who again were excellent field supervisors. Michael Finnegan continued his work on the charnel house burials and was assisted by Marilyn Saul and Lisa Ann Jurkoic. Kjell Sandved provided photographic documentation of the tomb excavations. Sami Rabadi was again the Jordanian Department of Antiquities representative and an archaeologist who provided advice and assistance on tomb excavation as well as continuing his role as liaison with the local workers from el-Mazra. Technical men from the Jordanian Department of Antiquities were Muhammad Darwish and Muhammad Jamrah.

1981 EXCAVATIONS

The 1981 excavations focused on evaluating the use of geophysical methods in identifying tomb locations and the amount of silting occurring in each chamber. Because of this, the field crew was kept to a minimum and the number of excavated tombs was significantly lower compared with 1977 and 1979. The second author was assisted by Christopher Albert in organizing and executing work during the field season. Albert acted as the field supervisor while the second author was visiting Bahrain and the Kingdom of Saudi Arabia at the end of the field season. As in 1977 and 1979 Sami Rabadi was the effective and helpful representative from the Jordanian Department of Antiquities. He also handled logistic supply problems and the hiring of a small local labor force. Bruce Bevan, an expert on electronic geophysical prospecting, visited the site for a few weeks to consult on the initial set-up and use of the EM-31 equipment. Edith Dietz, a Smithsonian conservator, provided advice on handling and preserving tomb contents. We also benefited from the assistance of several volunteers who helped both in excavation and with the subsequent conservation and registration of human remains and cultural objects. Volunteers included June Crowder, Jim Eighmey, Greta Kaltenbach, Hank Kaltenbach, Jeanette Olson, Marie Reilly, Ann Schelpert, and Tod Ziegler, most of whom split their time between the settlement excavations and the cemetery.

During the time when chambers were opened and the contents were recorded, we camped out next to the tombs to enhance the security of the site. Our stay at the site was greatly facilitated by support from the Arab Potash Company, which made one of the new apartments being constructed south of the site available to us for living purposes as well as for processing of the material. Of great value was our close collaboration with Walter E. Rast and R. Thomas Schaub and their team excavating the town site north of the cemetery. Their knowledge of the site as well as their extensive field experience there was a major asset in the work accomplished in 1981.

Subsequent curation and research on the EB I burials and the preparation of this book benefited from the assistance of several people. Evan Garofalo, Molly Zuckerman, and Liese Meier contributed to the careful sorting of the burials and were crucial particularly in identifying evidence of skeletal abnormalities during the detailed survey of every skeletal element for evidence of disease. Agnes Stix, who for thirty years was the first author's research assistant and, after retirement, continues as a volunteer research assistant, deserves recognition for her bibliographical research and the entry of references into a database system. Janet Beck is another volunteer research assistant who provided expertise in the development of data bases relevant to the research. Marcia Bakry, scientific illustrator, Department of Anthropology, Smithsonian Institution, was responsible for preparing the illustrations for publication.

Throughout the research and manuscript development for this book, Dr. R. Thomas Schaub and Dr. Walter E. Rast provided encouragement and expert advice on archaeological issues that emerged as we conducted the analysis of the burials. Sadly, Dr. Rast died before this book was completed, but both Dr. Schaub and Dr. Rast deserve our deepest appreciation for the vision they had in inviting our participation in the expedition and supporting our activities related to both the excavation and the subsequent analysis.

The authors wish to recognize with gratitude the staff members of AltaMira Press who were responsible for the publication of this book. They were a delight to work with and the speed with which the book was published after submission (less than one year) is a tribute to their effective efforts. Jack Meinhardt, Acquisitions Editor for Archaeology and Anthropology, enthusiastically supported publication of the book and piloted the book proposal through the review and approval process. During the early stages of getting the book manuscript into production Sarah J. Walker, Editorial Acquisitions, was a wonderful help in getting through the initial maze of contracts and other administrative details. After the book manuscript was submitted for publication, Jehanne Schweitzer, Senior Production Editor, coordinated and oversaw the production process with its many complicating factors, aided by copyeditor David Compton and proofreader Judy Fernow.

It is customary for authors to state what for us is an obvious truth, that those who have been acknowledged for their contributions should not be held responsible for any mistakes of either omission or commission that may exist in the published manuscript. To this the first author wishes to add an additional statement. During the many hours of writing in my home office, I was often accompanied by the presence of Scamp the household cat. As cats do, he took particular pleasure in leaping on my work table, swishing his tail in front of the computer screen of my notebook PC, striding across the keyboard, and nonchalantly sitting on top of the mouse. The latter two actions often had interesting results. All this leads the first author to issue the following caveat: any residual deficiencies encountered in this book are entirely the fault of Scamp.

CHAPTER 1

Introduction

WHY ARE HUMANS so interested in our biological and cultural origins? What is the social or biological payoff for the effort we spend in learning about these origins? This interest is virtually universal in world societies today and clearly has deep roots in past human societies. There is no easy answer to these questions, but certainly the past informs the present, and if we as individuals or human societies are willing to take our past seriously as we make decisions about the future, a knowledge of it can help us avoid at least some of the most serious mistakes revealed by our study of the past. This is, perhaps, an overly optimistic view of the practical benefit of history. Human societies do seem to repeat the mistakes of past generations. Nevertheless there are at least a few situations in modern human history where past experience does help in avoiding at least some of the egregious hazards that now confront human society. The persistent horror of the destruction of Hiroshima and Nagasaki, Japan, during World War II surely must be an important factor in the resolve of most nations of the world today to avoid another such disaster in the present.

Beyond any practical value knowledge of our biological and cultural origins may have, there does seem to be embedded in our individual and social psyches a desire to know about our past. Nelson Glueck, the legendary Old World archaeologist, writes of the tradition of oral history in the Middle East in which "there are old men in the tents of Arabia who can recite the history of their ancestors for forty generations" (Glueck 1946). It is somehow reassuring that people and societies in the past did confront major problems with varying degrees of success and that, despite the occasional less-than-ideal strategy, the human odyssey continues. Reconstructing this odyssey is the major focus of both archaeology and biological anthropology. One of the objectives of the Expedition to the Dead Sea Plain, Jordan, is to use the data obtained from excavations, particularly at the Early Bronze Age (EB) site of Bâb edh-Dhrâ' in the modern country, of Jordan to illuminate important aspects of this odyssey.

A BRIEF HISTORY OF EXCAVATIONS AT BÂB EDH-DHRÂ', JORDAN

More detailed accounts regarding the history of archaeological surveys and excavation of Bâb edh-Dhrâ' are available in other sources (Lapp 1966; Schaub and Rast 1989; Rast and Schaub 2003). The following few paragraphs are intended to provide a brief summary of these earlier analyses of the site as they relate to the field excavations conducted in 1977, 1979, and 1981.

The presence of a large walled structure at Bâb edh-Dhrâ' is mentioned in reports of surveys conducted by William F. Albright and colleagues (Albright 1924, 1924–1925). More recent knowledge of the site began to emerge when a very distinctive type of pottery began to be sold by the antiquities dealers in Jerusalem (Lapp 1966; Schaub and Rast 1989:18; Saller 1964–1965). The source of this pottery was ultimately traced to Bâb edh-Dhrâ' by Paul Lapp, and he conducted surveys and preliminary excavations that resulted in his three seasons of excavations between 1965 and 1967 (Schaub and Rast 1989:18; Rast and Schaub 2003:1). Walter E. Rast and R. Thomas Schaub participated in Paul Lapp's excavations in 1967 and took over the responsibility for subsequent excavations and publications following Lapp's untimely death in 1970. Schaub and Rast (1989) published a careful analysis of the materials excavated in 1965 and 1967. They conducted additional surveys and preliminary excavations in 1973 and 1975 that highlighted the significance of the site and further defined the major phases of occupation (Rast and Schaub 1974; Rast and Schaub 1981). These surveys and excavations provided the basis for a major effort to characterize the contents and significance of the site, including the large walled town and the cemetery located to the south of the town walls (Rast and Schaub 1974; Rast and Schaub 1981; Schaub and Rast 1987; Rast and Schaub 2003).

The first author of this book was invited to join the Expedition to the Dead Sea Plain in 1973 as the biological anthropologist with primary responsibility to conduct research on the human remains that would be excavated. It is a tribute to the vision of the directors of the expedition, Drs. Walter E. Rast and R. Thomas Schaub, that they recruited a biological anthropologist as an active member of the expedition team. They recognized the potential scientific value of a substantial and carefully documented human skeletal sample from the cemetery that would provide data to complement observations based on cultural materials recovered in earlier excavations in the cemetery and excavations elsewhere at the site. This research had the potential to provide a bioarchaeological base for the excavations of the town and other architectural features of the site.

THE EXCAVATIONS IN 1977, 1979, AND 1981

Excavations undertaken in field seasons of 1977, 1979, and 1981 were directed by R. Thomas Schaub and the late Walter E. Rast. General objectives were to clarify further the architecture and social significance of the large walled city ruin dating to the EB II–III time period (Table 1.1). Excavations of the cemetery area to the south of the city ruin had several objectives. Burial traditions are an important component of human society and tell us much about relationships among various components of a society. Clarification of burial traditions both within and between Early Bronze Age archaeological phases at Bâb edh-Dhrâ' was one objective. We were also interested in obtaining a better definition of the extent of the cemetery as a window into the size of the human population utilizing the site during the phases of the Early Bronze Age. However, the major objective was to obtain skeletal samples representative of the EB I–III phases of occupation to clarify biological relationships of the occupants of Bâb edh-Dhrâ' in each of the EB phases and the relationship to synchronous populations in other areas of the Middle East and Northern Africa.

Earlier excavations conducted by Paul Lapp as well as Schaub and Rast defined at least the broad outlines of the burial tradition in various phases of the occupation of the site during the Early Bronze Age. In the earliest phase of the Bronze Age (EB IA), burials were secondary and were placed in below-ground chambers associated with shafts, thus the term *shaft tomb burials*. The typical shaft tomb contained the central shaft and as many as five chambers radiating from the base of the shaft, creating a clover-leaf arrangement in plan view. The shafts were circular and usually a bit less than one meter in diameter. The depth of the shaft below the ground varied, but was about 1.5 meters. The dome-shaped chambers were about two meters in diameter and somewhat less than one meter high at the peak of the chamber ceiling (Schaub and Rast 1989:182). The contents of most shaft tomb burial chambers contained multiple secondary burials as well as varying types and amounts of cultural artifacts, including pottery, basalt vessels, clay female figurines, wooden bowls, beads, alabaster mace heads, textiles, reed matting, implements, and remnants of food gifts. Which of these types of artifact were placed in a given chamber varied, except that pottery was included in all chambers. Current evidence supports the observation that these contents were usually placed in the chamber during a single burial event. The typical arrangement within the chamber was skulls carefully arranged in a row to the left of the chamber entryway, postcranial bones in a pile in the center of the chamber, and tomb gifts, mostly pottery, arranged to the right of the postcranial bone pile (Schaub and Rast 1989).

It is possible, if not probable, that not all the chambers of a tomb were utilized during a single burial event. The location of and length of time interred in the primary burial site remain unknown variables, although our current data, based primarily on the variable preservation and completeness of the human remains within and between chambers, indicate that the length of primary burial was highly variable, probably ranging from a few days to several years (Frohlich and Ortner 1982:264).

In EB IB some use of shaft tombs continued, but above-ground charnel houses usually constructed with stone entryways and mudbrick walls became the typical place of interment. With smaller charnel houses, construction could be entirely of mudbrick with a stone-lined entryway. The walls were coni-

Table 1.1. Archaeological Chronology and Burial Types at the Site of Bâb edh-Dhrâ', Jordan

Period	Dates	Types
EB IA	3300–3200 BCE	Shaft tombs and secondary burials
EB IB	3200–3100 BCE	Shaft tombs and secondary burials + charnel houses with primary and secondary burials
EB II	3100–2750 BCE	Charnel houses
EB III	2750–2350 BCE	Charnel houses
EB IV	2350–2000 BCE	Cyst tombs

cal, creating a structure shaped like a beehive. Larger charnel houses also had stone-lined entryways and mudbrick walls. Wooden pole beams lay across the walls and were covered with reed matting and mud to form the roof. Some of the larger charnel houses were circular, but a more common floor plan was square or rectangular. Charnel house burial became the exclusive burial pattern in EB II and continued to be used through the EB III period.

Because of the association of charnel house burial patterns with increasing evidence of permanent settlement of Bâb edh-Dhrâ', an early hypothesis was that burial in the charnel houses occurred at the time of death; that is, burials were primary. Because, unlike burials in shaft chambers, multiple burial events were associated with charnel houses, burials within the house were disturbed as more recent burials were added to the contents of the house. This was further complicated by vandalism in antiquity that disturbed the burials. As will become apparent in the description of the burials in the G1 Charnel House dated to the EB IB period discussed in Chapter 7, it is highly likely, on the basis of variation in the prevalence of various skeletal elements, that at least some secondary burial continued to be practiced in the EB IB period.

The responsibility of the authors of this book was to oversee and direct the excavations of the cemetery in consultation and collaboration with Schaub and Rast. The skeletal samples excavated from each of the phases in the cemetery area would provide data on biological differences that might exist between the phases. For example, were the people of EB IA, thought to be proto-urban, biologically continuous with the early-urban people of EB IB? If so, this would support the hypothesis advanced by Schaub and Rast (1989) that cultural changes apparent during the Early Bronze Age at Bâb edh-Dhrâ' were the result of cultural evolution within a continuous human society and not caused by one or more invasions of new people.

Obtaining skeletal sample sizes needed to achieve this objective requires excavating substantial numbers of human remains to ensure reasonable delineation of the biological populations represented by the skeletons. The biological variation that occurs in skeletal samples, predominantly age and sex differences in the anatomy of skeletons, is one of the major factors creating the need for large sample sizes. Another factor is the loss of data that occurs in archaeological skeletal remains because of cultural burial traditions and taphonomic processes. To address these factors, sample sizes of at least three hundred burials were needed for each phase in the occupation of the Bâb edh-Dhrâ' site.

The limitations of sample size are particularly relevant in the analysis of skeletal evidence of disease. Only about 10% of a typical archaeological skeletal sample will show evidence of significant skeletal disorder (Ortner 2003). This 10% is distributed through many types of disease, although the most prevalent disorders are trauma, infection, and arthritis, which usually constitute about 85% of the skeletal disorders encountered in an archaeological skeletal sample. This means that the prevalence of many skeletal disorders will be low, and large skeletal samples are needed to draw relevant conclusions about the significance of these disorders relative to the health of an archaeological skeletal population.

It is also crucial to clarify as much as possible just how representative the burials were of the living population, since it is the living population whose history, biology, and health we want to reconstruct. Although everyone living in a society eventually dies, a sample of the dead is not the same as a sample of living people, since in a living population one has a range of people from those who are within days or hours of death to those who have a long life ahead of them (Wood et al. 1992). The burials in a cemetery context reflect only the biology of those who have died. There has been and continues to be a very vigorous debate about the extent to which one can make inferences about living populations on the basis of data from a skeletal sample of that population. Almost certainly there are some limitations in the accuracy of such inferences, and these need to be kept in mind as we interpret the data we obtain from archaeological skeletal samples (Ortner 2003). Nevertheless, the EB IA skeletal sample is substantial, with 578 burials. These include ages from fetal to old age and an equal distribution of males and females. Furthermore, rigorous field and laboratory methods ensured that virtually every skeletal element that had been placed in the tomb chambers or charnel house was recovered and included in the analysis. The age and sex distributions of burials within the chambers indicate that the EB IA society interred most if not all its members in the chambers following temporary interment at a primary burial site. Careful excavation and curation of the sample provides a skeletal sample that is as representative of the living population as any archaeological skeletal sample can be.

The burials recovered from the EB IB G1 Charnel House included very few fetuses, infants, and young children. The minimum number of individuals as-

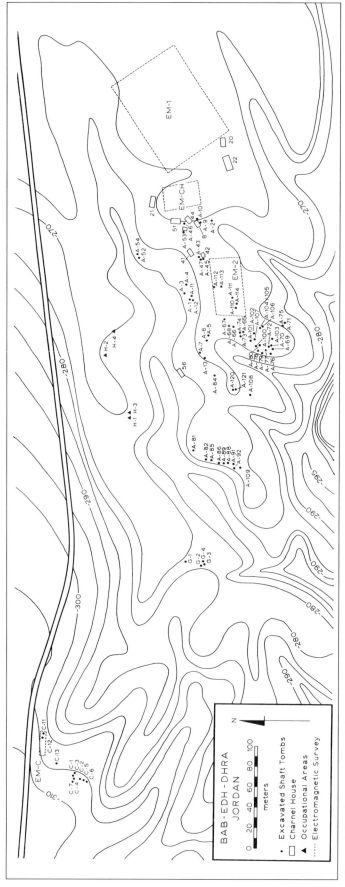

Figure 1.1. Contour map of the cemetery area, with shaft tombs and charnel houses indicated. Three areas designated EM1, EM2, and EM3 were surveyed using the Geonics EM-31 conductivity meter.

sociated with the charnel house is 115, which is much smaller than the EB IA sample. Furthermore, it was not possible to associate skeletal elements into discrete burials as we had been able to do with most of the EB IA shaft tomb burials. This means that the G1 burials are not as representative of the living population as is the case with the EB IA burials. Comparisons of the two samples are possible but need to be done carefully and with full awareness of the limitations of the GI sample.

It is also important to note that field methods for the cemetery excavations evolved, particularly between the 1977–1979 and the 1981 field seasons. In general, the quality of the human bone was fragile, most likely the result of the time since burial, the temperature, and the high sodium chloride content in the chamber burial environment. The ideal shaft tomb chamber was one that had remained as it had been created and used 5,000 years ago, with walls and ceiling intact and the chamber sealed, following interment, by blocking stones that were important in limiting or preventing one type of silting of the chamber. However, varying amounts of silting was a common complication in excavating the tomb chambers. In some chambers it is clear that silt flowed into the chamber from what was probably an open shaft where the blocking stone or stones were inadequate. However, there is evidence that some silt came into the chambers through fissures or holes in the walls and the ceiling of the chamber. There is also evidence of tomb robbing in antiquity that could have created a pathway for silting to occur.

In some cases silt may have provided protection to the human remains, particularly if ceiling fall occurred in the chamber following the deposit of silt. However, much depended on the quality of the silt. In some cases the silt solidified into a very hard matrix and, in that situation, removal of human remains and cultural artifacts was challenging to say the least.

Recovery of what may have been intact but fragile human remains was limited to fragments of skeletal elements in situations where the silt was very hard. In these situations, the probability is that one or more individuals buried in the chamber were represented in the human remains recovered. When recovery of skeletal elements was difficult, we also relied on observations regarding the burials made during excavation. Fortunately, this problem was encountered in very few chambers.

The difficulties of excavating silted tomb chambers and the associated loss of human bone led to an exploration of field methods that would identify the presence of significant silting within the chambers before excavation of a shaft tomb. In the 1970s the EM-31 ground conductivity meter was developed by Geonics Ltd., a Canadian electromagnetic geophysical equipment manufacturing company. The equipment is useful for evaluating soil conductivity to a depth of as much as five meters. Spatial resolution for the method was in some cases adequate to clarify the presence of intact versus silted chambers before excavation began (Frohlich and Ortner 1982; Frohlich and Lancaster 1986). Although identification of silted versus nonsilted chambers could not always be predicted accurately, we were 100% successful in identifying the location of the upper part of the shaft to within 0.5 meters. This significantly lowered the time used to identify the location of the shafts leading down to the burial chambers. This equipment and methodology was used only in the 1981 season. Tombs that showed significant evidence of silting in all the associated chambers were identified but not excavated. This field methodology greatly facilitated excavation, since tombs with significant silting could be avoided. A more detailed review of this method is presented in Chapter 3.

In the 1977 and 1979 field seasons the shaft tombs identified and excavated were relatively close to seasonal wadi channels, which seem to have been the preferred location of these tombs. In 1981 the area in the A cemetery that was in the flatter areas between the seasonal wadi channels was chosen for surveying and excavating (Figure 1.1). Some of the burials recovered were found with most of the bones in correct anatomical position. This indicates that the time between primary and secondary burial was very short compared with the EB IA burials excavated in the 1977 and 1979 field seasons. The significance of this difference is discussed in greater detail in Chapter 9, but there is a good possibility that the burials excavated in 1981 represent the late stage of the EB IA phase. This part of the A cemetery may have been used for shaft tombs as sites became scarce in locations closer to the seasonal wadi channels.

Most of the tomb chambers excavated in the 1977 and 1979 seasons were not significantly affected by silt, so samples recovered in each of the field seasons are comparable for the analytical procedures used. Variation in the data from tombs in different locations within the cemetery can be attributed to variation in the mortuary practices and not to factors associated with excavation and curation of the burials.

SCIENTIFIC VALUE OF THE BÂB EDH-DHRÂ' SITE

Many of the social and cultural developments that characterize modern human society have their earliest expression in the Middle East. Agriculture, and the associated domestication of many plant and animal food resources, is first seen in the Middle East. Sedentism and the emergence of urbanism with its large concentrations of people living in close proximity have their earliest expression in the Middle East. These cultural and social innovations have changed forever the relationships human beings have with each other and with the environment in which they live.

Throughout much of the evolutionary history of humankind, global population densities increased very slowly (Hassan 1983). The emergence of agricultural economies from hunting/gathering economies provided the food resources that made greatly increased population densities possible. The downside of this is that human health and longevity probably declined in agricultural societies (Cohen and Armelagos 1984) until the past 150 years, with the increased awareness of public health and understanding of the factors that influence it. The linkage of pathogenic bacteria to specific diseases by Robert Koch in the latter half of the 19th century provided the scientific basis for improving hygiene and had a significant impact on public health, but this is a very recent event in the context of the development of agriculture-based economies.

Human skeletal samples permit us to examine some aspects of health and morbidity in past human populations. There are potential problems in making inferences about living populations on the basis of skeletal data (Wood et al. 1992; Ortner 2003). Nevertheless, there is reason for optimism that these issues can be resolved as we understand these biases better and explore appropriate correction factors for them. Recent research based on very large archaeological skeletal samples indicates that important trends and patterns of morbidity can be identified in past human groups (Steckel and Rose 2002).

One of the observations that has emerged from our analysis of the Bâb edh-Dhrâ' EB I skeletal remains is the importance of careful excavation and recovery of human remains. Taphonomic factors may inject biases into excavated skeletal samples. The protein component of the EB I burials from Bâb edh-Dhrâ' had been destroyed by the conditions of burial, making the bone elements very fragile. Nevertheless, even very small fetal bones were recovered. In the EB IA phase at Bâb edh-Dhrâ' there seems to be a high probability that most if not all the people who died had at least some of their human remains placed in a tomb chamber.

As shown in Chapter 7, the EB IB skeletal elements offer a somewhat different picture in that very few fetal or newborn burials were recovered. This is remarkable especially when considering that the preservation of skeletal elements in the EB IB charnel houses appears to be much better than that of the burials from the EB IA period. This means that there was a change in the burial tradition and that the EB IB burials are not completely representative of all the individuals who died.

Tombs and their contents also provide a rich lode of information on the society represented by the tombs and burials. Differences in the type and amount of burial gifts may provide insight into status differences within the society. Who gets buried in the tombs can provide information on whether the society was hierarchical or egalitarian. The presence of infants testifies to the value the society places on life from birth to old age. The treatment of female burials may suggest the relative value of women to the society. The burial practices themselves are of interest in defining the type of society (Frohlich and Ortner 2000). Was it an urban complex society or a nomadic, pastoral society? Do the tomb gifts reveal any connections with other societies in the region? The high quality of the Bâb edh-Dhrâ' skeletal sample excavated and retrieved from 1977 to 1981 becomes an important component enabling us to explore and answer some of these questions.

CHAPTER 2

The Ecological and Social Context of the EB I People

ONE OBJECTIVE of this chapter is to provide a brief summary of what is known about the physical and biological environment of Bâb edh-Dhrâ' and the region that surrounds the site during the initial period of the Early Bronze Age (Table 1.1). It is this environment that provides the resources, shelter, food, and water on which the people linked to the site of Bâb edh-Dhrâ' relied for their existence. However, in human societies the social mechanisms that define the relationships between people within and between societies as well as the cultural strategies used to extract fundamental resources from the natural environment are equally, if not more, important in providing the context for the human society that lived at the site of Bâb edh-Dhrâ'. For this reason, we will also summarize some of the major ideas regarding what little is known about the social milieu of the Bâb edh-Dhrâ' people. Although the archaeological evidence is subject to future reinterpretation, a plausible case can be made for cultural influence of both the societies to the north and of Egypt to the south on the southern Levant, which includes Bâb edh-Dhrâ'.

However, it is not our intention to provide a comprehensive review of either the natural or the cultural context of the Bâb edh-Dhrâ' people. Much of what is known about the regional culture in the Early Bronze I and the Chalcolithic Period that preceded it rests on what can at best be described as a less-than-ideal base of data. Although considerable research has been done on the geology of the southern Levant, there remains much that is not presently known about the physical conditions at Bâb edh-Dhrâ' during the EB I period. For more comprehensive treatments of what is known about both the culture and the physical context, the reader should consult other sources (e.g., MacDonald et al. 2001; Ehrich 1992; Levy 1986; Donahue 2003).

THE PHYSICAL ENVIRONMENT OF THE SOUTHERN LEVANT

The geographic region that provides the physical setting for the EB I people of Bâb edh-Dhrâ' is the southern Levant, and more specifically the Jordan Valley, with Palestine to the west and the Jordanian highlands to the east. The northern boundary of the Jordan Valley is the Sea of Galilee, which feeds into the Jordan River and terminates in the Dead Sea, which today is about 1,200 feet (approximately 400 meters) below sea level. In the Early Bronze Age, the banks of the Dead Sea were somewhat higher and closer to the western margin of the site of Bâb edh-Dhrâ' (Neev and Emery 1967).

The Jordan Valley is near the northern end of the Great Rift Valley, which extends 4,000 miles beginning in the southern Levant, passing along the Red Sea, and continuing a southerly path through eastern Africa to Mozambique. The valley is in the Dead Sea–Jordan transform fault, which includes several north-south faults between which the Dead Sea trough developed. Several cross-faults, including the Kerak, Numeira, and Safi, connect the eastern escarpment with the Dead Sea and resulted in the development of the wadis and extensive alluvial fans. The Dead Sea area is one of the world's most tectonically active regions. This, combined with the movements in the faults, creates an unstable environment in which earthquakes are common events (Donahue 2003).

The site of Bâb edh-Dhrâ' is in the southeastern Ghor, about three kilometers east of the modern southeastern shore of the Dead Sea and almost directly east of the Lisan Peninsula, which juts out from the southern shore of the Dead Sea. The site is located on an alluvial plain between the Dead Sea and the base of the eastern escarpment, which marks the beginning of the steep ascent to the Ammon (eastern) plateau and the modern town of Kerak, which has a history extending back centuries and today contains the ruins of a massive Crusader castle. This plain occupies a substantial area of relatively flat land that supports agriculture today. In the Early Bronze Age, the climate of the Jordan Valley may have been slightly wetter than it is today (Zarins 1992:53; Rosen 2003), but, as in much of the rest of the southern Levant, irrigation was an important, indeed essential, component of effective farming. As Macumber (2001) emphasizes, the

Jordan Valley is an arid or semiarid region, and water from streams and springs is a requirement for human occupation.

In the late Pleistocene, Lake Lisan, the ancient precursor of the Dead Sea, covered the site of Bâb edh-Dhrâ'. Annual lake bottom deposits formed the clay/chalk layers of Lisan marl, which provided the best medium for the shaft tombs of the EB I period. This marl was relatively easy to excavate for the shaft and the tomb chambers and was remarkably stable. However, the distribution of marl was not continuous throughout the cemetery area, and some of the tombs were cut into other deposits, including consolidated sands and gravels, which were less stable and provided a challenge to the EB I tomb excavators. The locations of tombs and the numbers of chambers may be related to the presence of easily workable Lisan marl.

Today, as it did in antiquity, the east-west axis of Wadi al-Kerak passes through the site on its way to the Dead Sea and provides a continuous supply of water that is crucial to human existence at the site. The Wadi al-Kerak is one of several streams that begin in the highlands to the east of the Jordan Valley and flow down into the valley. They provided an essential resource for the societies utilizing the valley for either temporary or permanent settlements. Because of arable land and a favorable water supply for irrigation, the area around the site of Bâb edh-Dhrâ' would have been a valuable location for agriculture, which was the economic base for the Levant as well as the complex societies of Mesopotamia to the northeast and Egypt to the south.

HUMAN SOCIETIES OF THE CHALCOLITHIC PERIOD

Many of the cultural innovations that became widespread in the Early Bronze Age have their roots in developments occurring during the preceding Chalcolithic Period. In Mesopotamia, urbanism was flourishing by the Late Chalcolithic, and the societies there were developing a highly stratified social structure. Initially the social structure may have been defined by the religious leadership, and the temple was the focus of activity in the communities (Levy 2003). Later in the Chalcolithic the religious leaders were superceded by secular kings; palaces displaced temples as the foci of the society. Irrigation became a common method of providing crucial water to developing crops, and this innovation became even more important as settled communities developed in the Jordan Valley during the Early Bronze Age.

In the southern Levant there is no evidence of highly stratified societies during the Chalcolithic Period. The communities were small compared with the cities in Mesopotamia, and they were more self-contained. Levy (1986) argues that the best characterization of the social structure in Palestine during the Chalcolithic was as a chiefdom with close ties to kinship groups and distinct regional cultures. Equally important, Levy (1986:86) supports the hypothesis that these regional societies maintained their identity in the transition between the Chalcolithic and Early Bronze Age.

There is evidence of at least some human utilization of the resources at Bâb edh-Dhrâ' during the Chalcolithic Period (Bourke 2001:118). However, major use of the site begins in the Early Bronze Age I with tombs and burials at Bâb edh-Dhrâ' that are dated between 3300 and 3100 BC (Table 1.1). The Early Bronze Age was preceded by the Chalcolithic or protohistoric period, which extended from about 4500 to 3200 BC (Levy 1986). There is some variation in estimates for the end of the Chalcolithic (e.g., Stager 1992:27; Dever 2003). However, for the moment at least, we have assumed that 3300 BC is a reasonable estimate for the start of the Bronze Age phase at Bâb edh-Dhrâ'.

At Bâb edh-Dhrâ' the Early Bronze Age I is divided into two subphases: Early Bronze Age IA (3300–3200 BC) and Early Bronze Age IB (3200–3100 BC; Table 1.1). This division marks the transition between a less settled utilization of the site and the early stages in the emergence of a settled town with protective walls and the eventual construction of a massive walled town in the Early Bronze Age II–III (Rast and Schaub 2003).

During the Chalcolithic in the Levant and Mesopotamia, copper metallurgy was present, along with a characteristic flint-tool industry. Large more-or-less independent agricultural settlements were found in several areas within Palestine and the Jordan Valley (Levy 1986). Although there are clear cultural linkages between Palestine and both Mesopotamia and Egypt, the settlements in the southern Levant tended to have a distinctive pottery tradition that suggests culture change through time was a local and relatively independent development rather than the result of cultural diffusion from major urban centers in Mesopotamia or Egypt or invasion and replacement of local societies by those from the northeast or south.

The Chalcolithic is a time when urbanism was developing as a major social innovation in human society (Adams 1970). To the northeast of Bâb edh-Dhrâ', the great urban centers of Mesopotamia were thriving, complete with complex religious traditions, a stratified social structure, political control over agricultural resources basic to the survival of the cities, and complex economic relationships with other cities and city-states (Adams 1970).

By the Early Bronze Age I settled farming communities were becoming a common element of the social environment of the southern Levant. These communities were relatively small compared with the major urban centers in Mesopotamia, and they probably were not as highly stratified as the cities of Mesopotamia.

FUNERARY TRADITIONS IN THE LATE CHALCOLITHIC AND EARLY BRONZE AGE

Funerary traditions provide important insight into a human society and its values as reflected in the care of the dead, the structure of the tomb, and the contents placed in the tomb along with the human remains. Careful treatment of the dead has a very long history in human societies, and the specifics of this treatment provide an important window into the culture of past human societies.

Cave burials are in wide use during the Chalcolithic. During Early Bronze I, in the southern Levant, caves were a common place of burial for people living in settlements (Bloch-Smith 2003:105). At least by the Late Chalcolithic above ground stone tombs known as *nawamis* were found in the southern Levant (Zarins 1992:54; Bloch-Smith 2003:106).

Early Bronze IA is associated with the use at Bâb edh-Dhrâ' of shaft tombs with architecture of the southern Levant that appears to be unique at the site. These tombs, with a circular shaft and one or more tomb chambers radiating from the base of the shaft, appear to reflect the utility of Lisan marl in creating stable underground chambers that were relatively easy to dig.

EGYPTIAN SOCIETIES OF THE LATE PREDYNASTIC AND EARLY DYNASTIC PERIODS

Kantor (1992) provides a review of the Egyptian chronology and a brief account of cultural links to societies in the southern Levant. Unless otherwise referenced, the following summary is based on Kantor's review. Connections with the Levant were present in the Chalcolithic and continued in the Early Bronze Age. By the beginning of the First Dynasty, Upper and Lower Egypt had been united under a single king. The Gerzean period marks the end of the Predynastic phase in Egypt and corresponds roughly with the EB I period in the southern Levant. Between the Gerzean period and the First Dynasty there is an intermediate transitional period designated as Dynasty 0.

There is archaeological evidence of cultural influences moving in both directions between the southern Levant and Egypt in the Late Predynastic and Early Dynastic Periods. Kantor (1992) notes the prevalent assumption that domesticated grains and most domestic animals arrived in Egypt from western Asia, presumably through the Levant. Distinctive pottery as well as raw materials from Palestine are also found in Egypt. There is also plausible evidence of caravan trade between Egypt and Palestine along the north coast of the Sinai.

In the Late Predynastic Period cultural influence from Palestine to Egypt predominates. However, in the Early Dynastic Period this movement seems to be reversed (Stager 1992). The probable coastal trade route between the delta region of Egypt through the Sinai desert in the Early Bronze I has been identified on the basis of camp sites dated to that period that are located between the Egyptian delta and the coastal towns just north of the desert (Stager 1992).

Hoffman's review (1979) of the Predynastic Period in Egypt provides a brief reinterpretation of the significance of three late Predynastic sites labeled Omari and located near the modern town of Helwan, situated about 23 kilometers south of Cairo. The sites were excavated in the 1920s using field methods that leave many questions unresolved. Nevertheless, approximate dates for the sites range between 4000 and 3000 BC. The most recent of these sites Hoffman has designated as Omari C and dates to 3300–3100 BC (Late Gerzean or Protodynastic Period), which corresponds with the EB I period at Bâb edh-Dhrâ'.

According to Hoffman, the three Omari settlements are linked and provide a view of the evolution of a regional subculture over a one-thousand-year range of time. The social structure had a focus on the nuclear family, and each family was a relatively independent economic unit. The communities were egalitarian, like the other towns in Predynastic Lower Egypt. This is in contrast with the more hierarchical social structure of Upper Egypt at the same time. The stone tool technology at least hints at a link with Palestine with an emphasis on flake and blade traditions rather than bifacial tools.

The burial tradition of Omari A, the earliest of the settlements, is most clearly defined on the basis of the excavation reports. The dead were primary flexed burials in round pits within the settlement. The bodies were wrapped in cloth, matting, or animal skins and placed on their left sides with heads to the south and facing west. Grave offerings were minimal, and very limited evidence of any social stratification was in the grave contents. The Omari C burials were in shallow pits covered with circular mounds of stones. The bodies were in a flexed position with hands placed on the face. Unlike the Omari A burials, the bodies were not oriented in any specific direction.

Before the Gerzean period the presence of raw materials from sources east of Egypt provide evidence of active trade with the southern Levant (Kantor 1992:12). A line of small settlements in the Sinai following the Mediterranean coast contains both Palestinian EB I and Egyptian pottery, suggesting a trade route between Egypt and population centers to the east and north. In the Gerzean period itself, Mesopotamian cultural features are found in Egyptian pottery (Kantor 1992:14). Distinctive pottery types found in both Egypt and Palestine dated to the Late Chalcolithic/Early Bronze I periods also support links between the two regions (Zarins 1992:53).

MESOPOTAMIAN SOCIETIES OF THE PROTO-URBAN AND EARLY URBAN PERIODS

Mesopotamia, the "cradle of civilization," refers to the land located around and between the Euphrates and Tigris rivers. It includes land now defined as Iraq, eastern Syria, southeastern Turkey, and western Iran. The Mesopotamian region is the site of one of four major river systems where highly advanced civilizations, known for their use of written language, developed. These river systems are the Nile River, the Euphrates and Tigris rivers, the Indus River, and the Yellow River in China.

Mesopotamian history spans from the fourth millennium BC to the arrival of Alexander the Great around 330 BC. During a span of more than 4,000 years, four major city-states developed and flourished: (1) Uruk (3900 BC to 2900 BC), (2) the Sumerian city-state in the south (3200 BC to 2750 BC), (3) the Akkadian city-state in central Mesopotamia (2300 BC to 2100 BC), and (4) the Assyrian and Babylonian city-states, encompassing all of Mesopotamia (2300 BC to 500 BC). Of interest to us are the Uruk Period and Predynastic and early Dynastic Sumerian periods. These represent the Mesopotamian periods contemporary with the EB I eras (Adams and Nissen 1972; Kramer 1963; Matthews 2003; Pollock 1999; Postgate 1992; Potts 1994).

The Uruk Period is named after the city of Uruk in the southern part of Mesopotamia; it is one of the oldest known cities. It may have included more than 50,000 people within a walled structure, and it became one of the many cities that developed into the Sumerian city-state. The Sumerian city-state is the earliest known civilization of the ancient Middle East. It was centered in southern Mesopotamia and most likely coexisted over time with later city-states and empires such as those of the Akkadians in central Mesopotamia and the Assyrians in northern Mesopotamia.

Little is known about Uruk and Predynastic Sumerian trade and population movements to other geographical areas. However, Sumeria's access to the northern coast of the Arabian Gulf and the development of advanced shipping technologies resulted in trade activities reaching other states along the Gulf and further to the east. This would have included Dilmon, which is associated with present-day Bahrain and the Eastern Province of Saudi Arabia, and possibly places as far east as the Indus Valley, where Sumeria would have connected with the Harappan civilization (3300 BC to 1700 BC; Tosi 1986). More recently, some trade connections existed between Predynastic Egypt and Sumeria (Potts 1994). This most likely would have been along the coast, either following the southern route around present Yemen territories or across land to the northwestern part of Mesopotamia and then, using boats, along the eastern Mediterranean coast to the lower Nile Delta (Potts 1994). The latter possibility is supported by research by Stager (1992). We argue that any significant biological or trade relationship between the Uruk and Sumerian city-states and the people of Bâb edh-Dhrâ' during the EB I periods is unlikely.

At the beginning of the third millennium, the Sumerian city-state was being invaded by the Akkadians, about whom very little is known. The Akkadian empire is known to the world through written records and very minimally through the archaeological record (Matthews 2003). The Akkadian city-state developed into the first known empire, expanding rapidly during the middle of the third millennium to include all of Mesopotamia, northern Syria, and southern Turkey (Anatolia). During the time of the Uruk and Sumerian city-states, the vast lands between the Egyptian territories, the coastal areas of the Levant, and Mesopotamia were inhabited by Semitic-speaking nomadic people. These populations may have been the ancestors to the people who became the Akkadians. This possibility creates a potential research project to explore the bio-archaeological links between the EB I Bâb edh-Dhrâ' people and the early precursors of the Akkadians. Because of the proximity of the EB I Bâb edh-Dhrâ' people to the nomadic ancestors of the Akkadians, a plausible hypothesis is that the closest biological affinity of the Bâb edh-Dhrâ' people was with the Akkadian people in central Mesopotamia rather than with the contemporary Uruk and Sumerian populations.

Uruk and Sumerian artifacts have been identified some distance from southern Mesopotamia, with connections to Predynastic Egypt to the west, Anatolia to the northwest, and sites in Iran, including Shar-i Skohta, to the east. These links suggest the presence of a complex trading system during the Uruk and Sumerian periods. However, such trade may not have influenced the EB I Bâb edh-Dhrâ' people significantly.

As discussed in Chapter 10, the use of human skeletal remains in establishing degrees of affinities between contemporary and noncontemporary populations in the Middle East has limitations. At this time, this is the result of a combination of poorly documented skeletal samples, small sample sizes, and, in some cases, a lack of well-preserved and carefully excavated human remains. We hope that future excavations and improved documentation of human remains will provide the basis for exploring these biological relationships in future research.

CHAPTER 3

Methods in the Recovery of and Research on the EB I People

THIS CHAPTER PROVIDES a brief account of the methods used in excavating and conducting research on the EB I burials from the cemetery at Bâb edh-Dhrâʿ. This will include both the methods for locating tombs in the cemetery area and the conservation of the human remains once they were removed from the tombs or charnel houses. We also will review the methods used in data collection and analysis both during excavation and in the laboratory.

Fieldwork by Paul Lapp (1968) and Walter E. Rast and R. Thomas Schaub (Rast and Schaub 1974, 2003; Schaub and Rast 1989) had identified the location of the cemetery area a short distance south of the ruin of the walled EB II–III town. What was not known was the extent of the cemetery and the number of individuals that were buried there. The primary objective of excavations in the cemetery conducted in 1977, 1979, and 1981 was to obtain adequate skeletal samples from the EB I to EB III phases for analysis of the biological characteristics of the people. However, other objectives included defining the boundaries of the cemetery and estimating the size of the society that used the cemetery area at Bâb edh-Dhrâʿ to bury its members.

Earlier fieldwork had identified various areas within the cemetery, and these had been labeled with letters on the map of the site (see Figure 1.1). The greatest concentration of EB IA shaft tombs was in the A Cemetery area. In order to obtain the maximum number of burials, this location was the focus of our excavations in all three field seasons, although other areas of the cemetery were explored as well.

EXCAVATION METHODS

Three basic burial structures are present in the Bâb edh-Dhrâʿ cemetery: (1) the early shaft tombs, (2) charnel houses, and (3) cairns (Lapp 1968). Cairns are associated with the final EB IV phase; none of these tombs were excavated in 1977, 1979, or 1981, and they will not be discussed further. The earliest of the tombs are below-ground shaft tombs associated with the EB IA and EB IB periods. This type of tomb was used exclusively during the EB IA phase. In EB IB, shaft tombs continued to be used, but above-ground charnel houses were introduced and became the only burial structure used during EB II–III.

Shaft Tomb Excavation

During the field seasons of 1977 and 1979, test trenches were dug to identify the presence and location of a tomb shaft (Figure 3.1). These trenches varied in size and were square or rectangular. In most cases, if a shaft tomb was present, a more or less round area of soil that was different in type and color from the surrounding area identified the top of the shaft and became apparent within a few centimeters below the modern ground surface (Figure 3.2). Excavation of the shaft was done by one of the Jordanian technical men with the assistance of laborers from the nearby town of el-Mazra. The shaft fill was carefully screened by workers, and any cultural materials or skeletal elements identified were documented and retained for subsequent study (Figure 3.3). Student volunteers kept careful record of all stages in the excavation process using standard forms developed in previous fieldwork. The depth of the shaft varied from about one meter to three meters. It seems likely that a major factor in determining the shaft depth was the location of Lisan marl, which tends to be more stable than the sand/gravel layers. Lisan marl is a relatively soft combination of chalk and clay. Lisan marl is easy to dig into, which facilitated creation of the dome-shaped tomb chambers. Furthermore, it was less likely to collapse during and after the original excavation of the chamber.

Usually, one or more blocking stones and mud mortar were used to seal off the entryway to the tomb chambers. The base of the shaft was below the bottom of the blocking stone(s) and mud sealant. When the shaft fill had been removed to the base of the shaft, one of the senior excavation staff removed the blocking stones and conducted the initial survey of the tomb chamber. Burial chamber condition varied depending primarily on two variables, ceiling fall and silting. The matrix in which the chambers were cut ranged from

Figure 3.1. Test trench for Tomb A79 excavated in 1977. The early stage of excavating the test trench for Tomb A78 is apparent in the distance.

Figure 3.2. Initial stage in excavating the shaft for Tomb A78.

Figure 3.3. Two student volunteers, B. Prill and C. Larson, sifting shaft fill for skeletal elements and cultural artifacts.

consolidated sand/gravel to Lisan marl, the latter of which resulted from the cycle of annual deposits when the cemetery area was the lake bed of the much larger body of water, Lake Lisan, that preceded what is now the Dead Sea (Neev and Emery 1967; Donahue 2003).

In Chapter 1 we discussed the factors that led to silting of some tomb chambers, including possible flow of soil through fissures and holes in the chamber dome and silt flow through the entryway of the chamber when the shaft was still open. The amount of silt in a chamber varied from very little to filling virtually the entire chamber.

In most cases the silt was removed easily. However, in a few chambers the silt was almost like concrete and removal of chamber contents was difficult and time consuming, and damage to contents, particularly the fragile skeletal remains, was unavoidable. Ceiling fall does not seem to have been a major factor affecting the preservation of tomb contents. Even when ceiling fall was extensive, damage to skeletal elements was minor or nonexistent. In situations where ceiling fall occurred soon after contents were placed in the chamber, the bones would likely have been relatively strong and able to withstand the effect of marl falling from the dome. However, by the time of excavation the burials were very fragile because the protein component had degraded during the 5,000 years since the remains had been placed in the chamber. Ceiling fall on skeletal elements at this stage of preservation would likely have resulted in noticeable damage, but there was little evidence of this, so it seems likely that ceiling fall occurred early in the history of the tomb chambers.

Following the 1979 field season, the first author explored various options for avoiding shaft tomb chambers that had been heavily silted, since these required extra time to excavate and in some chambers resulted in serious damage to the chamber contents. In the 1970s geological survey equipment was being developed, and after consultations with various manufacturers and colleagues, the first author arranged to purchase the Geonics EM-31 conductivity meter manufactured by Geonics Ltd. in Ontario, Canada. This equipment had the potential to detect differences in soil conductivity as much as five meters below the ground surface. Since air does not conduct, chambers that were not silted would have a less conductive profile than those that were silted.

The archaeological application of electromagnetic surveying methods has previously been described and is widely used in different archaeological settings (McNeill 1979; Bevan 1983; American Geological Institute 1985; Frohlich and Ortner 1982; Frohlich et al. 1986; Frohlich and Lancaster 1986). Our use was initiated at Bâb edh-Dhrâ' in 1981 and proved very effective in identifying both the location of tombs and the presence of significant silting in the tomb chambers (Fig-

Figure 3.4. Second author and assistant recording EM-31 conductivity measurements during survey.

ure 3.4). The EM-31 consists of a single unit powered by eight C cell batteries and carried by one person. It operates by inducing an alternating electrical eddy current into the ground from the equipment's transmitter coil located at the end of a long tube connected to the meter box. This current creates a secondary magnetic field that is sensed at the equipment's receiver coil located at the end of another long tube attached to the opposing side of the meter. The signal recorded from the receiver coil measures the ground's cumulative conductivity between the surface and five meters down. The strength of the signal is read in millisiemens per meter (Frohlich and Lancaster 1986). Since the signal is a function of the ground's cumulative conductivity at the spot where the meter is read, different strata with different densities and conductivities cannot be identified in the normal mode of operation. However, applying two different modes of surveying, that is, by turning the equipment sideways, such levels can be differentiated, and the EM-31 is an effective tool in identifying multilayered soils and quantifying their electrical parameters (McNeill 1979; Bevan 1983; Frohlich et al. 1986; Frohlich and Lancaster 1986).

The specific response of the EM-31 to the presence of shaft tombs was not known prior to our field testing. We hypothesized that the equipment would differentiate between the shaft leading down to the chambers based on the probability that, relative to the undisturbed soil, the 5,000-year-old shaft fill would be softer and more porous and therefore contain more moisture, which would result in an increase of the field's electrical conductivity. This contrast with the soil surrounding the shaft is due to the greater density of the lake bottom deposits from several thousand years before the tombs were dug (Donahue 2003). The undisturbed, denser soil results in a detectable variation between disturbed and undisturbed soils (Frohlich and Ortner 1982; Frohlich and Lancaster 1986). In the shaft tomb area of the cemetery, a one to two square meter area with an isolated and significantly higher millisiemens/m value (high conductivity) is most likely a shaft.

The basic conductivity pattern of a shaft is relatively high conductivity surrounded by slightly lower conductivity associated with the less moist, higher density, undisturbed soil surrounding the shaft (Figure 3.5). The conductivity pattern of the shaft tomb chambers is significantly lower, but this varies with the unsilted chambers having low conductivity and silted chambers intermediate between the conductivity of the shaft and the undisturbed surrounding soil. Although this is the general conductivity pattern, there are complicating factors. One of these is that a combination of silted, partly silted, and unsilted chambers can result in confusing conductivity patterns (Figure 3.5; Frohlich and Ortner 1982; Frohlich et al. 1986; Frohlich and Lancaster 1986).

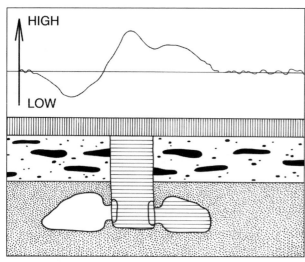

Figure 3.5. EM-31 plot of ms/m reading taken while passing over a shaft tomb.

a vertical shaft that was usually between 30 and 60 centimeters below the current ground surface. Furthermore, a shaft that had been opened in recent time, presumably by tomb robbers, had higher conductivity when compared to the conductivity of a shaft associated with undisturbed chambers. There was little point in excavating robbed tombs, and this difference in shaft conductivity made it possible to avoid such tombs (Frohlich and Lancaster 1986; Frohlich and Ortner 1982).

Our use of the EM-31 was highly successful. In all tested locations with high conductivity readings, we identified the vertical shaft leading to between one and four chambers. Variation in readings for conductivity surrounding the shaft made it possible to identify the presence of chambers and, before excavating, distinguish between those that had been silted and those that had not (Figure 3.6). However, this was not always certain, because of the relatively minor conductivity differences between silted, partly silted, and unsilted tombs and the surrounding soil layers (Frohlich and Ortner 1982; Frohlich et al. 1985; Frohlich and Lancaster 1986).

The assumptions we made about the conductivity patterns proved to be correct. In all cases, a concentrated high conductivity reading was associated with

Charnel House Excavation

Charnel house tombs vary in size and construction. Mudbricks are a major feature of all charnel houses. These houses may be circular, square, or rectangular, although the latter seems the most common design. Larger charnel houses may have had two levels (Lapp 1968). Particularly in smaller charnel houses such as Tomb A56, which dates to EB II and was excavated in 1977, mudbricks may have been the only material used in wall construction (Figure 3.7), aside from the stones used to line entryways and as blocking stones. In this type of charnel house each course of bricks was placed slightly overhanging the preceding course and formed a beehive-like wall structure.

The G1 Charnel House dated to EB IB and excavated in 1977 was also circular (see Figure 7.2). However, it was much larger than A56. The entryway was stone lined, and wooden cross-beams lay on top of the wall. These cross-beams were covered with reed matting and mud to form the roof.

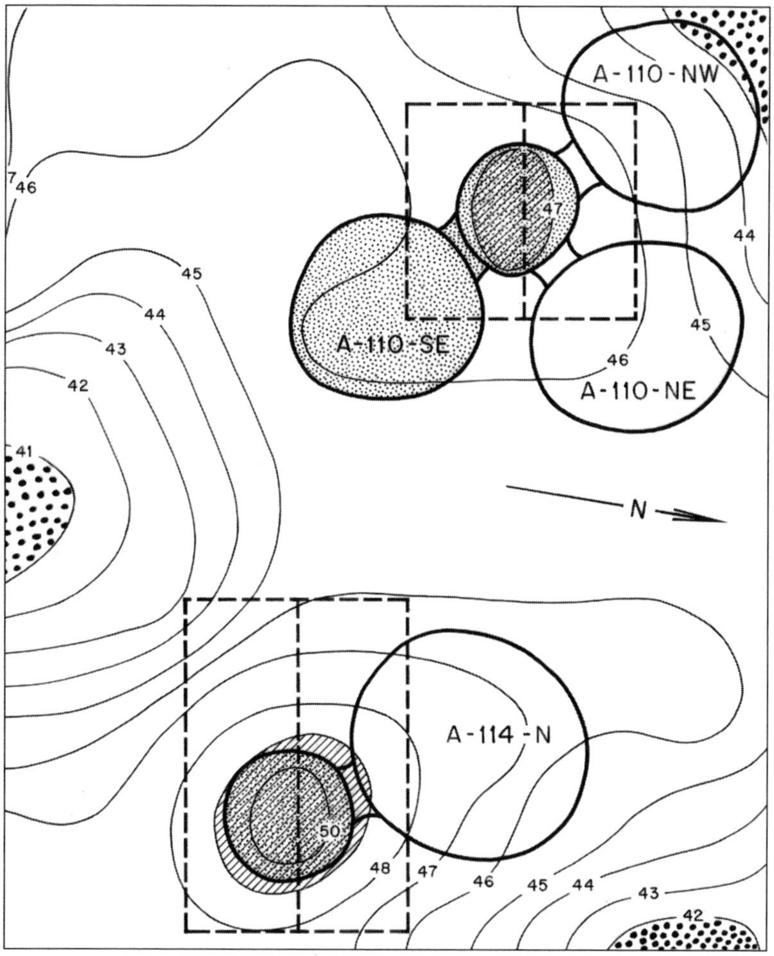

Figure 3.6. Conductivity plots for Tombs A110 and A114 indicating the locations of both the shafts, one silted tomb chamber (A110SE), and unsilted chambers.

Methods in the Recovery of and Research on the EB I People 17

Figure 3.7. Mudbrick charnel house construction for Tomb A56 dated to EB II.

Modern military construction had exposed the G1 Charnel House. It had been heavily damaged by destruction at the end of the EB IB period as well as recent robbing of some of the cultural artifacts. Modern testing by tomb robbers to locate cultural objects identified the location of some shaft tombs and EB charnel houses dated to EB II–III, including A55 excavated in 1977 and 1979 and the A22 Charnel House excavated during the 1979 field season. Unfortunately, there is a lucrative market for pottery and other cultural artifacts, and modern tomb robbers recovered some of these, which presumably were sold in the antiquities market. In some cases, the mudbrick walls of a charnel house such as A56 were identified during careful digging of test trenches.

The disruption of the original location of human remains and cultural materials in the G1 Charnel House caused by disturbances both in antiquity and in recent times limits the information that can be provided by systematic excavation of the remaining contents. For this reason, excavation was a salvage endeavor, and field methods involved careful exposure and removal of skeletal elements and cultural materials for subsequent analysis in the laboratory. Limited attention was paid to the location of contents within the charnel house. Standard systematic archaeological excavation methods were applied to the excavation of all other charnel houses. These included careful recording of the location of both cultural and human skeletal elements.

FIELD CONSERVATION METHODS

In some shaft tomb chambers there was no silting or roof fall. All chambers and their contents, including architectural features, human remains, and cultural features, were recorded carefully with top plan drawings and photography. Where appropriate, we also collected soil samples for subsequent testing and analysis. If only ceiling fall occurred in the chamber, damage to contents was usually minimal and careful removal of the roof fall, usually various sized pieces of Lisan marl, permitted routine recording and removal of chamber contents. If silting had occurred, the silt was carefully removed to completely expose both skeletal material and cultural objects that had been placed in the chamber. As in unsilted chambers, the location of all chamber contents was carefully recorded and photographed before the contents were

Figure 3.8. Second author applying PVA preservative in situ.

removed for registration and/or conservation. If the silt was hard, removal of contents was challenging and undesirable damage to the human remains was often unavoidable. In this situation, it was necessary to record the location of the chamber contents and remove them as they were identified and exposed. This posed serious challenges in recovering the contents. Nevertheless, we were remarkably successful in recovering the cultural materials and enough of the human remains to provide basic demographic data about most of the individuals interred in these chambers.

Regardless of the condition of the chamber, all human remains from the shaft tombs were extremely fragile, and it was necessary to treat the bones with preservative to have any chance of keeping the remains in good condition during shipping and subsequent analysis. At the time of excavation, the preferred preservative for bone was polyvinyl acetate (PVA), and we used a solution of 10% PVA-AYAT, a somewhat harder variant of PVA, in acetone or alcohol (ethanol) to treat the bones. In situations where the human remains were especially fragile, we applied PVA while the remains were in situ (Figure 3.8). Alcohol solvent was less of an irritant for the people involved in recovering chamber contents when these were conserved in situ.

In both 1977 and 1979 most of the bones were immersed in the PVA solution. However, during that time we discovered that some deformation occurred when complete immersion was used. This was most noticeable in mandibles; the postcranial bones were not at all or only slightly affected. In the 1981 field season, using a brush, PVA was applied routinely in situ to all skulls. Additional conservation with immersion in the PVA solution was done for particularly fragile bones. We found that initial in situ treatment was beneficial in preserving extremely fragile skeletal elements and some wood artifacts. It had the further advantage of securing teeth to the mandible and maxilla during subsequent processing. Brushing PVA solution in situ minimized distortion of bone but required several time-consuming applications of PVA to stabilize and preserve the bones.

The use of PVA to conserve the bones precludes the use of bone samples for dating the burials by carbon-14 dating. This was a troublesome result, although some bone samples were collected for dating purposes prior to the application of the PVA solution. Subsequent research revealed that virtually all the collagen had been destroyed, so this component would not have been available for dating. Carbon is present in the mineral phase of bone. However, the mineral phase of the bone undergoes recrystallization in the burial environment. During recrystallization, elements from the soil often substitute for elements originally part of the living tissue. Because of these substitutions, one cannot assume that the limited amount of carbon associated with the mineral component of bone represents the original tissue. This means that radiometric dating depending on carbon in bone mineral may not reflect the true age of the bone sample.

The burial environment of the Bâb edh-Dhrâʿ cemetery contains substantial amounts of sodium chloride. In the time since burial this salt has infiltrated into the bones and cultural artifacts. In material excavated in 1977 and 1979 we did not fully appreciate the potential this salt had to recrystallize and damage the chamber contents. Also the human remains were so fragile that we were reluctant to immerse the bone in water to remove the salt. With burials in the G1 Charnel House that had been burnt in antiquity, the stability of bone was much better and soaking in water was possible; this was not done until the remains were in the laboratory.

This procedure was changed during the 1981 field season, when most skeletal elements and cultural artifacts were treated with three to five changes of water to remove most of the salt. We took this risk to the human remains because exposure to a different environment after the removal from the chambers could result in damage to both skeletal and cultural materials as salt recrystallized and caused spalling of the outer layer of both bones and pottery. In some cases, the fragile condition of objects precluded treatment for salt removal. For example, water was not applied to wooden artifacts. Such objects were repeatedly treated with an alcohol-based solution of PVA while still in their original positions in the chambers. When applicable and if conditions allowed, further removal of salt for some tomb contents was done later in the laboratory, where the procedure could be done more carefully and effectively.

LABORATORY CONSERVATION METHODS

Once the human remains arrived at the Smithsonian laboratory, each burial was evaluated for any needed additional conservation. In some cases, we used additional immersion or brush application of PVA. Whenever possible, breaks in the bone, whether they occurred in antiquity, during excavation, or subsequently, were carefully glued in correct anatomical location and position using a thicker concentration of PVA in acetone. This repair was needed to obtain metric osteological data but also made evaluation of the association between cranial and postcranial skeletal elements easier and more accurate. It also enhanced evaluation of any evidence of skeletal abnormalities.

All the burials in the shaft tombs were secondary, and associating the postcranial skeletal elements with a specific skull was not possible in all cases. Fortunately, the average number of burials in a tomb chamber was slightly more than nine individuals. Because of the age, size, and sex differences of individuals interred in the chambers we were, in most cases, able to make reasonable linkages between cranial and postcranial skeletal elements. However, in situations where there were two individuals of similar age, size, and sex, associations of cranial and postcranial elements are less certain. Fortunately, this was an uncommon occurrence.

The human remains in the G1 Charnel House had been burned by vandals at the time the structure was destroyed in antiquity. In some cases, the bones had shrunk and been badly deformed by the heat (Figure 3.9). Generally, the bones, although fragmented, were better preserved than bones from the shaft tombs, presumably because of the stabilizing effect of high heat. As noted above, high concentration of salt in the burial environment resulted in high concentrations of salt on the bone fragments, and this tended to recrystallize on the bone surface in the more humid laboratory environment. Because the G1 bone was in better condition than the shaft tomb burials, it was possible to soak the bone in water to remove most of the salt.

The human remains from the G1 Charnel House were badly disturbed both in antiquity and in more recent times. However, as explained in Chapter 7, it seems likely that at least some of the burials were secondary as well. Unlike in the shaft tombs, many more individuals were interred in the charnel houses, and association of skeletal elements was not possible in most cases. We were able to reconstruct partially a few skulls from G1. Very few long bones from the same burial were found together during excavation.

ANALYTICAL LABORATORY METHODS

Basic to any bioarchaeological analysis of human burials is the objective to reconstruct the biological qualities of the living people represented by the skeletons. One goal of our research is to compare our Bâb edh-Dhrâʿ EB I sample with other archaeological skeletal

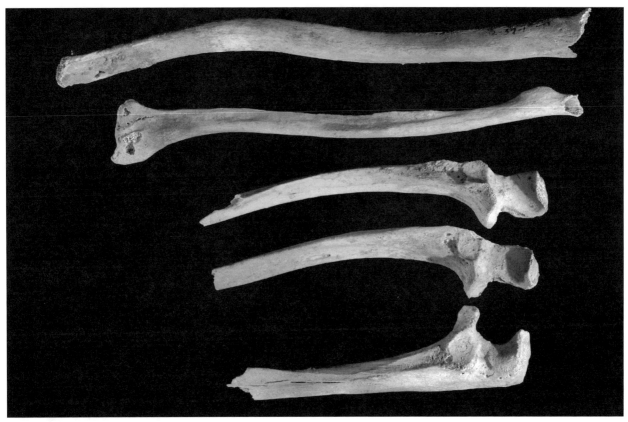

Figure 3.9. Heat-deformed long bones from the G1 Charnel House.

samples from potentially related regions. The objective of this comparison is to clarify possible biological and cultural relationships with contemporary populations represented by these samples. In doing this, we need to determine or assume that other skeletal samples are as representative of the populations from which they are taken as the sample from Bâb edh-Dhrâʿ is of the population it represents. Making inferences about many characteristics of a former living human population on the basis of skeletal data will always be a challenging exercise subject to the inherent limitations of the data. Nevertheless, the extensive research that has been done on human skeletal samples where much information is known about the people represented provides a baseline for interpreting skeletal data in human skeletal samples where this information is not known.

Three basic types of data are available to attempt this reconstruction: (1) metric data, (2) nonmetric observations and dental analysis, and (3) skeletal evidence of various disorders, including those that are most commonly expressed in the skeleton (trauma, arthritis, and infection (Ortner 2003). The metric data set was chosen on the basis of several years of experience in osteological research and long before the more recent efforts to standardize data protocols for human osteological research (Buikstra and Ubelaker 1994). However, there is considerable overlap between our data set and the variables included in the recommendations made by Buikstra and Ubelaker.

There is variation in the landmarks used to take some metric measurements that can create problems for comparing data gathered by different observers. There are additional problems in interobserver variability in recording observations of nonmetric variables. Our metric measurement methods are congruent with the definitions of landmarks and variables defined by Rudolph Martin (1928), Ales Hrdlička (Stewart 1952), William Howells (1973), and William Bass (1995).

Both authors measured all the human remains, and these measurements were compared statistically; with the exception of the interorbital width measurement, there were no significant differences between measurements made by the two observers. However, if the measurement difference exceeded more than one millimeter, the measurement was checked and corrected.

Evidence of skeletal disorder was evaluated by the first author with the assistance of three collaborators, Evan Garofalo, Liese Meier, and Molly Zuckerman.

Garofalo had completed a M.Sc. degree in skeletal paleopathology at the University of Bradford in the United Kingdom. Meier and Zuckerman both had experience in conducting research on human skeletal remains and received additional instruction from the first author before and during the inventory of the burials. All three observers demonstrated an ability to identify abnormal skeletal elements, even when these elements were broken and/or small. All evidence of skeletal disorder was evaluated by the first author regardless of who made the initial observation.

Fundamental to our research objectives was the need to determine that our skeletal sample is representative of the original living population to the extent that this is possible with an archaeological skeletal sample. The age at death and sex distribution in the Bâb edh-Dhrâʿ skeletal sample is what is expected in a skeletal sample that is representative of all who died in a living population. A limiting factor in our research is that the burials are secondary, making the association of the skull and postcranial bones challenging. In addition, some skeletal elements were not transferred from the primary site to the tomb chambers or charnel houses. This is further complicated because of breakage to some skeletal elements in which part of the element may be missing or sufficiently damaged so that measurement or observation of a variable is not possible. The result of these complications is that data sets for some burials are incomplete. Fortunately, in most cases the bones that were transferred to the shaft tombs were the skull and the major long bones, which provide the majority of our measurements and observations.

Our statistical analysis was conducted only on the metric data. We used univariate analysis to provide a basic and descriptive understanding of variables, including cranial measurements, in both male and female skeletons and various measures of postcranial size and proportion. We used multivariate analysis to compare the Bâb edh-Dhrâʿ sample with other skeletal samples to examine the biological relationship between groups.

We used SAS and Systat statistical computer software to analyze our data. The use of two independent statistical programs provides a useful strategy to ensure the integrity of the basic metric data. Analysis with each statistical package resulted in no discrepancies and provided added confidence regarding the quality of our data and the procedures used in analysis. Descriptive statistics provided tests for normality. This information guided us in determining whether parametric or nonparametric statistical tests were appropriate for a specific variable or combination of variables. If the distribution of a specific variable was normally distributed, we used parametric statistics for our group comparisons; if the distribution of data was not normal, we used nonparametric statistics. The test for normality also became an important tool in checking our data for recording mistakes and data entry mistakes.

Comparison of samples using multivariate statistics makes some assumptions about the data that may not be true. Although we are very aware of the strengths and weaknesses of the Bâb edh-Dhrâʿ skeletal sample, we rarely have comparable knowledge of data from skeletal samples studied or published by other researchers. For example, we usually do not know what the biases are relative to what was originally excavated and preserved for analysis. Is the skeletal sample representative of those who died? Were all skeletal remains recovered during excavation? Were all the burials from a sample retained by the institution that currently curates the skeletal sample?

A common problem is that we often don't have a complete data set for every burial in the sample. This limits the variables that could be used in the analysis to those that were most commonly present in the data set. In some situations we did use "missing data replacement" procedures to address this problem. To some degree, the selection of variables to be included in multivariate procedures became a function of the prevalence of missing data.

This created some limitations in the analyses that could be done. For example, in a large data set with many variables and no missing data we would conduct a pilot study using stepwise discriminatory analysis to determine which variables would be most responsible for between-group variation. These variables could then be selected for multivariate statistical analysis of all the skeletal samples and, at the same time, variables could be excluded that had little or no influence. However, in choosing variables for multivariate analysis we also had to consider the missing data points in the Bâb edh-Dhrâʿ sample, and the ideal from the standpoint of those with the least missing data variables in this sample were not necessarily the same as the best variables available in other skeletal samples.

The possibility exists that the variables we were able to use may not be the best discriminators in identifying multivariate differences between various skeletal samples and the populations they represent. Although in Chapter 10 we use multivariate statistical methods in comparing skeletal samples, the reader should exercise caution when evaluating our tentative conclusions.

Through the courtesy of the Jordanian Department of Antiquities, the skeletal sample from the EB I Bâb edh-Dhrâ' excavations is curated in the Department of Anthropology of the National Museum of Natural History, Smithsonian Institution, United States. Currently the human remains from the EB II–III charnel houses are being studied and curated in the Department of Anthropology, University of Notre Dame, South Bend, Indiana. Once this sample is analyzed it will be curated with the EB I sample at the Smithsonian Institution. All accessioned collections at the National Museum of Natural History are available for research by students and scientists.

CHAPTER 4
Cultural Artifacts of the EB I Tombs

R. THOMAS SCHAUB

THIS CHAPTER PROVIDES A BASIC interpretive summary of the cultural artifacts from the various areas of the cemetery in their archaeological and historical context. To the extent possible, study of the cultural artifacts is based on the stratigraphy. Since the tomb pottery has been extensively presented in previous publications, we include here only a sampling of the most widely used forms. Other cultural artifacts, such as stone jars, mace heads, figurines, and jewelry items, are briefly discussed.

In both quantity and quality the ceramic assemblage from the EB I tombs of the Bâb edh-Dhrâ' cemetery is unsurpassed for the southern Levant during this period. More than 2,500 vessels have been scientifically excavated by the Lapp expeditions in 1965–1967 and by the Expedition to the Dead Sea Plain (EDSP) from 1975 to 1981. The majority of the pots excavated are fine ware vessels with thin walls, carefully trimmed, elegant in form, and red-slipped and burnished. Together with other artifacts from the tombs, including basalt bowls, mace heads, clay figurines, and jewelry items, the pottery placed in the chambers offers an eloquent witness to the concern of the inhabitants for their dead.

An interpretive study of these cultural artifacts must consider several factors. Topographic location within the cemetery needs to be examined as the tomb types and cultural artifacts come from different areas. The A and C areas, for example, show marked differences in tomb architecture and grave goods. Within the different cemetery areas, a critical element of major chronological import is the existence of stratigraphical relationships between tomb chambers or types. Typological features of ceramic groups, based on the stratigraphical sequences, can offer further insights into the relationship between tomb groups and artifacts. Further analysis based on broad morphological features such as shape and especially size can be helpful in interpreting the function of vessels and for cross-cultural comparisons. Finally, comparative material from the local and regional contexts is important to relate the material to the recognized cultural periods of the area.

Each of these factors has been treated at great length in the previous publications of the EDSP. In the final publication of the tombs excavated by Paul W. Lapp in 1965–1967 (Schaub and Rast 1989), the basic stratigraphy and phasing of the tombs in the cemetery were treated in Part I. The tombs of EB I were described in detail in Part II by periods (EB IA and EB IB) and by cemetery area (A and C). Separate chapters included in-depth analysis of the pottery and tomb objects with abundant illustrations. Preliminary publication of the 1977–1979 excavations of the cemetery by the EDSP and the Smithsonian (Rast and Schaub 1980, 1981) included extensive treatments of the burial patterns (Schaub 1981b) and ceramic sequences of the EB I tombs groups illustrated with abundant figures and descriptions (Schaub 1981a). Publication of the well-preserved tombs excavated in 1981 (Frohlich and Ortner 1982) included tables of the cultural artifacts.

TERMINOLOGY

In harmony with the earlier publications of the EDSP, we continue to use the terms EB IA and EB IB to denote the earlier and later uses of the EB I period at Bâb edh-Dhrâ'. Recent studies on this period have expressed concern about this terminology. Two concerns especially have been raised. First, new C14 dates have indicated that the time period between the Late Chalcolithic and EB II is much longer than previously recognized. The longer time period appears to call for a more complex division than EB IA and IB. Second, it is quite clear that there are major regional differences in the various sites of the EB I period, and it is not a simple matter to find consistent parallels in the regional assemblages. Because of these two factors new schemes for dividing the period have emerged, but no consensus has been reached. Some commentators now prefer the generic division of Early EB I and Late EB I (Braun 2001; Philip 2001:174). At Bâb edh-Dhrâ', because of earlier publications and the local regional factors, we still prefer the terms EB IA and EB IB to denote the uses of the cemetery.

TOMB LOCATIONS AND TYPES

Paul Lapp excavated EB I tombs in two separate areas of the Bâb edh-Dhrâʻ cemetery during the 1965–1967 field seasons. The A Cemetery, located to the southwest of the town site at a lower elevation, was the largest and most extensively explored. In this area Lapp excavated 28 shaft tombs with 48 chambers associated with the EB IA period. Three tombs—a surface burial, a shaft tomb, and a round charnel house—based on stratigraphic relationships and distinctive ceramic types, were assigned to the EB IB period. A second area, called the C Cemetery, was located on the western edge of the limestone bluffs. Here Lapp discovered that the shafts for these tombs had been cut into the slopes diagonally and yielded distinct ceramic groups from the A cemetery. All six of the "shaft" tombs excavated by Lapp from the early period in this area were single-chamber tombs. Their positioning on slopes did not lend itself to obvious shafts. The openings to the chambers were often closed off with stone walls. Based on the relationship of one of the "shaft" tombs to an EB II–III burial house and ceramic parallels, these tombs were assigned to the EB IA period.

The EDSP excavated additional EB I tombs in the A and C cemeteries as well as in a new area, the G Cemetery. In the A Cemetery, 22 shafts with 63 chambers of the EB IA period contained skeletal and artifactual material. Five chambers in A112(4) and A111S were empty of artifacts and probably had been robbed. Three tombs in the A Cemetery—a surface chamber, a shaft tomb with a single chamber, and a chamber off the same shaft with three EB IA chambers—contained EB IB artifacts. In the G Cemetery, located between A and C, three EB IA chambers were excavated along with a round burial house and a single-chamber tomb of EB IB. In the C Cemetery, the EDSP explored four EB IA tombs—three chambers and a surface burial. Two of the chambers (C10E, C10W) located on the crest of a hill were cut off of a deep shaft, similar to those found in the A Cemetery. At a third chamber, located nearby C9, the shaft was not located and the roof of the tomb chamber had collapsed.

STRATIGRAPHICAL RELATIONSHIP OF TOMBS

The distinction between early (EB IA) and later tombs (EB IB) is based on the discovery of a number of superimposed tombs with different ceramic groups. Lapp uncovered Tomb A13, a surface chamber with spouted bowls and line-group painting, directly over the shaft of Tomb A7, which included two chambers with the distinctive pottery associated with the majority of the shaft tombs. A similar superimposition was found by the EDSP with Tomb A104, a surface burial with wares and types associated with the line-group pottery, directly over the shaft of Tomb A105, a four-chamber tomb with the typical EB IA pottery. Another outstanding example is in the relationship of Tomb G5 and Tomb G1. The latter is a round burial building excavated in 1977 with dominant EB IB pottery types. The floor of this house was cleared and a trench cut down the middle to make certain that the floor had been reached. In the off season, a staff member, David McCreery, visited the site after a heavy rainstorm and discovered an opening in the trench floor that led to the shaft and two-chamber tomb of G5 with distinctive EB IA pottery types (Schaub 1981a: Fig. 11). The shaft of Tomb G5 had been cut and its two chambers built before the floor and walls of the burial house G1 had been built. Another example is Lapp's Tomb A43, which contained a large group of line-group pottery types (Schaub and Rast 1989:205–208). The shaft of A43 reused an earlier shaft, of EB IA Tomb A45, and was also stratigraphically beneath an EB II rectangular burial house, A42. The consistent sequential relationship of the line-group pottery types occurring in chambers stratigraphically above the tombs with dominant early EB IA types is utilized below in developing the ceramic types.

Other common features of the EB IB tombs should be noted. The two round charnel houses, A53 and G1, both have formal stone-lined entryways with standing monolith for jambs, a threshold stone, and a large slab blocking the entryway. This is also a common feature of all of the later EB II–III charnel houses. It is also a feature associated with three of the four earlier EB IB shaft tomb chambers, A43, A88L, and G2. The fourth chamber with EB IB pottery, A100N, is distinctive in having a slab-lined floor along with a series of standing stones partially along the chamber wall. These fairly large flat stones lining the doorway appear in only one (A6N) of the 111 excavated EB IA chambers in the A Cemetery. The six EB IB chamber tombs, including the two surface burials, are all single chambers, in contrast to the more normal pattern of multiple chambers in the EB IA shaft tombs.

The differences in tomb architecture noted above could be used to argue for a major cultural shift in burial patterns. Single-chamber tombs with stone-lined entryways giving way to round burial houses certainly indicate that major changes are taking place. The surface chambers of A13 and A104 over earlier shafts suggest that the EB IB groups may not have been aware of the placement of the earlier tombs.

The same may be said of A43 and A88L, which cut into the shafts of earlier tombs. On the other hand these circumstances may be an indication of how concentrated the tombs had become in the A Cemetery. Wherever one tried to place a new tomb, older tombs would be encountered. This in turn could have led to the use of single-chamber tombs and eventually the move to above-ground burial houses. A more gradual shift seems to be better supported by the comparative study of the ceramics from EB IA and EB IB and the stratigraphy of the nearby town site.

STRATIGRAPHY IN CONTEMPORARY OCCUPATIONAL AREAS

A similar sequential relationship between EB IA pottery forms and EB IB line-group family forms occurs in the Stratum V–IV sequences excavated in and around the town site of Bâb edh-Dhrâʿ (Schaub and Rast 2000:73–90; Rast and Schaub 2003:62–155). Three areas associated with Stratum V were excavated by the EDSP. Area F, slightly to the west of the later walled town site, contained several tombs with typical EB IA pottery: F1, F2, and F4 (A, B, and C). The three tombs of F4 were stratigraphically beneath mudbrick debris and a thick layer of ash containing Stratum IV material, including line-group ware forms. Area H, on the northern ridge of the A Cemetery, was excavated during the 1977 season with several trenches. H1 and H3 yielded evidence for campsite occupation, with midden deposits containing sequential loci of animal bone, wood and ash, pebbles, and a great deal of pottery. The pottery assemblage (Rast and Schaub 2003:74–101) contains typical EB IA tomb forms along with domestic forms of hole-mouth storage jars and cooking pots. The third area, J, located approximately 650 meters southwest of the town site and 30 meters below the northern ridge of the cemetery, yielded two superimposed structures. Both structures had lower walls constructed of small to medium-size boulders. The lower structure was associated with Stratum V material. It may represent the first signs of permanent location at the site. The upper structure is clearly identified with Stratum IV artifacts. The stratigraphic evidence from the occupational areas north of the cemetery clearly confirm and demonstrate the sequential relationship of the early EB I cultural material (EB IA) and the later EB I cultural material (EB IB).

CULTURAL ARTIFACTS FROM THE EB I TOMBS

Artifacts deposited in the EB I tombs included pottery, basalt bowls, mace heads, unbaked clay figurines, jewelry, bone and wood objects, and textile remnants. Clay pots were by far the most frequently found artifact in the chambers. The 48 EB IA chambers excavated in 1965–1967 in the A Cemetery yielded 1,132 vessels, an average of 23.5 per chamber. A closely similar average, 23.6 pots per chamber, was associated with the 1,322 vessels from the 56 chambers with cultural artifacts that were excavated in the A Cemetery in 1977–1981 (see Table 4.1). The C Cemetery, with 4 chambers and some surface finds, had 94 pots (Table 4.2). Lapp's C Cemetery had 6 tombs with 50 pots. The G Cemetery had three EB IA tombs with 29 pots (Table 4.3). The five EB IB tombs from the A and G cemeteries yielded 224 pots (Table 4.4). Three EB IB tombs in the Lapp excavations had 123 vessels. Other artifacts did not occur with the same regularity. Basalt bowls were placed in 23 of the 54 Lapp EB IA excavated tombs, and 27 of 64 of the EDSP chambers. Mace heads were even less common (11 of 54 and 16 of 64). Jewelry items, including beads and shell bracelets, turned up in smaller overall proportion (13 of 54 and 7 of 64). Unbaked clay figurines were found in only three of the chambers excavated by Lapp and in five of the EDSP chambers. Remnants of perishable items, including straw mats, textiles, and wood objects, survived in only a few chambers, but impressions of mats discerned in silted layers underneath many of the disarticulated bone piles suggest that perishable straw mats were probably originally in most of the tombs, if not all.

The Pottery

The large quantity of complete pottery forms from the EB IA shaft tombs of the cemetery has presented an unusual opportunity to develop new methodological and statistical approaches to examine the potter's craft of this period. Two professional potters examined the construction techniques of the forms and helped to develop methods for recording variations in the techniques. A computer program was devised to record all of the size and formation techniques of the individual pots along with profiles for statistical analysis. Petrographic studies of thin sections of the typical wares were also completed by the staff geologist to assist in determining basic ware fabrics. All of these data were used in developing several different classification systems (basic form, specific type, functional types, and ware families) to enhance our understanding of the EB IA potters of Bâb edh-Dhrâʿ.

BASIC FORMS. The purpose of a basic form classification system is to stress the function of the vessels by closely relating overall form attributes (closed,

Table 4.1. EBIA Cultural Material of A Cemetery Excavated by the EDSP

Tomb	Pots	Basalt Bowls	Ceramic Imitations	Mace Heads	Figurines	Wood	Jewelry	Bone Objects
A78NE	28						1	
A78SE	29			1				
A78NW	21					1		
A78SW	21			1	3			
A79N	13	1						
A79E	27							
A79S	14	1					1	
A79W	41	2						
A80N	27							
A80E	6							
A80S	21							1
A80W	53							
A86NE	20		1					
A86SE	7		1					
A86SW	17		2					
A89NE	32	1						
A89NW	39			1			78	
A89SE	46							
A91	30	2		1				
A92	22	1		1			2	1
A100E	48				2	1		
A100S	8							
A100W	23							
A101S								
A102E	26		1					
A102S	27	1	1					
A102W	17		1					
A102NE	12		1					
A103	22							
A105SE	25						1	
A105NE	11							
A105NW	37							
A105SW	10							
A106	29						1	
A107N	35							
A107E	52							
A107S	45							
A107W	48	2						
A108N	6							
A108NE	30							
A108NW	25							
A108SE	22							
A108SW	24							
A109N	22							
A109S	12							
A109E	6							
A110NE	24	1			1			
A110NW	22	1			2	1		1
A110SE	30	2		2			67	
A111E	15	1			2			
A111W	14	1		1		1		1
A111N	49	1						
A113	3							
A114N	23	1		1		10		
A120S	5			1				
A121N	1			1				
Totals	1,322	19	8	11	10	14	151	4

Table 4.2. EB IA Cultural Material of C Cemetery Excavated by the EDSP

Tomb	Pots	Basalt Bowls	Ceramic Imitations	Mace Heads	Figurines	Wood	Jewelry	Bone Objects
C9	19						2	
C10W	18							1
C10E	45	1		2			2	2
C11	8	1		1				
C Surface	4			2				
Totals	94	2		5			4	3

Table 4.3. EB IA Cultural Material of G Cemetery Excavated by the EDSP

Tomb	Pots	Basalt Bowls	Ceramic Imitations	Mace Heads	Figurines	Wood
G4	15					
G5E	7	1				
G5W	7	1				
Totals	29	2				

Table 4.4. EB IB Cultural Material Excavated by the EDSP

Tomb	Pots	Basalt Bowls	Ceramic Imitations	Mace Heads	Figurines	Bone Objects
ABBL	89	2		3		5
A100N	34	1				
A104	31					
G1	40					
G2	30				2	
Totals	224	3		3	2	5

necked, and open) with size dimensions, especially volume. Abstracting from site-specific attributes such as rim and handle formations at this stage also allows for better cross-cultural comparison based on simple morphological features such as shape and size. To determine the forms, each pot excavated by the EDSP was drawn at 1:1 and then reduced to 1:4 for scanning and arrangement on plates. A computer program was devised to trace the drawings while recording the profile image along with the variables for 16 attributes. The size attributes recorded included maximum width, mouth width, height, neck height for jars, wall thickness, base thickness, and ratios of these attributes with one another. The program also computed the volume for each vessel. Other attributes recorded in the database included types of neck joins and base type. The basic forms were determined by statistically ranging pots with similar size attributes, including volume and ratios, into basic form groups. For example, medium-large jars more than 21 centimeters in height with a volume of over 1,800 cc comprised one large group. Subtypes within this range were distinguished by tall or short necks and mid to low tangent of the profile. Vessels from the C Cemetery tombs, with their distinctive carinated profile and different volume ranges, stood apart from the A Cemetery forms. The dominant basic forms of the EB IA groups are illustrated in Figure 4.1. EB IB basic forms are illustrated in Figure 4.2. The volume ranges make it possible to suggest the intended function of these vessels (Schaub and Rast 1989:248). The ranges also suggest that the potters were well aware of controlling the size of different groups. In the EB IA bowls, for example, each size range from small (S) to medium-small (MS) to medium (M) to medium-large (ML) to large (L) tends to be either quadruple or double the size of the closest smaller range.

CONSTRUCTION TECHNIQUES. The vessels are all handmade. Finger grooves on the interior and exterior walls of the large and medium bowls demonstrate that the bowls were coil-formed built up and smoothed from flat pancake bases. Two techniques were involved in forming the necks of jars. In the fine ware vessels the neck was formed separately and then joined to the body, forming a sharp turning point. Necks tended to be tall and cylindrical. In the plain ware vessels and in the EB IB groups, the more common technique is to draw the neck up from the body of the vessel, resulting in a fairly smooth curve. Necks are generally shorter and more flaring. Small bowls in the fine ware EB IA

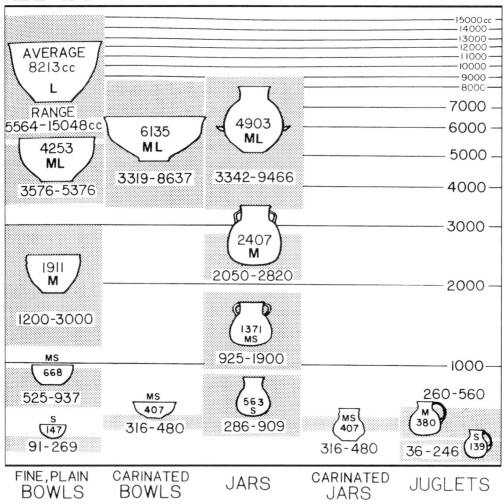

Figure 4.1. Volume ranges and averages of EB IA basic forms.

groups may have ring or disk bases. The EB IB small bowls often have a pushed-up (omphalos) center in the base. Handles are attached in a different manner in the various ware groups. The fine ware vessels normally have the handle attached at or slightly below the rim. In some plain wares and especially in the EB IB groups the handles are wider and flatter and attached from above.

The most striking feature, especially to professional potters who have examined the pottery, is the extraordinary care that was used in forming and finishing the fine ware vessels. Excess clay was carefully trimmed, forming thin walls and elegant profiles at the base and rims. The decorative feature of raised bands on the circumference of both bowls and jars was often carried out with precision using a tool to form regular and even delicate patterns in the bands. One potter estimated the time involved in constructing and finishing a group of 20 pots to be 24 hours (Rast and Schaub 1981:79).

WARE FAMILIES. In addition to basic forms and construction techniques, further analysis of these data compared features of the pots such as fabrics and surface decoration to determine the major ware families. Wall thickness emerged as an important attribute from discriminate analysis of the database. It also was noted above in the professional potters' examination of the vessels as an important clue to finishing techniques. Thin section analysis of the fabrics noted a distinction between coarse and fine wadi sand used as temper in the vessels. Surface decoration features, including the use of slip and burnishing and incised or raised decoration, added a further level of data to consider. Analysis of these data led to the deter-

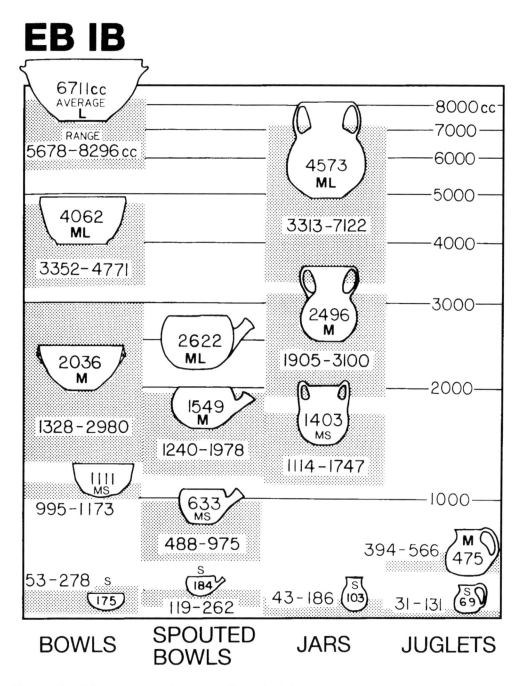

Figure 4.2. Volume ranges and averages of EB IB basic forms.

mination of two basic ware families of the EB IA A Cemetery assemblage: fine and plain. The fine wares, as described above, were produced with excess clay trimmed and had thin walls. Rims were elegantly formed. In surface finish, they were usually slipped and burnished. Surface decoration, if used, was often carefully done, producing regular lines of punctuate or more often raised bands, which were artistically slashed. Plain wares had thicker walls without the signs of finishing techniques of the fine wares. Fabrics of the plain ware used a coarser wadi sand. In form they were often asymmetrical, with slumping rims or walls. Most of the plain wares did not use slip or burnish and the decorative techniques, mostly punctate, were uneven in their application. The dominant features of the fine and plain ware families are summarized and compared to the carinated and line-group ware families in Table 4.5.

Table 4.5. Distinctive Features of Ware Families

	Fine	Plain	LGW	Carinated
Basic forms construction finishing	All vessels trimmed to thin walls	Thick walls and base	Medium thick walls	
	Deep, lightly closed bowls	Deep, open bowls	Medium bowls, lightly closed	Medium shallow open bowls, carinated
	Corner points on necked vessels	Inflected points on necked vessels	Narrow necks, spouts	very wide, tall necks, flat bases on jars and juglets
Most frequent forms	Large bowls	Medium bowls	Large bowls and spouted vessels	Large and medium carinated bowls
Surface treatment	Slip and/or burnish common	Slip uncommon, burnish rare	Slip fairly common, burnish infrequent, line-group design on some vessels	All slipped and burnished
Relief features	Frequent, Raised bands and regular punctate	Fairly common—irregular punctate	Some irregular punctate, many incised designs	Frequent added bands at carination
Fabrics, temper	Fine, fine wadi sand	Medium coarse, coarse wadi sand	Medium coarse, Z temper	Coarse wadi sand
Specific diagnostic features	Everted bowl rims, handles below rim, ring bases, thin ledge handles, small pierced lugs	Simple, direct rim bowls, handles at rim, thick ledge handle, large knobs	Simple. Direct rims handles above rim, omphalos bases, duckbill ledge handles, large knobs, column handles, ear handles	Everted bowl rims, small loop handles, no ledge handles, added dots

The pottery from the C Cemetery stands apart as a separate ware family. In basic form, the bowls from the C Cemetery tombs are usually carinated in profile with an angle in the wall at about the midpoint of the vessel. Jars also may have a carination. In ware fabric, the vessels from the C tombs are also distinctive. The temper tends to be coarser than that of the fine ware A Cemetery vessels and the fabric is generally tan in color rather than the orange-red of the A Cemetery vessels. In surface treatment all of the C Cemetery vessels have a red burnished slip and frequently have short, raised bands on the surface at the carination angle.

The vessels of the line-group ware family of the EB IB tombs have different fabrics and distinctive forms and surface treatment. Inclusions in this pottery consist of rounded limestone particles with varying quantities of subangular to subrounded quartz, occasional smaller micritic limestone, shell fragments, and tabular red grains, possibly grog. The fabrics, especially in the smaller vessels, have a bright orange color and are smooth and powdery. They are described in the town site volume as orange chalky wares (Rast and Schaub 2003:135). Distinctive forms, long associated with later EB I pottery in the southern Levant, include a range of spouted bowls, column handle jars, omphalos bases on small bowls, and narrow-necked amphoriskoi. The term *line-group ware* (LGW) is derived from the distinctive painted style that appears on some of these vessels.

SPECIFIC TYPES. Specific types were determined by the regular association of morphological features of handles, rims, and bases among the basic form groups and ware families. Specific types of jars included ledge-handled, loop-handled, and those without handles. Subtypes incorporated size and ware categories into groupings such as ledge-handled jars that were medium-size to large and made of fine ware. Rim formation was a key factor in the bowl classifications. Everted rims comprised a separate specific form category distinguished from direct rim bowls. Again, different size ranges and ware fabrics were used for subtypes. Everted rim bowls, lightly closed in profile, large, and made of fine ware, were distinguished as a type from direct rim bowls, which were neutral to open in form, large, and made of plain ware. The dominant types in the EB I ware families (fine, plain, carinated, and line-group ware) are illustrated in Figures 4.3 and 4.4).

TESTING THE DATA. One of the primary goals of adopting statistical approaches to the classification of the Bâb edh-Dhrâʿ cemetery pottery was to confront the problem of subjectivity in ceramic typologies. With the focus on size and ratios, we hoped to determine

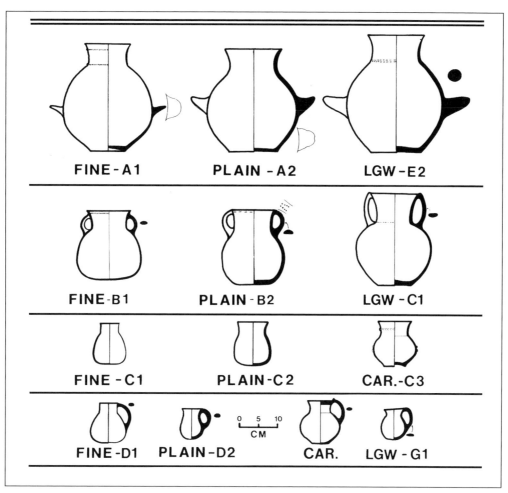

Figure 4.3. Dominant types in EB I ware families—necked vessels.

consistent, nonrandom basic form groups that would reveal the behavioral patterns of the EB I potters. The methodology was first applied to the large group of pottery excavated by Lapp in the 1965–1967 excavations (Schaub and Rast 1989). The EDSP excavations produced a similar body of data that may be considered as a randomly selected control group. When the two basic form classifications are compared, the basic distribution of the forms is remarkably similar (see Table 4.6). The total number of pots and number of chambers from the EB IA assemblage of the A Cemetery is very close: 1,132 pots from 48 chambers excavated in 1965–1967 and 1,254 pots from 56 chambers excavated in 1977–1981. The total number here differs from the total number of pots in Table 4.1 because some forms were not given type categories; it differs from the total number in Table 4.10 because that figure did not include Tombs A113, A120, and A121. Comparable percentages for large to medium-large jars (7.5% to 8.3%), medium jars (7.9% to 10.0%), small jars (8.2% to 5.5%), juglets (22.8% to 25.8%), large to medium bowls (31% to 31.9%), and small bowls (19% to 17.2%) in the two databases (further reflected in the subtypes) would appear to confirm the validity of the size ranges. A further analysis of the basic forms distributed in the chambers offers additional confirmation of the size ranges. In the Lapp database, for example, in the 38 EB IA tomb chambers with three or more jars:

28 of 38 had a small jar
35 of 38 had a medium-small jar
36 of 38 had a medium to medium-large jar
35 of 38 had both a medium-small and medium to medium-large jar
23 of 38 had a small, medium-small, and medium to medium-large jar

In the EDSP database of the 39 chambers with three or more jars:

31 of 39 had a small jar
38 of 39 had a medium-small jar

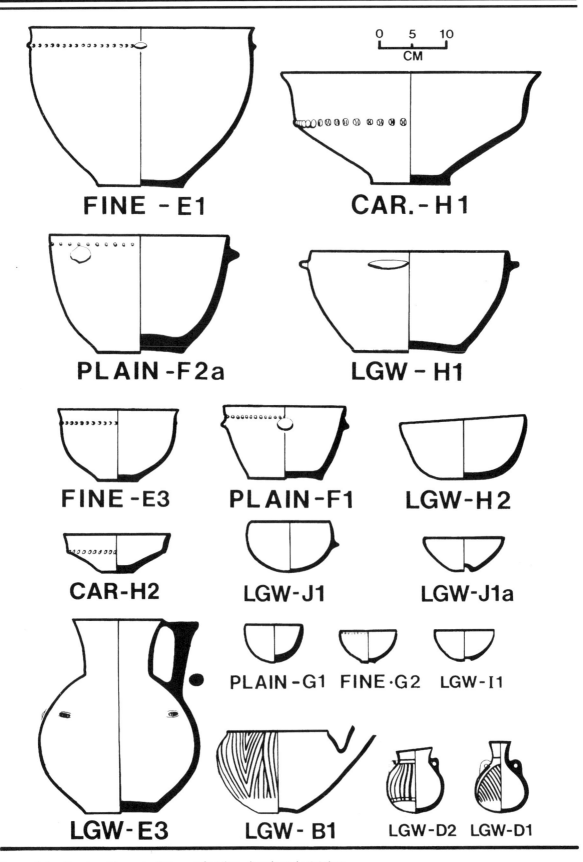

Figure 4.4. Dominant types in EB I ware families—bowls and LGW jars.

34 Chapter 4

32 of 39 had a medium to medium-large jar
33 of 39 had a medium-small and medium-large jar
25 of 39 had a small, medium-small, and medium to medium-large jar

Similar distribution of size ranges in the chambers is associated with the juglet and bowl basic forms. The consistent distribution of the basic form-size ranges in individual chambers suggests that the EB IA potters had certain standard size ranges in mind. If the size ranges have been correctly determined, then it also should be possible to relate these size ranges to possible intended functions of the vessels (Schaub and Rast 1989:248).

CHRONOLOGICAL RANGE OF THE POTTERY. In ware fabrics, especially orange chalky ware (OCW), basic forms and specific types of the EB IB tomb pottery duplicate many of the same features in the pottery assemblage of Stratum IV of the Bâb edh-Dhrâ' town site. OCW combines a distinct temper, fabric texture, and surface feel that make it easily recognizable. The ware color is reddish-yellow and the surface feel is smooth and chalky. It is most often associated with smaller vessels such as the medium-small wide bowls with incurved profile, spouted bowls, amphoriskoi, juglets, and omphalos bowls. In surface treatment, the forms may have line-group decoration, but they also may be slipped or plain. Numerous examples of this ware from the Stratum IV loci are listed and illustrated in the town site volume (Rast and Schaub 2003:135; Figs. 7.1:15; 7.2:15–17; 7.3:7–10, 14, 20–25). Examples of parallels in form include the typical bowl forms with ledge handles below the rim (cp. Fig. 4.4, LGW-H1, to Rast and Schaub 2003:Fig. 7.1:3); spouted bowls (cp. Fig. 4.4, LGW-B1, to Fig. 7.1:15; 7.2:20); amphoriskoi (cp. Fig. 4.4, LGW-D1 and D2, to Fig. 7.2:15,17; large jars with inflected curves at the necks (cp. Fig.4.3, LGW-E2, to Fig. 7.2:3); small juglets with handles attached above the rims (cp. Fig.4.3, LGW-G1, to Fig. 7.2:160) and small bowls with omphalos bases (cp. Fig. 4.4, LGW-J1A, to Fig. 7.3:10). In addition, a common ledge-handled form in the Stratum IV pottery is a "duck bill" ledge handle (Rast and Schaub 2003:Fig. 7.2:21, 22), which is one of the distinctive features of the large jars in the EB IB tombs.

Comparative material from Stratum IV for other EB IB settlement sites in ancient Palestine is listed in Rast and Schaub (2003:Table 7.7). Among the sites from which EB IB parallels are cited are Arad IV, Jericho Tell PU layers, Tell Halif, Tel Yarmuth, Tel Erani, and 'Ai. Extensive parallels for the EB IB tomb pottery of Bâb edh-Dhrâ' with that from Jericho, Tell el-Far'ah and other sites are also listed in Schaub (1981b:74–77).

Establishing a relative sequence among the EB IB tombs is difficult because tombs A104 and G1 were disturbed and their pottery assemblages are incomplete. In addition, tomb chamber A88L had a very large group of pottery. The basic forms and specific types of the EB IB tomb pottery illustrated in Figures 4.2, 4.3, and 4.4 above occur in different quantities in Tombs G2, A100N, A104, A88L, and G1 (Table 4.7). Large bowls (Figure 4.4, LGW-H1) with small ledge handles below the rim are the most dominant type. They occur in all of the tombs and in large numbers in A88L and G1. Other types that occur in all five

Table 4.6. Distribution of Basic Forms in the Lapp and EDSP Excavated Tombs

	Lapp		EDSP	
Type	n	%	n	%
Kraters	24	2.1	11	0.9
Jars, large to medium large	94	8.3	104	8.3
Jars, medium	90	8.0	126	10.1
Jars, small	93	8.2	69	5.5
Juglets	259	22.9	324	25.8
Bowls, large to medium large	352	31.1	400	31.9
Bowls, medium-small to small	216	19.1	216	17.2
Special	4	0.4	4	0.3
Total	1,132	100.1	1,254	100.0

chambers are the medium bowls with incurved profiles (Figure 4.4, LGW-H2) and the small juglets with inflected curves at the necks and handles attached from above the rims (Figure 4.3, LGW; Tomb G1, Figure 4.5:2). The distinctive medium-large jars with tall necks and handles attached from above the rims (Figure 4.3, LGW-C1; Tomb G1, Figure 4.5:1) appeared in all of the tombs except G2. Since G2 has no forms with line-group painting or incised designs and has the fewest of the other specific types, it probably should be placed early in the sequence of these five tombs. The forms that most often have line-group painted designs are the small amphoriskoi with ear handles and the bowls with trumpet spouts. Wide-necked amphoriskoi appear in A100N and A88L, and spouted bowls appear in all except G1. Charnel House G1, however, has several amphoriskoi with narrow necks (cf. Figure 4.5:3). Small bowls with pushed-up bases (omphalos type, Figure 4.4, LGW-J1a) occur in Tombs A88L and G1 (Figure 4.5:5).

G1 also has several distinctive forms not found in the other EB IB tombs, which support placing it last in the sequence of these tombs. One is a tall, narrow-necked jug, highly polished, similar in form to the so-called Abydos jugs (Figure 4.5:8). There are also three carinated platter forms (Figure 4.5:9–11) and two jars with tall necks and pronounced lug handles on the shoulders (cf. Figure 4.5:7). A good parallel to the latter forms appears in tomb chamber A114A from Jericho (Kenyon 1960: Fig. 18.12), a tomb that has many other parallels to EB IB types (Schaub 1981b:76). Similar forms also occur at Tel Erani, assigned to EB IB1 by Yekutielli (2000: Fig. 8.7.1), and at Phase I (EB IB) of Abu al Kharaz (Fischer 2000:12.2.6). Phase I of the same site also produced a tall-necked jug very close in form to the jug from G1. Fischer calls it early metallic burnished ware, a harbinger of the later classic EB II forms. The dominant range of the carinated platter bowls is EB II, but numerous earlier examples from EB IB strata occur at 'Ai, Tel Dali, and Arad and at the Stratum IV (EB IB) town site at Bâb edh-Dhrâ' (Schaub 2000:461). Despite these early parallels, one could argue that the G1 Charnel House should be seen as a transitional tomb between EB IB and EB II or even dated to EB II from its latest ceramic forms. Against this late dating are the occurrences of thorough burning throughout the G1 House, which duplicates the burn found also in round Charnel House A53 and in the destruction layers found throughout the town site closing off Stratum IV. The destruction of Stratum IV of the town site can be securely dated to EB IB by extensive comparative material and by carbon-14 dates (see below).

Establishing a seriation for the EB IA tombs is much more problematic. Tackling this problem in earlier publications, we proposed that "the best candidates for the latest groups are chambers that reflect typological features associated with EB IB tomb architecture, burial practices and also include pottery and artifacts closest to EB IB types" (Schaub and Rast 1989:27). The two chambers of A6, for example, had stone-lined entryways and pottery groups with features similar to the EB IB groups. They were placed in the proposed latest group of EB IA tombs along with other tomb groups that had similar ceramic features (Schaub and Rast 2003:29). None of the EB IA tombs excavated by the EDSP had stone-lined entryways, so this criterion is not applicable. If ceramic seriation is applied with the dominant plain ware groups as later (see Table 4.8), only a few tombs fall into a later group (A79W, A89SE, A107N, and A108SW). A second group has a fairly even distribution of plain and fine wares (A79E, A80E, A102NE, A107E and A107W, A108N and G4). The remaining groups all have predominant fine ware. These two ceramic traditions are basically contemporary, and any seriation remains problematic.

The basic relative chronological position of the carinated ware groups is clear, but some questions remain. This group is at least partly contemporary with the fine and plain ware groups. Fine ware vessels occur in C Cemetery tombs. Tomb C11 had three fine ware vessels: a typical large bowl with everted rim, a juglet, and a small bowl. Tomb chamber C10W had two plain ware vessels, while C10E contained five plain ware vessels and two fine ware pots. In the A Cemetery, tomb chamber A109S included two carinated ware juglets, and in the G Cemetery, Tomb G5 had two carinated ware bowls. It is difficult to determine, however, exactly where the C Cemetery tombs fit in with the areas to the east. Parallels to the red-slipped and burnished carinated ware vessels with raised bands have been cited from Jericho VIII, Meser Stratum II, and Afula (cf. Schaub and Rast 1989:273), but they are not particularly helpful in focusing the chronological range. Since the C tombs are at least partly contemporary with the A and G fine and plain ware vessels, perhaps the most that can be said at this point is that the carinated ware family vessels belong to the latter third of the fourth millennium. Fischer has mentioned rare bowl sherds similar to gray burnished ware in Phase I at Abu al Kharaz and commented that, "'Grey' can be a misleading color description because vessels falling into this category may also

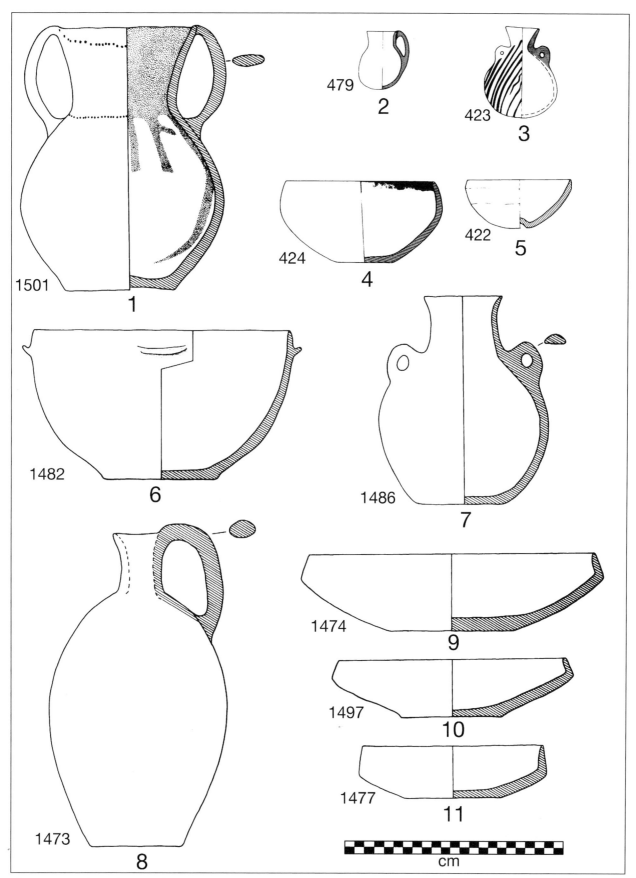

Figure 4.5. Selected pottery from Tomb G1.

Cultural Artifacts of the EB I Tombs 37

Table 4.7. Common Forms and Features of the EB IB Tomb Groups

	Tombs				
	G 2	A100N	A104	A88L	G1
Forms					
Large bowls—H1	4	4	2	17	11
Medium large jars—handle above rim—C1	0	2	2	2	1
Large jars—inflected curved, no handle	1	2	1	2	0
Spouted bowls—B1	1	4	1	13	0
Amphoriskoi—narrow neck—D1	0	0	0	0	4
Amphoriskoi—wide neck—D2	0	2	0	1	0
Omphalos bowls—J1A	0	0	0	6	3
Medium bowls incurved profile—H2	2	1	1	2	1
Juglets—handle above rim—G1	1	7	5	5	2
Column handle jars—E3	0	0	0	1	0
Features					
Line-group painting	0	2	1	8	2
Incised reed marks	2	5	0	0	4
Incised designs	0	1	1	14	4

have yellow, reddish-brown and black surfaces" (Fischer 2000:204). He also recognizes a problem with including "grey burnished ware," usually placed in early EB I in his Phase I, and regards them as probably residual. Phase I at Abu al Kharaz is dated to Late Early Bronze Age I. The emphasis on gray burnished ware carinated bowls has often meant that red-slipped and burnished vessels, even if carinated, are overlooked.

Absolute Chronology

Three carbon-14 dates from Stratum IV of the town site are discussed by James Weinstein (Rast and Schaub 2003:641–43). The calibrated dates with the highest probability at the 1 Sigma level are 3520–3410 cal. BC (Beta 134011), 3100–2910 cal. BC 9 (Beta 134012), and 3340–3210 cal. BC (Beta 134013). After discussing recent theories and synchronisms with Egyptian historical data, Weinstein concluded that "the two latest EB IB dates from Bâb edh-Dhrâʻ are not inconsistent with the radiocarbon evidence from Palestine or the archaeological and historical evidence from the Nile Valley" (Schaub and Rast 2003:643). All three of the Stratum IV dates were from charcoal samples. A more recent short-lived sample of charred seeds from Stratum IV at a 1 Sigma–calibrated a range of 3310–3230 BC and 3100–3010 BC (Beta 221164) confirms the date of the Bâb edh-Dhrâʻ EB IB stratum to the last quarter of the fourth millennium. An earlier sample from EB IA tomb chamber A100E offered a calibrated date of 3545–3345 BC (Weinstein 1984:337). Current theories have tended to move back the beginning of the EB I period to 3500 BC, with some researchers proposing even earlier dates (Rast and Schaub 2003:641). A100E would fit comfortably in that range.

Basalt Bowls

The dark green to black basalt stone bowls are among the most distinctive items placed in the EB I chambers. These bowls offer a striking contrast to the dominant presence of the numerous tan and red pottery vessels. In form, the basalt vessels have thick bases with graceful, slightly flaring wall profiles ending in tapered rims. Some of the examples are decorated with one or two raised bands, which may be plain or notched and provide rope-like effects.

In size, the basalt bowls range in diameter from 24 centimeters to 12.5 centimeters. In height, the tallest is 20 centimeters and the shortest is 8.5 centimeters. Base width is frequently similar in size to the height of the vessels, varying from 18 to 9.5 centimeters. Overall, the bowls average 18 centimeters in diameter and 13 centimeters in height, with a base width of 13 centimeters.

There is substantial evidence that these bowls were considered to be prestige items. Their limited occurrence compared to the numbers of pottery vessels is one line of evidence. The total number of bowls recorded from the excavations in all areas of the cem-

Table 4.8. Distribution of Fine and Plain Wares in EDSP Tombs

Tomb	Fine Ware	Plain Ware	Not Classified
A78NE	25	3	
A78NW	19		2
A78SW	13	7	1
A78SE	22	6	1
A79E	15	12	
A79N	12		1
A79S	9	5	
A79W	17	24	
A80E	4	2	
A80N	14	13	
A80S	17	4	
A80W	47	6	
A86NE	19	1	
A86SE	7		
A86SW	17		
A89NE	24	8	
A89NW	30	9	
A89SE	22	24	
A91	23	7	
A92	15	7	
A100W	15	5	3
A100S	5	2	
A100E	26	14	
A102E	20	6	
A102NE	7	5	
A102S	16	11	
A102W	11	6	
A103	5	17	
A105NE	10	1	
A105NW	26	11	
A105SE	18	7	
A105SW	8	2	
A106	7	22	
A107E	28	24	
A107N	17	18	
A107S	29	16	
A107W	24	24	
A108N	3	3	
A108NE	20	10	
A108NW	16	5	4
A108SE	20	2	
A108SW	9	15	
A109E	3		3
A109N	22	1	
A109S	5		2 Carinated
A110NW	16	9	
A110SE	28	2	
A110NE	18	6	
A111E	10	5	
A111N	34	15	
A111W	14	1	
A114	9	2	
G4	6	7	

etery was 60, with 28 from 23 chambers recorded by the Lapp excavations (Schaub and Rast 1989:294–302) and 32 from 27 chambers recorded by the EDSP (Tables 4.1, 4.2, and 4.3). Six of Lapp's chambers had two bowls, as did five of the chambers excavated by the EDSP. These small numbers present a strong contrast to the large number of pots (on average over 23) found in the tomb chambers of the Bâb edh-Dhrâʿ cemetery. Further evidence may be adduced from the demanding craftsmanship, including decoration on some of the bowls (Schaub in press). In addition, evidence for the special role of these basalt bowls comes from the presence in some chambers of ceramic imitations of the stone bowls. In 13 chambers without stone basalt bowls, ceramic imitations in similar form were placed, apparently as substitutes. Two chambers (A67N and A102S) had both stone and ceramic imitations.

Attempts to find significant patterns in the placement of the stone bowls in the chambers have met with limited success. The stone bowls are placed in the chambers in different positions. Most frequently they are together with pottery groups, sometimes to the left of the central bone piles and other times to the right of them. In only a few instances are they isolated and placed to the left of the doorway. The bowls also occur in chambers of all areas of the main parts of the cemetery from A to C to G. Often, however, they do appear in several chambers of the same tomb and in tombs in the general locale. All of the A68 chambers, for example, had at least one basalt bowl, and three of the chambers had two (Schaub and Rast 1989: Fig. 169). Nearby tombs—A65, A66, A67, A69, A70, and A72—all had chambers with at least one basalt vessel. All three chambers of Tombs A110 and of A111 had at least one basalt bowl (Table 4.1). It is possible that there was a peak production of the bowls when these tombs were fashioned. An interesting circumstance also can be observed with the eight imitation stone bowls excavated by the EDSP. Each of the chambers of Tombs A102 and A86 has one imitation vessel, and A86SE has two. Perhaps the market for the basalt bowls had dried up. A possible association of the basalt bowls with female skeletal material is explored below.

Although a local stone production workshop has not been discovered in the Bâb edh-Dhrâʿ region, previous studies have strongly suggested that this region may have been a center of production (Braun 1990; Philip and Williams-Thorpe 1993, 2000, 2001). Braun compiled a corpus of 80 basalt bowls of the EB I horizon in the southern Levant before the publication of the Lapp excavations. The majority of his examples represent his Type IB, "upright walls with externally concave profiles and tapering rims" (Braun 1990:87). This description perfectly fits the Bâb edh-Dhrâʿ examples. No other region in Braun's survey has the sheer quantity of bowls found in the Bâb edh-Dhrâʿ tombs. Philip and Williams-Thorpe analyzed sixteen archaeological samples, including nine basalt bowls from Bâb edh-Dhrâʿ and two from the es-Safi cemetery, together with twenty geological samples. In their conclusions, they state "seven vessels from Bâb edh-Dhrâʿ and one from Safi are of basalt from the same outcrop, probably in the Kerak area. A further vessel from Safi and one from Bâb edh-Dhrâʿ are from a different source, perhaps in Wadi Mujib, and are identical to each other" (1993:59). The remaining Bâb edh-Dhrâʿ sample is from an unidentified source in the Dead Sea area. Commenting on the possible different sources, the authors also suggest that wadi boulders may have been used: "The simplest method would have been to collect blocks from the vicinity of flows, or water-redeposited rocks from any of the various wadis which border the flows in this region" (1993:60).

Since some of the larger bowls would have required boulders at least 30 centimeters in diameter and would have weighed approximately 50 pounds, gathering the raw material would not have been an easy task. The effort to collect the material, drill out the center of the bowls, and fashion the profile of the bowls, at times with decoration, testifies again to the major commitment the inhabitants had to furnishing the tombs with quality objects.

The regional parallels and the chronological range of finely worked basalt vessels, extending from Late Chalcolithic through EB I, have been well documented in previous studies (see Amiran and Porat 1984; Epstein 1975, 1988; Schaub and Rast 1989:299; Braun 1990). The basalt bowls of the Bâb edh-Dhrâʿ cemetery may well represent the last stages of this tradition.

Mace Heads

The stone mace heads did not demand the same labor effort as the basalt bowls, but they also appear to have been regarded as special items to place in the tombs. In the Lapp excavations, mace heads were found in only 11 of the 47 chambers of the A Cemetery (Schaub and Rast 1989:289–93). Only 10 chambers excavated by the EDSP in the A Cemetery yielded mace heads. Mace heads occurred in the C Cemetery with more regularity. They were found in two of the four chambers excavated by the EDSP, and two more were found in a bowl in a surface find. In form (pyriform, rounded, spheroidal, elongated) and material (limestone, chalkstone, alabaster) the EDSP examples duplicate those found in the Lapp excavations. Most of the mace heads had a hole drilled from both ends, which often had curved profiles, indicating that they had not been hafted. It seems likely that they were symbolic gifts rather than functional weapons. A discussion of the significance of the mace heads, including other Middle Eastern contexts and parallels, is found in Schaub and Rast (1989:292–94). The possible correlation of the mace heads with males buried in the tombs is discussed below.

Figurines

Some of the most interesting artifacts found in the chambers are unfired anthropomorphic figurines. Fourteen examples were registered by the EDSP, doubling the number recorded by the Lapp excavations (Schaub and Rast 1989:274–89). The figurines tended to occur in groups. The seven Lapp figurines were from two tombs: A5E (four) and A7S (three). A similar pattern of multiple examples appeared in the EDSP group: A78S (two), A100E (three), A110NW (two), A111E (two), and G2 (three). Only two chambers, F2 and A110NE, had single examples. In height, the EDSP figurines range from 8 to 16 centimeters and average 13 centimeters. Thickness at the waist is usually around 4 centimeters, and the width at the extension varies from 6 to 9 centimeters.

Features associated with the figurines vary and are affected by the state of preservation. Several have broken arms or cracks in the torsos. The most consistent features include rounded heads with flanges and pierced holes. Facial features include beak-like noses, small holes for the eyes and nostrils, and slashes for the mouths. Arms are upraised at approximately 45 degrees. In some instances they seem to form a crescent. Most have rounded-to-pointed breasts and pinched lumps at the abdomen and on the back; the back lumps usually have a slash representing the buttocks, while a gouge below the abdominal lump represents the vulva.

There is considerable homogeneity among all of the figurines with regard to technical elements. All are hand modeled from rough clay containing grits of limestone. They are not kiln fired but merely sun baked, and the levigation is generally quite rough. The color of the clay tends either toward a light browning gray (e.g., Munsell 2.5Y 6/2) or toward a brighter tan with a pink quality (e.g., Munsell 7.5YR 7.5/4). Construction is based on a cylindrical slab of clay out of which the arms, legs, and head features are molded; breasts, abdominal lump, and buttocks are sometimes molded and sometimes formed with added lumps applied to the torso.

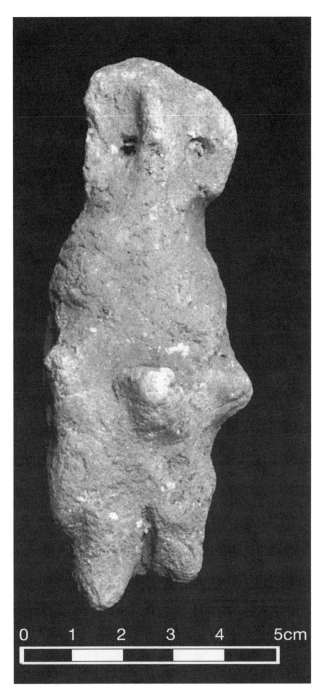

Figure 4.6. Figurine from Tomb G2, Reg. # 629.

Figure 4.7. Figurine from Tomb A78SW, Reg. # 777.

In general, the figurines are quite crudely modeled, particularly in comparison with the finely made ceramic ware from the same EB IA tombs. Indications of carelessness, haste, or lack of skill on the part of the makers are quite apparent: breasts stuck on carelessly, facial features punched out of place, perforations of the head-flanges that sometimes do not go all the way through, and so forth. There is some variation in the quality of craftsmanship from tomb to tomb. A figurine from Tomb G2 (Figure 4.6), for example, is very crudely made, with arms and face suggested simply by a pinching of the clay; a figurine from tomb chamber A78W (Figure 4.7), by contrast, shows rather careful, symmetrical modeling of the features.

In the absence of any written texts and without close parallels for the late fourth millennium from nearby regions (see the discussion in Schaub and Rast 1989:286–88), the meaning and symbolism behind the placement of the figurines in the Bâb edh-Dhrâ' tombs remains undetermined. In principle they would

Cultural Artifacts of the EB I Tombs 41

appear to be part of a widespread tradition of representing the gestating female.

Jewelry and Perishable Items

Items of jewelry were not common in the tombs. Carnelian beads were found in pottery vessels in tombs A79, A89NW, A110SW, and C8. Other beads were made of bone (A89NW, C9, and C10E). Some of the largest groups were formed of talc (A79S, 289 beads; A89NW, 40 beads). Two distinctive shell "bracelets" from Tomb 92 are similar to one published from tomb chamber A76E (Schaub and Rast 1989: Figure 183). They have been described as "bangles" in the literature and by David Reese (pers. comm.), who has identified the source as a large Red Sea gastropod, *Lambis truncata sebae* (the Giant Spider Conch), or a scorpion shell. Parallels come from Tell el-Farah (N), Jericho, T. el-Ghassul, and many sites on the Sinai Peninsula (see Schaub and Rast 1989:311). Other parallels and the significance of these items are discussed in Schaub and Rast (1989:310–12). A more complete treatment of the role and significance of shells in early Sinai societies may be found in Bar-Yosef (1999).

A few perishable items are still being studied. They include remnants of wooden bowls (A110NW, A111N), fabric (A110NE), a piece of string (A111N), and ten pieces of worked wood (A114). Mat impressions, similar to those previously published by Yedlowski and Adovasio (Schaub and Rast 1989:531), were found in several tombs.

DISTRIBUTION OF ARTIFACTS RELATED TO THE INDIVIDUALS REPRESENTED BY THE SKELETAL MATERIAL

Since most of the chambers have both male and female adult skeletal material, it is not easy to correlate the types of the nonceramic artifacts with the sexes of individuals in the chambers. Of the 63 chambers with skeletal material for the EB IA period, 59 chambers (92%) had a least one male adult represented. The exceptions were A78NE and NW, A79W, and A109N and S. Similar figures are associated with female adults in the chambers. Fifty-eight of the 63 chambers had at least one female adult. The exceptions were A78S, A79E, A80N, A87, and A110NW.

In some chambers, a multiple occurrence of similar items may be used to argue that, by tradition, the items were associated with either males or females. The occurrence of more than one basalt bowl in some chambers may be associated with female skeletons. Seventeen chambers had at least one basalt bowl and three of these chambers had two basalt bowls. Fifteen of the seventeen chambers had both male and female skeletal material. One of the exceptions is A79W, which had no males and two female adults. A79W is one of the three chambers with two basalt bowls. The other two chambers with two basalt bowls are A91 and A107W. Both of these chambers have multiple adult female burials: two in A91 and three in A107W. Ceramic imitations of basalt bowls are also found in seven chambers. All seven had both male and female adult skeletal material. One chamber, however, creates a problem for this hypothesis. A basalt bowl was found in A110NW, a tomb in which there were two male and no female adults represented.

Chamber A100NW also creates a problem for an attempt to associate the unbaked clay figurines with female burials. These artifacts were found in five chambers, and four of the five had more than one figurine. All of the chambers had both male and female adults except A110NW.

Mace heads were found in 12 chambers. All 12 had both male and female adults. Two chambers, A110SE and C10E, had two mace heads. Both chambers had multiple male burials but also multiple female burials.

Early attempts to establish correlation of the number and types of pots to the number of individuals ended with negative results. If one looks only at the adult individuals represented in the chambers, there are many inconsistencies. In chamber A105NE, for example, there are nine adults but only eleven pots. A105SE has only two individuals and twenty-five vessels. A89NW has three individuals and thirty-nine pots. Many other absences of correlation could be cited. It would appear that the placement of the vessels in the chamber is unrelated to the number of individuals interred.

A more thorough comparative study, however, yields some interesting patterns. If one looks only at the single-chamber tombs, the number of pots correlated with adult individuals appears to be consistent (Table 4.9). The average number of pots per individual varies from 6.0 to 7.7. An overall comparison of all of the relevant chambers of the A Cemetery demonstrates

Table 4.9. Single Chamber Tombs

Tomb	Adult Individuals	Pots	Average number of pots per individual
A91	5	30	6.0
A92	3	22	7.3
A103	3	22	7.3
A106	5	29	5.8
A114	3	23	7.7

a similar pattern. In the 52 chambers (excluding the robbed chambers of A101, A113, A120, and A121), there are a total of 1,313 pots, or 25.3 pots per chamber. Adults represented in these chambers total 198, or 6.6 pots per individual (Table 4.10). Several multiple chamber tombs—A102, A107, and A111—vary widely in the number of pots per chamber, but the tomb as a unit has an average of seven pots per individual. Other multiple chamber groups vary in their averages from 4.2 to 9.8 pots per individual. It would appear from these figures that, although the placement of vessels in the individual chambers may have been random, the potters tended to produce the vessels for the tombs in similar numbers.

Table 4.10. Average Pots per Individual in Tomb Groups

Tomb	Adult Individuals	Pots	Average
A78	10	99	9.9
A79	11	95	8.6
A80	12	107	8.9
A86	10	44	4.4
A89	12	117	9.8
A91	5	30	6.0
A92	3	22	7.3
A100	19	79	4.2
A102	14	82	5.9
A103	3	22	7.3
A105	18	83	4.6
A106	5	29	5.8
A107	25	180	7.2
A108	22	107	4.9
A109	7	40	5.7
A110	8	76	9.5
A111	11	78	7.1
A114	3	23	7.7
Totals	**198**	**1,313**	**6.6**

CHAPTER 5

The Funerary Tradition of the EB I People

THE PRIMARY OBJECTIVE OF this chapter is to reconstruct, as much as possible, the funerary tradition of the EB I people of Bâb edh-Dhrâʿ and compare it with what is known about funerary traditions at other archaeological sites dating to approximately the same time period and located in Palestine, the Levant, and Egypt. This reconstruction is based on both biological and cultural evidence and the inferences that can be made from these two sources of data.

The investment made in the construction of both shaft tombs and charnel houses as well as the value of the objects placed within the tombs argues for the importance of the funerary traditions throughout the EB I period. Placement of food offerings, jewelry, and other personal items as well as the clay female figurines with a possible symbolism linked to fertility and/or rebirth imply a belief in the continuation of life beyond normal human existence.

Current knowledge about the human societies occupying various sites in Palestine and the southern Levant in the Late Chalcolithic and Early Bronze I periods indicates that they were relatively independent social entities (Hennessy 1967; Philip 2001). Undoubtedly these societies were influenced by and probably had economic ties with each other. There were also cultural influences from the major urban centers in Mesopotamia (Hennessy 1967) as well as the complex society of Late Predynastic and early Dynastic Egypt (Hennessy 1967; Miroschedji 2002). However, the archaeological evidence supports relatively independent development of their social structure and the specifics of their cultural traditions.

Funerary traditions of a society during a specific time period reflect multiple factors, including the culture of the ancestral society and the diffusion of cultural traits from other societies. These traditions were also influenced by the natural environment in which the societies developed. Sandy, unstable soils require above-ground burials or some method of preventing collapse of the walls of an underground grave, as one sees in the cairn and cist burials associated with some locations and time periods in the southern Levant.

Another example is the shaft tomb burial tradition of the EB I people of Bâb edh-Dhrâʿ, in which the presence of abundant, although not universal, layers of Lisan marl formed an ideal material for creating the distinctive underground shaft tomb architecture.

What is known about the development of the human society at Bâb edh-Dhrâʿ is that there is minimal evidence of permanent dwelling structures in the EB IA period in which the shaft tomb funerary tradition was exclusively present. The paucity of evidence for permanent occupation of the site changes in the EB IB phase with evidence of houses (Rast 2003). In EB II, the site would begin to develop into a substantial walled town supporting an agricultural economy.

The insignificant evidence for permanent structures in the EB IA period at Bâb edh-Dhrâʿ provides the basis for the hypothesis that the people utilizing the site were nomadic pastoralists. This does not imply that Bâb edh-Dhrâʿ was not important to the people who used the site in EB IA. The large area utilized for shaft tombs as well as the elaborate funerary structures and the ceremonies implied by the tomb contents highlight this importance. Furthermore, the spring water that passes through the site in what today is Wadi Kerak represents a vital resource that would almost certainly have been exploited by any society that knew about it and was in proximity to it during one or more periods in a year.

During the EB IA period at Bâb edh-Dhrâʿ, all burials were secondary; primary burial was at an unknown site and reburial typically but not always came after the soft tissue had completely decayed. As we will demonstrate more clearly in Chapter 7, at least some of the burials in the EB IB period were secondary as well, although there is evidence of at least one primary burial (see discussion of tomb chamber A100N in Chapter 6).

The secondary burial tradition would lend itself to a society that was nomadic but would nevertheless place great importance on careful treatment of its dead members. Secondary burial permits interment at or near the location where death took place and is a response to the socially defined need to provide eventually a central place of ceremonial significance for the deceased in a society whose members range over a large geographical area. Subsequent transfer and reburial at a site and with ceremonies more representative of the value the dead had to the society could then occur at a more convenient time.

It is not until EB IB that the society at Bâb edh-Dhrâʿ shows archaeological evidence for permanent occupation of the site. Since the time span from the beginning of EB IA to the end of EB IB was approximately 200 years, this transition probably took place during a relatively short time between the two periods. The archaeological evidence suggests that transition was an evolution of the local culture, not due to an intrusion of a different society (Rast 2003; Rast and Schaub 2003). Nevertheless, the changes in the funerary tradition between EB IA and EB IB are substantial. Among the more significant changes, the below-ground shaft tomb burials gave way to above-ground charnel houses that were, in general, much larger and contained many more burials. There was, however, continued although limited use of shaft tombs during the EB IB period.

The EB IB G1 Charnel House excavated in 1977 represents, at least in tomb architecture, a fairly substantial change in the funerary tradition of the Bâb edh-Dhrâʿ people. In Chapter 7 we present evidence of other changes in the burial tradition, including the greatly reduced prevalence of skeletal elements associated with late-term fetuses, infants, and young children. This is in contrast with the large number of skeletons from these developmental stages that are found in the EB IA shaft tombs. We also find some evidence of primary burial in one of the EB IB tomb chambers (A100N), but there is very convincing data that secondary burial continued to be a significant part of the burial tradition.

Analysis of dental data supports a close genetic relationship among at least some of the people interred in a shaft tomb (Bentley 1987; chapter 13 of this book). Thus far the research on the charnel house burials has not progressed to the point where genetic relatedness of the burials within a charnel house can be established. Although there are examples of relatively small charnel houses (e.g., A56) that could easily be for a family, most of these structures contained many more burials than could be accommodated in a shaft tomb complex. Chesson (1999) argues that charnel houses were for members of an extended family, implying more genetic relatedness of interments within a given charnel house than between two or more houses. However, the biological evidence for this hypothesis remains to be demonstrated.

Research to clarify the relatedness of individuals interred in a charnel house is likely to be a challenging exercise. Many of the charnel houses show evidence of intentional destruction by a hostile society at the end of the EB III period. Burning and vandalism within the tombs is extensive. Furthermore, most of the charnel houses contain pottery that extends from EB II through EB III, a time range of several hundred years. This range in the use of a charnel house is in contrast with the range for the shaft tomb tradition, which extended for about two hundred years but was the principle tomb type for about the first one hundred years. The length of time many charnel houses were used is certainly compatible with the hypothesis that the houses were used by extended family members both synchronically and diachronically, although other explanations are possible.

If we assume, for the moment, that the hypothesis regarding a more sedentary society emerging in EB IB is correct, one might expect the adoption of a primary burial tradition in which the deceased are placed in charnel houses immediately upon death. What the evidence suggests is that the EB IB burial tradition combined new architectural developments with the retention of some aspects of the old tradition, including secondary burial in at least some cases. However, in addition to the shift from underground tombs to much larger, above-ground mudbrick structures, the paucity of skeletons of fetuses, infants, and young children also indicates a fairly significant cultural change in the value ascribed to individuals in these young age categories.

SHAFT TOMB ARCHITECTURE

Underground shaft tombs were the exclusive burial structure at Bâb edh-Dhrâʿ in EB IA and continued to be used in EB IB, but above-ground charnel houses were introduced in EB IB and became the only type of burial structure used in EB II–III. The shaft tombs at Bâb edh-Dhrâʿ consist of a central, roughly circular shaft about one meter in diameter that extends about one to three meters below the ground surface. It seems likely that the workers constructing the shaft tombs used the shaft excavation to identify the presence and depth of a thick layer of Lisan marl, which was ideally suited to creation of the underground tomb chambers. Marl is easy to excavate and provided a relatively stable material for the tomb chamber dome.

In the Bâb edh-Dhrâ' cemetery area, Lisan marl occurs sporadically in one or more layers that vary in thickness. The ideal thickness for tomb chambers was at least one meter. When Lisan marl was present, shaft depth tended to extend near or below the bottom of the marl layer.

Dome-shaped chambers about two meters in diameter at the base and a bit less than one meter high at the peak of the dome were dug from the base of the shaft. The roughly square entryway into the chamber was about one-half meter on each side. The floor of the chamber was about twenty centimeters below the lower margin of the entryway. The number of chambers associated with each tomb varied from one, particularly in situations where Lisan marl was absent or occurred only in a thin layer, to as many as five chambers. In ideal areas of the cemetery, that is, where Lisan marl provided the optimum material for chamber excavation, shaft tombs were very dense, with walls between the chambers of adjacent tombs dug so close that breaks occasionally occurred into an adjacent chamber of another tomb (see Figure 6.64). In some cases, these breaks in the wall were repaired with stones and mud mortar (see Figure 6.29). Breaks that were not repaired probably occurred after the tomb chambers were sealed and not during construction of the tomb.

The precision in the placement of shaft tombs raises several questions about tomb construction. It is possible that, in anticipation of use, several tombs would be excavated, in which case the location of additional tombs being dug would be known. However, this would have required some method of preventing water and wind-blown silt from filling in the chambers and shafts at least partially. Although we do find evidence of silting, it all seems to have entered the tomb chamber following placement of human remains and cultural artifacts in the chamber (see Figure 6.7).

If tombs were linked to family units as some of the dental genetics data suggest (Chapter 13), shafts might have been dug and one or two chambers prepared because of deaths in the family, but spaces for other chambers in the same tomb might have been left for later creation of one or more chambers once additional deaths had occurred. In this situation, the shaft might have been left at least partially open and the location of existing tombs would have been well defined.

An alternative and more parsimonious possibility is that tombs were excavated as needed. Since all burials in the shaft tombs were secondary, the people using the cemetery would have known how many tomb chambers would be needed at any given time to contain all the burials to be returned for final interment in the shaft tomb. This seems the more plausible explanation of this aspect of the EB IA funerary tradition, but it requires knowledge of the location of existing tombs when new tombs were dug. Existing tomb locations would be apparent even in situations where all the chambers were utilized and sealed with blocking stones and the shaft was refilled, since the fill would settle somewhat and leave a temporary depression at ground level until wind- and water-borne sand had filled the depression. Some evidence of this possibility occurs in the profile of the partially excavated shaft of A78 (see Figure 3.2). The upper fill is funnel shaped, which would be a likely profile if the shaft fill had settled and the exposed shaft wall had been eroded by wind and rain but eventually filled in by wind or water-borne silt. The density of tombs and the fact that intrusion from one tomb to another is rare argues for some method of indicating the location of a shaft.

TOMB CHAMBER CONTENTS AND ARRANGEMENTS

Biological Contents

The principal contents of the tomb chambers were, of course, the human remains that had been recovered from the primary interment site and placed in the chamber. The time between primary burial and secondary interment varied from what was probably no more than a few days to sufficient time for the bone to be degraded to the point that it broke during the process of transfer from the primary to the secondary site. The speed of the degradation process is very much influenced by the burial environment at the primary site. It is not clear what conditions in the primary burial site would have degraded bone in this way this quickly. Bone is normally a very tough material, and breakage of the type that occurs in some of the human burials would not have occurred in the relatively short interval between primary and secondary interment without unusually rapid degradation of the tissue in the primary burial site. It is possible that high salt content in the soil of the primary burial site accelerated the breakdown of bone proteins, making the bone fragile. Another possibility is that a very dry burial environment or one in which the soil contains one or more natural desiccant salts, such as magnesium sulfate, denatured the bone (Tuross, pers. comm. 2007) and made it more vulnerable to breakage. Breakage of bone between primary and secondary burial sites is not a common condition of the EB IA human remains and primarily occurs in major long bones. We cannot rule out the possibility that these bones were intentionally broken to facilitate transportation to the secondary burial site.

Bones from late fetal through old age as well as male and female remains were placed in the tomb chambers. All the evidence suggests that no distinction based on either age or sex was made regarding who was buried in a given tomb chamber. In most cases not all the bones of a burial were recovered from the primary site. Completeness of a skeleton varied from one or two skeletal elements to virtually the entire skeleton being present. In general, the emphasis in the transfer to the secondary burial site was on the bones of the cranium and the major long bones. We argue that the variation in completeness is primarily due to the elapsed time between primary and secondary burial. All of the EB IA burials excavated in 1977 and 1979 were incomplete. The time between primary and secondary burials was sufficient for all the soft tissue to decay. A somewhat different condition was encountered in some of the tombs excavated in 1981. In some of these tombs, the relationship between most of the skeletal elements of a burial is in approximate anatomical position, an unlikely situation unless the bones had been held in place by soft tissue. In one burial, the bones were in approximate anatomical position except that the ends had been rotated so that the proximal ends of the bones were in the distal position. Clearly, most of the bones of this burial were still attached at the time of burial but some were not, and an attempt had been made to place these in the correct anatomical position.

Bones generally were arranged carefully, with skulls placed in a row to the left of the chamber entryway, postcranial bones placed in a central bone pile, and cultural artifacts placed to the right of the central bone pile and the entryway. This pattern was very consistent in the chambers. There was variation in the care with which bones were arranged in the central bone pile that ranged from very careful placement to the appearance of having been carelessly dumped in the pile. In some cases, smaller bone elements were placed in pottery (see Figure 6.35). If the base of the skull had been broken away, the small bone elements might have been placed in the inverted skull vault (see Figure 6.6).

Cultural Contents
A common feature in tomb chambers was a reed mat placed on the ground below the bones (see Figure 6.6). In some of the chambers, a cloth shroud had been placed over the bones. As the shroud oxidized over time it sometimes left a print of the cloth on the bone.

By far the most common cultural artifact placed in the tomb chambers was pottery. Other artifacts included large and heavy basalt vessels and wooden bowls. Some of the bowls contained clear evidence of food having been placed in the container at the time of interment. Food included grapes (see Figure 6.51), seeds of various types, and prepared food that left a residue on the walls of the pot (see Figure 6.52b). Analysis of the prepared food indicated the presence of amino acids (Ortner 1981).

Other artifacts included mace heads and wooden shafts that may have been used with the mace heads. Mace heads were not common, which raises the possibility that they were symbolic of a more prominent person interred in the chamber. Another fairly uncommon artifact was unfired clay female figurines (see Figure 6.12). Evidence of necklaces and amulets was also uncommon but did occur in some of the chambers. Some included gold pieces and semiprecious stones.

BURIAL TYPES
Earlier in this chapter we noted that the EB I burials were mostly secondary, that is, at the time of death the body was buried in a primary burial site, where it remained for a variable period of time. In some cases, the relationship between the bones was almost anatomically correct and virtually all the bones of the hands and feet were present. This burial condition would have been unlikely if the bones had not still been held together by soft tissue at the time of removal from the primary burial site and placement in the shaft tomb chamber or the charnel house. At the other extreme, some of the long bones were fractured during the transition from the primary burial site to the secondary site. Bone is a very tough combination of various proteins, primarily collagen, that constitute approximately 35% of the weight of bone. The remaining component is bone mineral, which consists primarily of hydroxyapatite, in which calcium and phosphorus are the major elements. The biological combination of protein and mineral forms a very strong material, as it must be able to withstand the various types of biomechanical stress the human skeleton absorbs during life.

In many burial environments, bone retains its strength long after the soft tissues of the body have decayed. The best burial environment is alkaline soil with good water drainage. Acidic soils tend to destroy bone, and the speed of this destruction depends on the degree of acidity and the amount of water in the soil (Von Endt and Ortner 1984). Bone recovered from acidic soils appears much eroded and crumbles easily. None of the burials from the EB I tombs at Bâb edh-Dhrâ' have this appearance, although some exhibit taphonomic deterioration probably resulting from the salty burial environment. This means that the bones probably were not exposed to acidic soils at either the primary or secondary burial site. Why, then, did some of the bones break so easily during the transfer from the primary to the secondary burial site?

One possibility is that there was some taphonomic process occurring at the primary burial site that degraded the bone to the point that some of them broke during transfer. Earlier in the chapter, we speculated on possible causes of this condition, but the answer to the question must await further research. Another question that remains for future research concerns the location of the primary burial site or sites. If the people were nomadic pastoralists, it seems that it certainly would have been easier to bury people near where they died until the soft tissue had decayed and then transport the bones to the final interment. This would have involved initial burial in many different burial environments and some method of marking the primary burial site so that the remains could be recovered at a later time.

CHARNEL HOUSE DEVELOPMENT IN EB IB

As we have indicated in Chapters 1 and 4, the current archaeological evidence suggests that the EB IA people at Bâb edh-Dhrâ' may not have been permanent residents of the site. However, by EB IB, the archaeological evidence (Rast 2003) supports the presence of a more sedentary society utilizing the resources of Bâb edh-Dhrâ'. This hypothesis is supported by a gradual change in the funerary architecture with the development of above-ground charnel houses in EB IB that may reflect the changing culture of a more sedentary society. Charnel houses in EB IB were round, perhaps reflecting a link with the round structure of the shaft tomb chambers. This contrasts with the square or rectangular charnel houses more typical of the EB II–III phases.

The EB IB G1 Charnel House excavated in 1977 had been badly damaged both in antiquity and in recent times, so caution is needed in interpreting data. Nevertheless, the damage relative to human remains is likely to be random and unlikely to affect interpretation relative to funerary tradition. Fetal, infant, and early childhood skeletons were very uncommon, in contrast to the EB IA tradition. Details of the architecture of the G1 tomb are provided in Chapter 7.

At least some of the burials in the G1 Charnel House were secondary. The justification of this conclusion is developed in Chapter 7, but the essential finding is that the prevalence of various skeletal elements recovered is not in agreement with the expected prevalence if all the bones of the bodies had been interred at the time of burial. The continuation of secondary burial would not, of course, preclude increased sedentism. It may simply reflect a continuation of one aspect of the burial tradition from the EB IA period in a different architectural burial structure.

The G1 Charnel House was a mudbrick structure with a roof and a single, small rectangular entryway including a stone threshold, stone orthostats, and lintel, as well as a single blocking stone. We estimate the minimum number of burials in this tomb to be 115 individuals, which means that remains were placed in the tomb at multiple times. In other charnel houses, the evidence suggests that more recent interments involved moving earlier burials and cultural artifacts to the periphery of the tomb; this movement would likely have damaged some of the tomb contents. In Chapter 7 we include a photograph (Figure 7.8) of a large bowl that had been broken before the charnel house had been vandalized in antiquity. The charring that occurred when the tomb was burned in antiquity did not align when the bowl was reconstructed. This type of evidence that breakage occurred before the burning event is not common, so considerable care probably was expended when new burials were placed in the tomb to avoid damage to burials and cultural artifacts that had been placed in the tomb earlier. Another important difference between the burial traditions of EB IA and EB IB is the paucity of bones from fetuses, infants, and young children in the G1 tomb.

BURIAL TRADITIONS OF THE EB I SOUTHERN LEVANT

Because of the relative independence of human societies in Palestine and the southern Levant during the late Chalcolithic and Early Bronze Age, it is not surprising that the cultures of these societies varied, and with that, the funerary traditions varied as well. In the southern Levant this variability has been summarized (Bloch-Smith 2003), and it indicates that the funerary tradition at Bâb edh-Dhrâ' was relatively unique. On the Transjordan plateau to the east of Bâb edh-Dhrâ', the escarpment between the Dead Sea plain and the plateau provided innumerable caves that were used for tombs by the societies of the plateau. In the southern Sinai, above-ground circular stone buildings with corbelled roofs known as *nawamis* were the typical burial structures. In the southern Levant, the EB IA shaft tombs at Bâb edh-Dhrâ' were unique for that time period (Bloch-Smith 2003), although the principle of an underground burial chamber excavated from the base of a shaft does occur in other places and time periods.

It is important to recognize the multiple factors that influence burial traditions. One of these factors is the immediate natural environment in which a given society exists. Cave burials become a reasonable possibility if the caves are conveniently located and accessible. Societies whose operations are based on the plain are more likely to depend on structures that can be created by the members of the society rather than to utilize natural geological features such as caves. Bloch-Smith (2003) cautions that the association between a type of society (e.g., nomadic versus sedentary) and the burial tradition does not fit into neat typological boxes.

CHAPTER 6

The Tombs and Burials of the EB I People: Shaft Tombs Excavated in 1977

DONALD J. ORTNER, EVAN M. GAROFALO, AND BRUNO FROHLICH

THE PRIMARY OBJECTIVE of this chapter is to provide a detailed statement about the human remains found in each of the EB I shaft tombs and tomb chambers. The human remains from the EB IB charnel house also excavated during the field season of 1977 is described in Chapter 7. Burials from tombs excavated in 1979 and 1981 are described in Chapters 8 and 9, respectively. The major emphasis in these chapters is on the evidence of abnormality in the bones. Chapter 12 provides a more comprehensive and integrated treatment of skeletal disorder. Although mention is made there of some of the dental pathology, a careful analysis of this aspect of the human burials is presented in Chapter 13. Chapter 10 reviews the metric and nonmetric osteological data about the human remains. This chapter, as well as Chapters 7 through 9, describes the contents of each chamber with a particular emphasis on (1) conditions of burial; (2) the age, sex, and minimum number of individuals (MNI) recovered from the chamber; (3) evidence of skeletal disorder; and (4) what can be observed about the burial traditions on the basis of the tomb chamber contents, particularly skeletal evidence.

CONDITIONS OF BURIAL

Burials in the EB IA tomb chambers at Bâb edh-Dhrâ' were all secondary, after a variable time at the primary burial site. We know very little about the primary burial conditions. The presence of soil constituents in some of the cavities or marrow spaces of bone in some tomb chambers that had not been affected by either ceiling fall or siltation implies primary burial below ground level in those cases, but this does not mean that all primary burials received the same treatment. The variation in the completeness and fragility of the bones at the time of secondary interment in the tomb chambers suggests a fairly broad range of time at the primary burial site from a few days or weeks to two or more years. We see evidence of breakage of both long bones and skulls that argues for burial at the primary site long enough to result, in some burials, in reduction of the natural strength the bone had at the time of death. It is also possible that conditions at the primary burial site led to relatively rapid taphonomic degradation of the bones, making them more vulnerable to breakage during the transfer to the secondary burial site. However, some burials, particularly some of those excavated in the 1981 field season, were relatively complete, with some bones in the correct anatomical position; this indicates that soft tissue was present, holding the bones in place at the time of secondary burial.

The preservation of bone following secondary interment was affected by several factors. The warm environment and high salt (sodium chloride) content of the burial environment contributed to breakdown of protein molecules such as collagen that are important to the biomechanical strength of bone. By the time of excavation, all the bones were fragile and required treatment with plastic infiltration to minimize damage during handling (see Chapter 3 for more details on conservation methods).

Three conditions within the tomb chambers also affected bone preservation. In some chambers, collapse of the domed ceiling may have damaged the underlying bone. Bâb edh-Dhrâ' is in the northern end of the Great Rift system, an area noted for its earthquakes. Perhaps the remarkable aspect of the EB IA tombs is that there is not more evidence of ceiling collapse in the chambers. Sand and gravel eroding from the chamber wall in some cases flowed around the bone. However, the most serious problem was siltation in which a soil/water mixture flowed into the chamber in some cases through the central shaft and, partially to completely, surrounded and encased the human remains as well as the tomb gifts placed in the chamber.

Within each tomb chamber the burials were numbered. In general, skulls were designated, beginning with the skull nearest to the entryway, as Burial 1, and continuing with ascending burial numbers until all relatively intact skulls had a unique burial number within the tomb chamber. In many cases, the postcranial bones could be sorted into discrete burials and associated with one of the skulls. When this was possible,

Figure 6.1. Early stage of excavation of Tomb A78. Note the funnel-shaped discontinuity of the shaft fill.

the postcranial bones were given the burial number of the associated skull. When there were cranial or postcranial skeletal fragments that were most likely not from one of the relatively complete burials, they were given a discrete burial number, usually beginning with 50 to ensure that these less-complete burials were easily identified in our records and not confused with the more complete skeletons. In some cases, the burial number assigned during excavation contained the bones of more than one individual. In that situation, the additional burials were differentiated with letters added to the original burial number. In some cases, higher burials numbers were used, and there are situations where the reasons for a given burial number are obscured by the gap between the assignment of a number many years ago and the preparation of the descriptions of the burials. To avoid potential confusion in the assignment of a burial number, the numbers assigned in the field or during initial laboratory analysis many years ago have been retained.

TOMB A78

During excavation, the location of the tomb was identified as a discontinuity in the soil exposed by a 2×10-meter test trench (Figure 6.1). The soil refilling the central shaft is clearly distinguishable from the undisturbed adjacent soil. At the base of the shaft, blocking stones sealed the four chambers (Figure 6.2). The tomb chambers were arranged approximately equidistant from each other (Figure 6.3), although the wall between the south and east chambers was unusually thin and minor collapse of part of the wall occurred. This defect in chamber construction is apparent in other tombs as well but it is remarkably uncommon.

Tomb A78, Chamber Northwest (North)

Sand erosion had affected the bones at the margins of this chamber. There was no evidence of silting, but ceiling fall was significant (Figure 6.4). However, most of this fall was on the central bone pile and did not affect the skulls. How much of the breakage apparent in the postcranial bones was caused by the ceiling fall and how much was due to conditions prior to secondary burial is not known, but ceiling fall seems to have had a minimal effect on preservation. Indeed, the reed matting placed on the chamber floor before interment, although very delicate after 5,000 years, was virtually intact once the ceiling fall had been cleared (Figure 6.5).

Figure 6.2. Shaft of Tomb A78 after removal of the shaft fill. Blocking stones have been removed from one of the chambers.

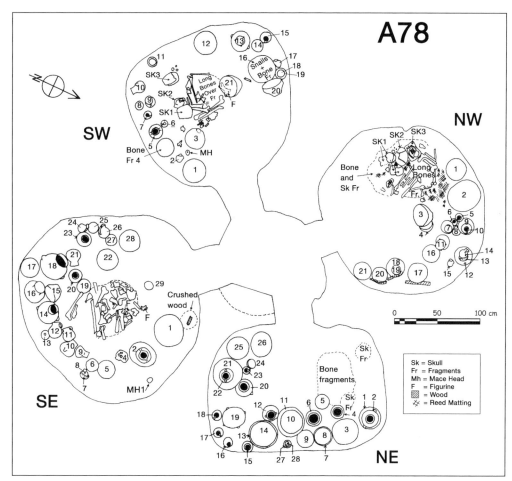

Figure 6.3. Plan drawing of Tomb A78.

Shaft Tombs Excavated in 1977 53

Figure 6.4. View of tomb chamber A78N immediately following removal of blocking stones. Note the extensive ceiling fall.

Figure 6.5. View of tomb chamber A78N after removal of ceiling fall. Note the careful placement of the skulls in a row to the left of the bone pile and the pottery arranged to the right of the long bones.

Figure 6.6. Detailed view of the skulls in chamber A78N. Note the use of the broken skull vault to hold some of the smaller bones and bone fragments.

The human remains were relatively well preserved (Figure 6.6), with completeness of the adult burials in excess of 50% of the skeletal elements for each burial. Evidence of subadult remains is largely represented by cranial elements. However, these remains included some late-term fetal, infant, and child postcranial elements. The MNI in this chamber is twelve, including three adults, one child, seven near-term or term fetuses or infants, and one infant about one year of age. Sex could be determined for two of the adults, both of which were female. There was one adult right parietal element that was not associated with any of the adult remains. This may reflect some mixing of human remains at the primary burial site or during the recovery and interment of the remains at the secondary site. Another possible explanation is that the chamber was cleared of most earlier burials except for the parietal fragment and was then reused for the remains we recovered.

Pathology apparent in the teeth includes enamel hypoplasia, which was present in most of the dental remains. This is indicative of at least one episode of biological stress at some point during early childhood. Each of the two adult skulls (Burials 1 and 3) has an area of abnormal bone deposition on the left side of the frontal. The lesion in Burial 3 may be secondary to a depressed fracture. The lesion on Burial 1 consists of a small area of bone deposition approximately five millimeters in diameter. The cause of this lesion is unknown.

Tomb A78, Chamber Northeast (East)

The skeletons from this chamber are both poorly preserved and incomplete. This is partially due to significant ceiling fall that damaged some of the skeletal remains. However, this does not entirely explain the incomplete recovery of skeletal elements. It seems likely that at least some of the missing skeletal elements were the result of inadequate recovery from the primary burial site. Some of the skeletal elements had soil imbedded within the cancellous bone and marrow spaces. This could have occurred from silting in the tomb chamber, but also may have resulted from conditions in the primary burial environment. The poor condition of the skeletal elements precludes obtaining any metric data.

Figure 6.7. View of tomb chamber A78S. Note the silting that had partially encased chamber contents.

The three burials recovered from the chamber include an adult approximately 40 years of age at the time of death, one four-year-old child, and an infant about nine months to one year of age at death. Reactive bone formation is present on some of the endocranial surfaces of the skull fragments of both subadult burials. Infection and scurvy are among the most common pathological conditions that cause this abnormal bone formation. Reactive bone formation is also apparent in unidentified fragments, including one of the sinuses. This is probably caused by chronic infection of this sinus.

Tomb A78, Chamber Southeast (South)
The human remains in this chamber were poorly preserved and incomplete, in part due to significant silting that raised the floor of the chamber about twenty centimeters and encased much of the skeletal remains (Figure 6.7). The remaining evidence indicates that the burials may have been incomplete and in poor condition at the time of secondary interment (Figures 6.8 and 6.9). Skeletal elements of two individuals, an adult male with an estimated age at death of 25–35 years and a subadult 10–14 years, were recovered.

There is no evidence of skeletal pathology in any of the remains.

Tomb A78, Chamber Southwest (West)
There was some sand erosion from the chamber walls onto the floor and around the bones and artifacts near the margin of the chamber but no evidence of siltation (Figure 6.10). Two artifacts from this chamber are of particular interest. The first of these is the broken end of a stone tool (Figure 6.11). The tool end is rounded and abraded from use and is approximately the size associated with the grooves apparent in the ceiling and walls of the shaft tomb chambers. It may have been the broken end of the tool used in carving the tomb chamber. The other artifact is a clay female figurine (Figure 6.12). Several of these figurines were recovered during excavations in other chambers.

A minimum of eleven individuals were buried in this chamber. Three of the burials were assigned burial numbers in the field based on the three skulls that were arranged to the left of the tomb chamber. The preservation and completeness varies between burials. Four of the burials were adult skeletons, two of which were probably female (including Burial 1) and

Figure 6.8. Central bone pile in A78S. The bones were in poor condition.

Figure 6.9. Detailed view of the broken long bones in A78S.

Figure 6.10. Contents of tomb chamber A78W. Some coarse sand had eroded from the base of the chamber wall.

Figure 6.11. Skulls and postcranial bones in A78W. A broken stone tool is seen in the lower right corner.

Figure 6.12. Clay female figurine in A78W.

two of which were probably male. One of the burials was an adolescent (Burial 2) between 11 and 17 years and two were children. The age estimate for one of the children (Burial 3) is 7.5 to 9.5 years, the other is 4.5 to 5.5 years. The bones of one infant and three late-term fetuses were also identified. Skeletal paleopathology included possible cribra orbitalia in Burial 3; one of the adult male skeletons had spondylolysis of the fifth lumbar vertebra. The femora and tibiae of the child burial (4.5–5.5 years) had periosteal woven bone (Figure 6.13), indicative of active pathology at the time of death. The involvement of multiple bones in the child's skeleton indicates a systemic condition as the cause with infection being most likely but other diagnostic options include cancer and metabolic diseases such as scurvy.

TOMB A79

Tomb A79 had the basic clover-leaf relationship between the tomb chambers (Figure 6.14). Two of the chambers (North and South) were sealed by blocking stones. Chambers east and west did not have blocking stones. All of the tomb chambers had significant silting. Why this occurred in the two chambers with blocking stones is unclear. It seems most likely that the silting occurred before the blocking stones were in place, suggesting a variable time gap between interment, final sealing of the tomb chambers, and refill of the shaft. Another possible scenario is that the blocking stones were ineffective in preventing the silting. In neither case it is likely that significant silting could have occurred if the central shaft had been completely refilled. In either scenario it is likely that the central shaft remained unfilled long enough for exposure to at least one annual rainy season.

Tomb A79, Chamber North

Preservation of burials in all the chambers in Tomb A79 was much less than ideal. Silting varied from a few centimeters to around a half meter in depth. The silting in A79N covered most of the central postcranial bone pile (Figure 6.15). The problems created by silting were exacerbated by a limited amount of ceiling collapse. All of the burials in the chamber contained less than 50% of the skeletal elements, and most had less than 25% of the bones present. In some cases, a burial is represented by only one or two elements. How much of the poor recovery of bones can be

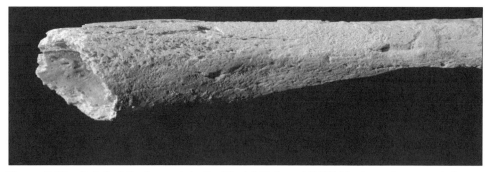

Figure 6.13. Detail of the femoral shaft of Burial 58 from A78SW. The bone fragment is from a child about five years of age. Multiple bones were affected in this burial.

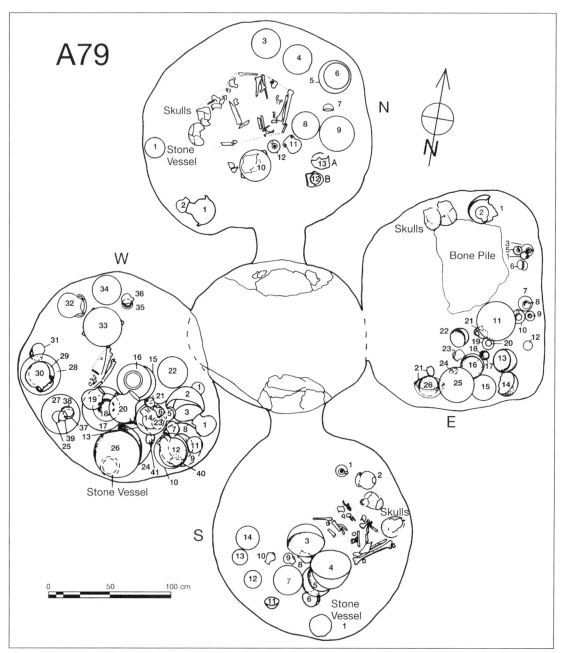

Figure 6.14. Plan drawing of Tomb A79.

attributed to silting and ceiling collapse is not known. The preservation of bones that were present in the chamber was typical of bone preservation in other tombs at the site where recovery of burials was more complete. Because of this, it seems likely that many of the skeletal elements were simply not recovered from the primary burial site and returned for secondary interment at Bâb edh- Dhrâʿ.

The MNI recovered from this chamber is 17. There were three adult burials, including one female and two male skeletons. Two of the burials, one possible male and one possible female, were in the late adolescent or young adult age range. There were seven burials between the ages of two and nine years. Four burials ranged in age from perinatal to about 1.5 years. There was one burial that was in the late fetal to perinatal age range.

Evidence of skeletal disease is limited. The single adult female burial (Burial 1) was about 45 years of age at death. The spine exhibits some evidence of osteoarthritis. One vertebral body fragment probably associated with this burial and most likely from the lower lumbar region has a porous destructive lesion of the anterior superior margin of the vertebral body. This is possibly another manifestation of osteoarthritis apparent in other vertebrae. However, this location as well as the morphology of the lesion raise the possibility of an infectious condition such as brucellosis, evidence of which is more conclusively demonstrated in other burials from the site (Tombs A108NW and A111N) that are described in Chapter 9. In this burial there is also a porosity of the left mandibular fossa that is indicative of breakdown of the articular cartilage and probably would have been associated with temporomandibular joint problems during life.

In Burial 1a, the endocranial surface of the two-to four-year-old child's skull exhibits multiple irregular depressions or convolutions (Figure 6.16) in which fine vascular channels are apparent. The depressions are indicative of an imbalance between the size of the skull and the brain. The convolutions are the result of either a pressure erosion by an abnormally enlarging brain or slow growth of the cranium associated with a failure of the bone to reach its normal thickness because of pressure by the brain. There are a number of conditions that could have caused this disorder, including the various manifestations of hydrocephalus in which there is an excessive amount of cerebrospinal fluid within the cranial compartment.

Figure 6.15. Tomb chamber A79N showing extensive silting.

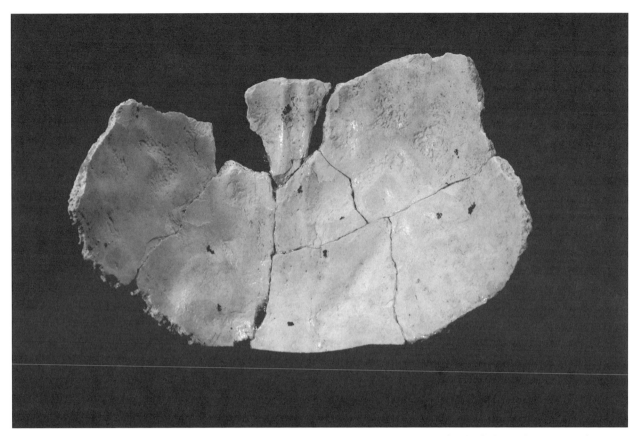

Figure 6.16. Endocranial surface of Burial 1a with abnormal depressions in the skull of a child about three years of age.

Burial 2 is the skeleton of an adult male that exhibits evidence of osteoarthritis in at least two joints; there is no other evidence of disease in this burial. In a bone fragment of a maxillary sinus that cannot be associated with a specific burial, there is evidence of reactive bone formation probably indicative of a chronic infection. Burial 50 is that of a child about eight years of age. The teeth of this burial show evidence of enamel hypoplasia (Figure 6.17). This condition is apparent in teeth associated with Burial 51, a six year old, and Burial 52, who was about five and one-half years of age. Enamel hypoplasias are also apparent in teeth that cannot be associated with a specific burial. There is no evidence of skeletal pathology in any of the remaining skeletal elements that were recovered during excavation.

Tomb A79, Chamber East
Substantial silting of this chamber had occurred in antiquity. This process resulted in embedding some of the skeletal remains in silt and also infiltrating soil into the medullary spaces of the bones. The burials were rather fragmented, and few long bones were sufficiently complete to permit measurement. However, at least some of the individuals buried in the chamber had 25% or more of the skeletal elements present.

The MNI buried in this chamber was 13. Only one of the burials was an adult that probably was male. An adolescent female skeleton about 16 years of age was present. There were two children's skeletons about eight years of age. One skeleton was from a six-year-old child and another from a two-to three-year-old. There were two infants less than one year of age, two neonatal

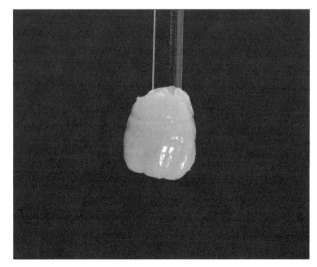

Figure 6.17. Linear enamel hypoplasia in the left maxillary central incisor in a child about eight years of age. Burial 50 from A79N.

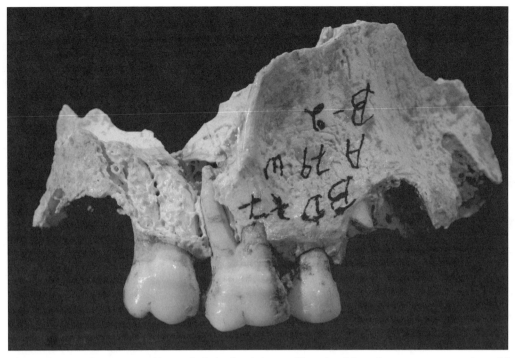

Figure 6.18. Linear enamel hypoplasia in the right maxillary dentition of a young woman about 20 years of age. Burial 2 from A79W.

skeletal elements, and three late term fetuses. The only evidence of skeletal pathology occurs on one of the two neonatal infant skeletons. In this skeleton there is reactive bone formation on one of the femora. There are several diagnostic possibilities for this disorder, including infection, metabolic disease, and cancer. The evidence is inadequate for a more specific diagnosis.

Tomb A79, Chamber Southeast
Major silting in this chamber resulted in poor preservation and incomplete recovery of the human remains. On the basis of the skeletal elements present there were at least eight individuals buried in the chamber. Of these, four were adults (two female and two male). There was one adolescent that probably was female and one child about two years of age. One of the burials was a newborn and the other was a late-term fetus.

Burial 1 is a female about 20 years of age at the time of death. This skeleton shows evidence of asymmetrical development of the skull in which the right side is underdeveloped. The metopic suture of the frontal bone remains. Generally, osteoarthritis is uncommon in the Bâb edh-Dhrâ' skeletal remains; this makes the marked development of joint problems in a young adult burial especially notable. Both femora have significant bone changes in the distal joint surfaces that are associated with osteoarthritis. There are subchondral bone deposits on the femoral articulation with the patella. Slight cystic development is also apparent in the subchondral bone. It is possible that the developmental anomaly apparent in the skull was associated with locomotor deficiencies in the postcranial skeleton that gave rise to the early onset of osteoarthritis.

Tomb A79, Chamber West
As in Chamber A79SE heavy silting affected both the preservation and recovery of human remains from the chamber. Field notes indicate the presence of five or six skulls, but only three are available for study. The fragmentary nature of the skeletal elements as well as the poor recovery limit the data that can be extracted. Age and sex determinations are made with the recognition that the potential error factor in these assignments is higher than in situations where the preservation and recovery is good to excellent. Nevertheless a minimum number of nine individuals were identified. Three of these were adults (two possible female skeletons and one of unknown sex). There is one adolescent skeleton, one six-year-old child and four late-term fetal or infant remains.

Burial 2 is the skeleton of a young woman about 20 years of age. The dentition is severely affected by poor enamel formation resulting in dental hypoplasia. The orbital roofs exhibit porous reactive bone formation known as cribra orbitalia. Both the linear enamel hypoplasia and cribra orbitalia (Figures 6.18 and 6.19) are indicative of at least one

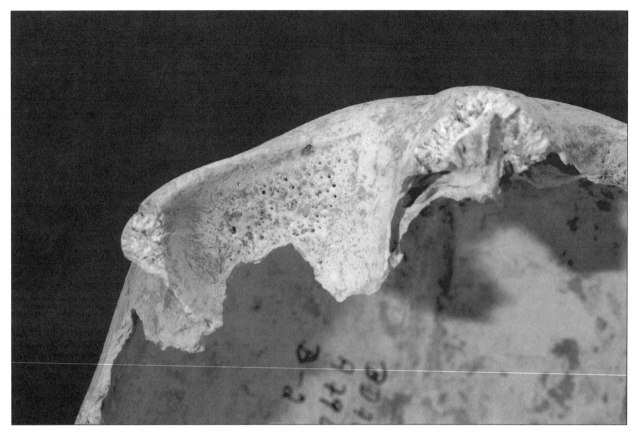

Figure 6.19. Cribra orbitalia in the right orbital roof of a young woman about 20 years of age. Burial 2 from A79W.

episode of serious abnormal stress. Dental hypoplasia occurs only during childhood development of teeth; cribra orbitalia could have occurred shortly before death.

In this chamber, there is other evidence of dental/skeletal disorders, but these cannot be linked to a specific burial. They include the presence of dental hypoplasia in a majority of the loose teeth recovered. A fragment of a right acetabulum has marginal osteophyte formation indicative of osteoarthritis that may have been related to the abnormally shallow depth of the acetabulum. This might have been associated with repeated partial dislocation (subluxation) of the hip. The absence of secondary joint formation (pseudarthrosis) makes more severe complete dislocation (luxation) unlikely. There is no additional evidence of skeletal disorders in the fragmentary and incomplete skeletal elements from this chamber.

TOMB A80

Tomb A80 had a central shaft and four chambers with entryways equidistant from each other and located at the base of the shaft (Figures 6.20, 6.21, and 6.22). All chambers had been sealed with blocking stones. As in Tomb A79, there was silting in all of the tomb chambers, although the amount of silting varied considerably from slight in A80E to substantial in A80W. The variation in the severity of silting between the chambers may reflect different episodes of silting for each of the chambers in which the entryway to the chambers was open to silting at different times, perhaps with an interval of more than one year for some of the silting events. It provides additional evidence that there may have been a gap of at least one year between interment in the chambers and placement of the blocking stones and refill of the shaft. The combination of silting with ceiling fall in each of the chambers adversely affected the preservation and recovery of the human remains.

Tomb A80, Chamber North

The silting that affected the contents of this chamber was exacerbated by moderate ceiling fall (Figure 6.23). The silt was not evenly distributed throughout the chamber and was concentrated at the entryway and the central area of the floor. This suggests that the slurry was thick and must have flowed slowly into the chamber through the entryway. The combination of a limited amount of thick slurry and slow flow pre-

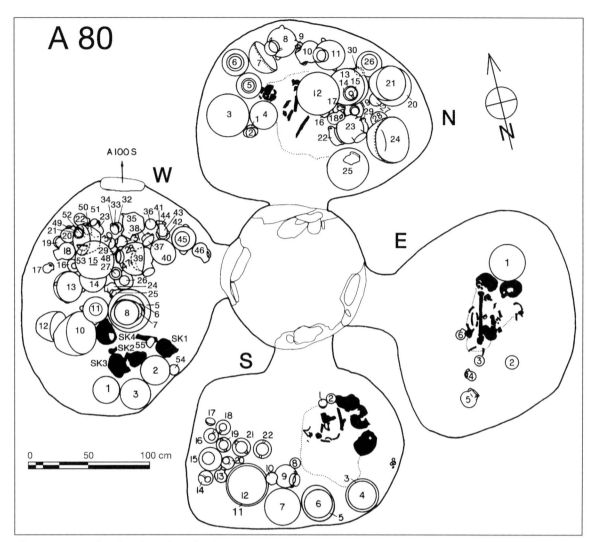

Figure 6.20. Plan drawing of Tomb A80.

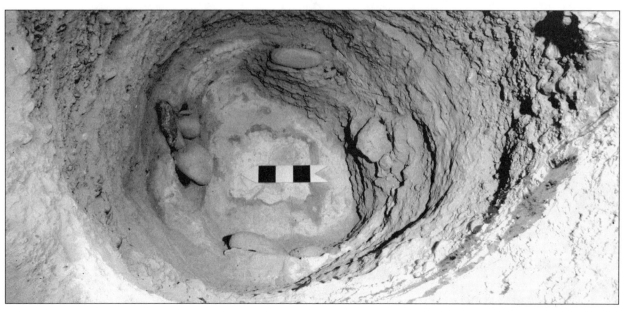

Figure 6.21. Excavated shaft of Tomb A80 with all blocking stones still in place.

Shaft Tombs Excavated in 1977 65

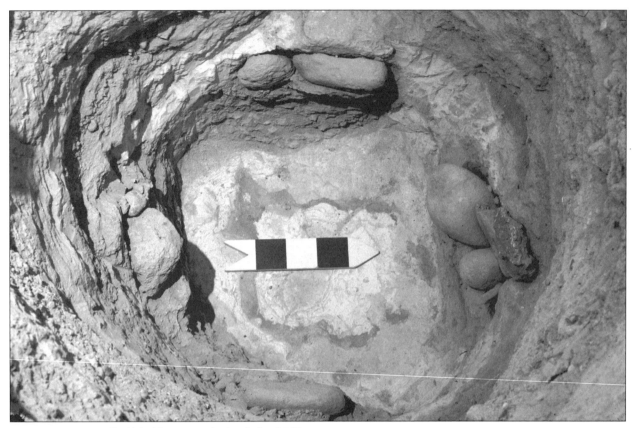

Figure 6.22. Detail of blocking stones in the shaft of A80.

Figure 6.23. Tomb chamber A80N, with extensive silting.

vented even distribution of the silt throughout the chamber. Most of the bones were embedded within the silt. The MNI interred in this chamber is three and includes two adults and an infant.

Burial 50 is the skeleton of an adult about the age of fifty at the time of death. The skeleton is gracile, but the anatomy of the pelvis indicates the probability that the individual was male. The left humeral head is abnormal (Figure 6.24) and was displaced inferiorly with the anatomical neck absent; this is a condition known in modern clinical practice as slipped epiphysis. The subchondral bone of the head is poorly formed and porous. These features are most likely associated with trauma in early childhood that caused detachment of the humeral head at the cartilaginous growth plate, with disruption of growth and subsequent reattachment in the abnormal inferior position. Marginal osteophytes of the humeral head indicate osteoarthritis associated with the abnormality. The corresponding articular surface of the scapula is fragmentary, due to postmortem breakage, preventing evaluation of abnormality in this component of the joint. Additional evidence of osteoarthritis is apparent in the left acetabulum and subchondral surfaces of the right elbow. The vertebrae also exhibit widespread evidence of osteoarthritis with both marginal osteophyte formation of the vertebral bodies and enthesopathies associated with tendon and ligament attachments.

Burial 51 is the skeleton of an adult about 35 years of age and most likely male, although the individual was gracile. There is evidence of osteoarthritis in the appendicular skeleton that is particularly apparent as subchondral bone damage in both knees and osteophyte formation at the subchondral margins of both elbows. The vertebrae also exhibit evidence of osteoarthritis as well as a possible compression fracture of the 11th thoracic vertebral body.

Burial 52 is the skeleton of an infant represented by only a fragment of the right temporal bone. Little can be said about this burial except that the temporal fragment is normal.

Other skeletal elements that cannot be associated specifically with any of the numbered burials show evidence of disease. The dentition of an adult maxilla has hypoplasia indicative of two stress events during childhood. A right fragment of a mandible with dentition contains evidence of enamel hypoplasia from three stress events. There is a mandible with considerable postdepositional tooth loss. However, antemortem tooth loss occurred with all molars of the left side and the right third molar. There is evidence of dental abscess associated with the left first molar that probably was associated with loss of the tooth.

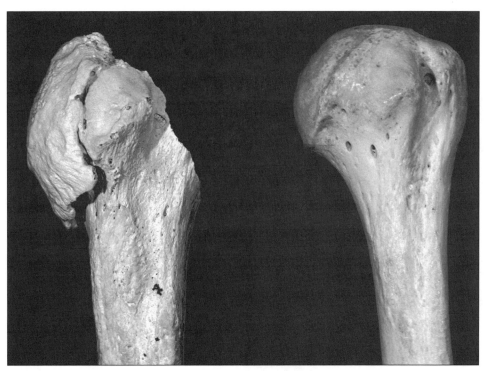

Figure 6.24. Abnormal development of the left humeral head (A80N, Burial 50) compared with a normal left humeral head. Burial 50 is the skeleton of an adult male.

Tomb A80, Chamber East
Silting was slight in this chamber, but considerable ceiling fall was present. Preservation of the human remains in the chamber was poor and very few post-cranial bones were recovered (Figure 6.25). Much of the information on the human remains is based on the skulls; two of the three recovered were in relatively good condition. The MNI recovered from this chamber is three, including an adult female and two adult males.

Burial 1 is the skull of a young adult about 20 years of age at the time of death. Anatomical features of the skull indicate that the individual was female. There is evidence of dental caries in the molar teeth. Burial 2 consists of the skull and C1–C3 vertebrae of a male individual about 35 years of age when he died. No evidence of disease was present in the skull. Burial 50 consists of the occipital fragment and complete vertebral column and pelvic girdle. The individual represented by these skeletal elements was a male about 30 years of age at death. There is a porous lesion of the anterior body of C3 that is probably due to disk degeneration. A left maxillary fragment that cannot be assigned to a specific burial has evidence of antemortem tooth loss (M^2) probably resulting from dental abscess. There also is evidence of reactive woven bone formation in the maxillary sinus, probably indicative of active infection at the time of death.

Tomb A80, Chamber South
In this chamber, ceiling fall was minimal but silting was substantial and embedded most of the bones (Figure 6.26). The bones are fragile and removing them from the silt resulted in additional damage. Despite these problems, preservation and recovery was good, with most of the burials being represented by more than half of the skeletal elements. One burial was limited to the bones of the hands and feet. There is evidence that some of the breakage occurred between primary interment and secondary burial in the tomb chamber. Bone is a remarkably tough tissue, often surviving intact long after death and burial. The breakage seen in the skeletal remains occurring between the primary and secondary burial sites requires some combination of time and/or a primary burial environment conducive to unusually rapid taphonomic degradation of the bone. Which of these two variables was more important is not known, but there is at least a possibility that burial at the primary site was long enough to be a significant factor in producing bone tissue sufficiently fragile to break when removed to the secondary site.

However, it also seems likely that primary burial was in conditions that degraded the protein component of bone more rapidly than normal. If the primary burial environment was not unusual, the length of time in the primary burial site for many burials is more likely to be measured in years rather than months.

The MNI recovered from this chamber is five, including two children and three adults. Two of the adults are male skeletons and one was female. Burial 1 is the skeleton of a child about 3.5 years of age. The skull of this burial has a disorder of the orbital roof known as cribra orbitalia (Figure 6.27a), in which woven bone that is very porous and poorly organized has been added to the normal compact bone surface of the orbit. The manifestation of this condition is associated with several disorders, including infection and scurvy. The presence of reactive woven bone on the alveolar process of the maxilla (Figure 6.27b) and in the bone surrounding the infraorbital foramina was probably caused by chronic bleeding. The two abnormalities are most likely caused by scurvy, although not all the classic features of scurvy of the skull (Ortner and Ericksen 1997) are present in this case. There are also lytic lesions in the maxillary sinuses that may reflect active sinus infection at the time of death. Children with scurvy have increased vulnerability to infectious diseases.

Burial 2 is the incomplete skeleton of a child about 4.5 years of age. This burial also has cribra orbitalia and reactive bone deposition on the occipital bone as well as in the bone surrounding the infraorbital foramina. There is a lesion on the left humeral shaft that may be the result of chronic bleeding and associated with the lesions seen in the bones of the cranium. Scurvy is the most likely diagnostic option for the disorders apparent in this case.

Burial 3 is the skeleton of a female about 27 years of age. In the cranium of this burial, the maxillary sinuses have both compact and woven reactive bone formation most likely indicative of chronic infection. Similar lesions are apparent in the sphenoid sinuses. Evidence of osteoarthritis is seen in the subchondral bone destruction of the right glenoid fossa and the right acetabulum. The latter abnormality is also associated with development of marginal osteophytes. Woven bone in the process of remodeling to compact bone is seen on the right tibia; the left is normal.

Burial 4 is a male skeleton about 40 years of age at the time of death. There are several abnormalities of the spine, including asymmetry of the lower lumbar vertebrae and sacrum. The fourth lumbar vertebral

Figure 6.25. Tomb chamber A80E showing the poor preservation of skeletal elements.

Figure 6.26. Skulls and postcranial skeletal elements from tomb chamber A80S.

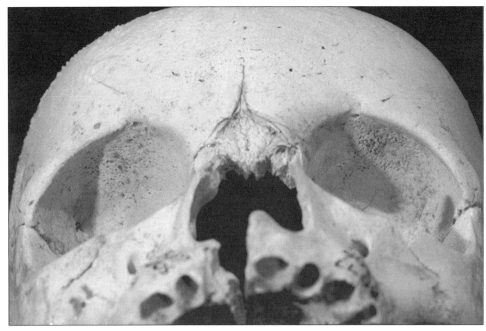

Figure 6.27a. Cribra orbitalia in the skull of a child about 3.5 years of age. Burial 1 from A80S.

body has anterior erosion on the superior margin that is associated with osteophyte development. The most likely cause of this disorder is trauma with anterior herniation of the vertebral disk. The fragment of the fifth lumbar vertebra shows similar lesions. Osteoarthritis is also apparent in some of the cervical vertebrae, including the vertebral bodies but also the diarthrodial joints. Additional evidence of osteoarthritis is seen in the left hip and in the bones of the feet that were available for study; the right hip is normal.

Burial 60 consists of the incomplete bones of the hands and feet and cannot be associated with any of the previous burials. The bones are from an adult and probably male skeleton. No skeletal disorders are present in the bones available for study.

Tomb A80, Chamber West
There was no ceiling fall in this chamber, but silting had raised the floor approximately 50 centimeters, or about half the distance from the original floor to the ceiling. Many of the pots included as tomb gifts floated on the slurry (Figure 6.28), but the bones were largely embedded in the silt and difficult to extract. Damage to the human remains during excavation and clearing of the chamber was considerable. One of the interesting features of this chamber is the opening into the south chamber of Tomb A100 (Figure 6.29). Stones to fill the defect had been placed in the opening in antiquity, apparently from Tomb A100, indicating that tomb chamber A80W may have been created and possibly used before the south chamber of Tomb A100. This is not the only tomb chamber with a

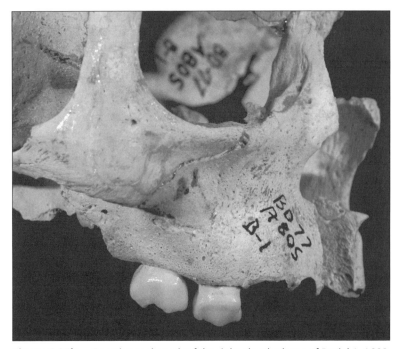

Figure 6.27b. Reactive periostosis of the right alveolar bone of Burial 1, A80S.

man-made defect in the wall, but it is remarkable, given the density of tombs in this portion of the A Cemetery, that more intrusions from one tomb chamber into another did not occur. Clearly, the tomb makers had a fairly clear idea where the other tombs were. This would not have been particularly challenging if the shafts remained partially to completely open for an extended period or if the shaft fill settled, leaving a depression indicating the presence of a tomb at that location.

The MNI interred in this tomb chamber is eleven. There were four adult skeletons, including a young adult female, a female about 40 years of age, and two adult males, both of which were 45 years of age or older at death. The remaining burials were subadults of unknown sex. One of these was a late adolescent about 15 years of age. One of the burials was a child about 10 years of age. There were two children's skeletons about 4 years of age. One skeleton was of a young child about 18 months of age. The burials included an infant about one year old and a late-term fetus.

Burial 1 is the skeleton of one of the adult males; a more specific age estimate is not possible. Both the size of the bones and the well-developed muscle attachment areas on the bone indicate a robust individual. The presence of osteophytes at the margins of several joints, as well as the abnormalities of some of the subchondral bone surfaces, are indicative of osteoarthritis and suggest a physically demanding life.

Burial 1a is the skeleton most likely of a female about 40 years of age when she died. The ectocranial surface of the skull has extensive abnormal porosity with fairly fine holes in the anterior portion of the parietals that become larger toward the posterior cranium, including the occipital bone. There is some abnormal thickening of the parietals, but this appears to affect the outer table and not the diploë. Fine bony spicules occur along the posterior sagittal suture. Infection or metabolic disease are the most likely causes of the lesions. Anemia does produce porosity in the cranium in some cases, but the lack of involvement of the diploë and the spicule formation argue against this diagnostic option.

Burial 1b consists only of the cranial fragments of a child about 10 years of age. The teeth exhibit hypoplastic lines of the permanent dentition near the cemento-enamel junction. The deciduous dentition that still remains has no evidence of dental hypoplasia. Dental hypoplasia in the permanent teeth implies probably more than one stress event associated with early childhood when the crowns of these teeth were forming. The endocranial surface of the frontal and right parietal has smooth-walled depressions associated in life with the convolutions of the brain. The left parietal is poorly preserved but likely had a similar abnormality. This condition is seen in hydrocephalus and craniostenosis; it also occurs with some cranial tumors such as pituitary adenoma. The cause of the abnormal endocranial surface in Burial 1b is not known, but certainly there was an imbalance between the growth of the brain and that of the cranium, with the brain being too large for the available intracranial space. The left orbital roof has a porotic lesion of the compact bone surface. There is no evidence of this condition on the right fragments of the orbital roof. This lesion is nonspecific but may reflect malnutrition that could have contributed to the other evidence of disorders in the cranium and dentition.

Burial 1c consists of cranial fragments that cannot be assigned to another skull. The remains are from a child between three and five years of age. Cribra orbitalia is present in the left orbital roof but is not present in the right orbit. Like Burial 1b, there are endocranial depressions in the fragments indicative of inadequate space for the growing brain. This could be the result of subnormal skull development or abnormal enlargement of the brain as occurs in hydrocephaly and some tumors of the brain compartment.

Burial 1d consists of the mandible only of an individual about 15 years of age at the time of death. Dental eruption is abnormal in this mandible. Both deciduous second molars have been retained long after they should have been lost. The right canine was delayed in its eruption as well. All of the permanent dentition exhibits dental hypoplasia, primarily near the cemento-enamel junction. This abnormality is indicative of at least one stress event during the development of the permanent teeth.

There are no skeletal elements designated Burial 2 in this chamber. Burial 3 is the skeleton of a robust adult male about 45 years of age. There is widespread evidence, including enthesopathies and osteoarthritis, of hard physical activity during life. The latter includes lipping at the margins of subchondral bone and bone deposits on some of the subchondral bone surfaces. Tooth wear is considerable and particularly severe in the first molars, with the pulp chamber exposed, creating a pathway for infection with abscess in the alveolar sockets that penetrated through the labial alveolar process (Figure 6.30). Chronic infection stimulated reactive woven bone formation that was remodeling into compact bone adjacent to the cloaca. This is indicative of a long-standing infectious process. There is a generalized

Figure 6.28. Severe silting in tomb chamber A80W. Some of the pottery floated on the silt, leaving the pottery near the ceiling.

Figure 6.29. Break in the wall of A80W into tomb chamber A100S. Note the attempt to repair the defect with stones and mud mortar.

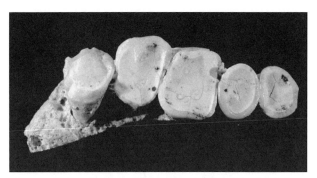

Figure 6.30. Right maxillary dental arch of Burial 3 from A80W, showing severe wear of the premolars and molars.

loss of alveolar bone with dental root exposure that reflects periodontal disease. In the appendicular skeleton, the only evidence of disease other than osteoarthritis is reactive woven bone formation of the left distal tibial crest that is probably the result of traumatic injury to the tibia. The woven bone is remodeling into compact bone, indicating that months if not years had elapsed since the onset of the abnormality.

Burial 50 is the skeleton of a late-term fetus. Given the high rate of morbidity and mortality associated with pregnancy, the most likely cause of death of this fetus is the death of the pregnant mother. In this chamber, the presence of at least two adult female skeletons, one of which was a young adult, raises the possibility of one of these women being pregnant at the time of death.

Burial 51 is the incomplete skeleton of an infant with no evidence of disease in the bones available for study. Burial 52 is the incomplete skeleton of a child about 18 months and, like Burial 51, shows no evidence of skeletal disease. Burial 53 is represented by the right and left petrous temporal bones of a child about 4 years of age. There is no evidence of disease apparent in this burial. Burial 54 consists of the left os coxa of a young adult female about 19 years of age. No evidence of skeletal disease occurs in the single bone of this burial.

There are two unassociated bones that show evidence of a skeletal disorder. A left greater wing of an adult sphenoid has a porous lytic lesion on the endocranial surface. A fractured rib was in the active healing stage at the time of death. Woven bone callus is in the process of remodeling into compact bone, so the elapsed time since the fracture event was probably a few months.

TOMB A86

Tomb A86 contained three burial chambers and is unusual in having an antechamber leading into the southwest chamber of the tomb (Figures 6.31 and 6.32). Skeletal preservation is generally poor and incomplete, presumably because of inadequate transfer of the remains from the primary burial site. Furthermore, some of the skeletal elements that were in the tomb have been lost or misplaced subsequent to excavation, making the task of providing an accurate description of all the burials more difficult. Silting varied greatly from minimal in A86SW to substantial in both A86NE and A86SE. Ceiling fall was not a major factor in the preservation or recovery of the burials. There was considerable variation in the preservation of bone within tomb chambers. A86SW had minimal silting; despite this, some of the long bones had soil infiltrating into the marrow spaces and cancellous bone that must have occurred during primary burial. Also, some of the bones in A86SW showed cracking while others did not (Figure 6.33). This probably reflects variation in taphonomic conditions during primary burial.

Tomb A86, Chamber Northeast

This chamber was completely infiltrated with silt; the tomb contents were completely covered. Because of the hardness of some of the silt, only crania remains associated with Burials 1 and 2 and some postcranial remains were recovered. At least four individuals were interred in the chamber. Two additional skulls, Burials 3 and 4, were left in situ. Burial 1 is the skull of a female about 21 years of age. This skull had bilateral cribra orbitalia, indicating some type of disorder. As stated in earlier descriptions, this condition can be caused by several disorders, including infection, scurvy, and anemia. Burial 2 is the cranium of a robust male 45 or more years of age at the time of death. No evidence of pathology was apparent in the bones of the skull.

Tomb A86, Chamber Southeast

Silting covered most of the skeletal remains in this chamber. Despite the silting, some of the skeletal remains were in fair condition at the time of excavation. However, the remains are fragmentary. Although the MNI interred in this chamber was three, one of the three (Burial 50) is represented by three fragmentary skeletal elements about which little can be said. Burial 1 is the skeleton of an adult female. Included in the burial are fragments of the skull and most of the major long bones. Marginal osteophyte formation in the right acetabulum indicates at least mild osteoarthritis of the hip. The left innominate is not available for study. Fragments of some vertebrae also show minor evidence of osteoarthritis, as do some of the bones

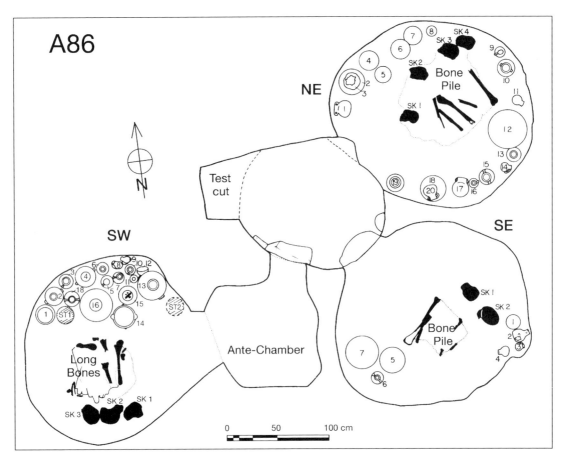

Figure 6.31. Plan drawing of Tomb A86.

Figure 6.32. First author inspecting tomb chamber A86SW from the antechamber.

Figure 6.33. Central bone pile in tomb chamber A86SW. Note the surface cracking of some long bones.

of the feet. This evidence of mild osteoarthritis most likely means that this individual lived past early adulthood.

Burial 2 is the skeleton of a robust adult male 40 or more years of age. Postmortem breakage of the skeletal elements is extensive, but many bones of this burial are represented. Multiple muscle attachment sites exhibit enthesopathies indicative of hard physical activity. This observation is further supported by evidence of osteoarthritis associated with many of the joints available for study. Antemortem tooth loss, probably resulting from complications of caries, is apparent in the first and second molars of both the maxilla and mandible. Some remodeling of alveolar bone has taken place, but dental sockets remain well defined, suggesting that loss of the molars had occurred within a year or two before death.

Burial 50 is represented by the right zygomatic bone and fragments of the first and second cervical vertebrae. The limited evidence from these bone fragments suggests that the individual was probably an adult male. The zygomatic fragment has reactive periosteal bone deposition on the external surface. Trauma is a possible cause, but infection is also.

Tomb A86, Chamber Southwest

The southwest chamber of Tomb A86 was entered through an antechamber that was completely empty. The antechamber was smaller than the burial chamber and was roughly square in floor plan (See Figure 6.31). The burial chamber had a slight amount of ceiling fall, and silt was present in the central portion of the floor extending from the entryway (Figure 6.34). As in other chambers, the skulls were arranged to the left of the entryway, postcranial bones were carefully placed in the center of the floor, and tomb gifts were placed on the right. Some of the smaller bone fragments had been placed in one of the bowls (Figure 6.35).

The MNI interred in the chamber is seven. These include two males, one about 19 years of age and the other 45 or more years, a probable adult male of unknown age, and two female skeletons, one of undetermined age and the other about 35 years of age. Two children were buried in the chamber, one about eleven years of age and the other about seven.

Burial 1 is poorly preserved but likely is an adult female, with no evidence of skeletal disease. Burial 2 is the relatively complete skeleton of a male about 19 years of age. There is a small lesion with woven bone remodeling

Figure 6.34. General view of tomb chamber A86SW, with some skeletal elements embedded in silt.

Figure 6.35. Detail of pot contents in A86SW, in which some small bones and bone fragments had been placed.

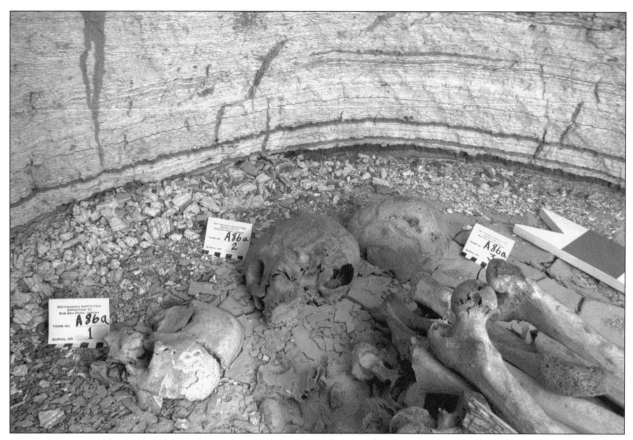

Figure 6.36. Skulls partly embedded in silt in A86SW. Note the surface erosion apparent in the skull vault of Burial 3, which is the skull closest to the far end of the chamber. This is the result of scalping with survival for several months following the trauma.

into compact bone on the left maxillary alveolar process superior to the socket for the third molar. The tooth is missing postmortem, but there is no indication of dental abscess that would account for the lesion, which is probably due to an inflammatory process of the gum. Dental hypoplasia is apparent in the dentition, indicative of a stress event during early childhood, and dental caries occurs in some of the molar teeth.

Burial 3 consists of the cranium, with probable postcranial bones associated. The individual was a male 40 years or older at the time of death. The skull vault contains an extensive destructive lesion with a clearly defined margin at the boundary between the lesion and normal bone (Figure 6.36). The outer table of the skull within the lesion was necrotic for several months before the death of the individual. The margin of the necrotic bone shows the undercutting characteristic of scalping cases. A report of this burial has been published (Ortner and Ribas 1997). Thus far this is the only case of scalping identified in the EB IA remains, and it seems likely that this case is the result of accidental trauma, possibly resulting from an encounter with a wild animal. The cause of death is very likely meningitis, a common complication in scalping cases (Ortner 2003). Other evidence of skeletal disorders in this burial includes bony fusion of several vertebrae associated with a condition known as DISH in the clinical literature (Figure 6.37). The cause of this disorder is not known, but it tends to be associated with evidence of tendon and ligament transformation into bone in other areas of the skeleton. A lesion of the subchondral bone of the lateral condyle, known as osteochondritis dissecans, occurs in this burial in the left femur (Figure 6.38). It usually involves traumatic damage to cartilage and underlying subchondral bone that typically results in a loose body within the joint capsule that may reattach in an abnormal position during the healing phase.

The remaining burials in this chamber consist of fragmentary skeletal elements about which little can be said beyond identifying the sex and approximate age is some cases. Burial 50 may be associated with Burial 1, but this cannot be established with certainty. The skeletal elements are that of a female 35 or more years of age. The cortical bone present is abnormally thin, which may be indicative of osteoporosis. Today in Western societies, osteoporosis is usually associated with the postmenopausal period in females, which would suggest a much older age estimate for this skeleton. However, other conditions, such as malnutrition, can result in osteoporosis as well, and this can occur at any age in either sex.

Burial 51 consists of cranial fragments of an adult female of unknown age. A calcaneus possibly associated with this burial has porous reactive bone formation on the medial surface that may be the result of a localized infection. Burial 52 is the fragmentary remains of a child about ten years of age. There is no evidence of skeletal disease in the fragments available for study. Burial 53 consists of very fragmented bones of what is probably an adult male. A right fibula, possibly associated with this burial, exhibits reactive bone formation on the distal diaphysis. The cause of this disorder is uncertain, but localized infection is the most likely factor. Burial 54 included a few fragments of a child about six years of age. There is no evidence of skeletal disease in the bones available for study.

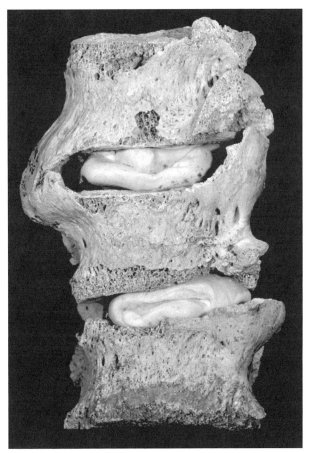

Figure 6.37. Abnormal bone formation with fusion between vertebrae apparent in Burial 3 from chamber A86SW. Today this disorder is known as DISH.

TOMB A87

Major silting occurred in this tomb, creating considerable difficulty in removing the contents. The human remains consist largely of very fragmented bones. Much of the fragmentation appears to be breaks during the excavation and postexcavation processing. Individual burials are represented by only a few elements. Only one chamber is associated with this tomb. The condition of the bones is varied. In some skeletal elements, the outer layer of the bone cortex is flaky with cracks occurring both longitudinally and transversely. However, other bones in the same chamber do not have this condition. This taphonomic variation seems to be linked to specific burials, with most of the bones of some burials affected and other burials unaffected. It seems likely that this difference in taphonomic condition of the bones reflects variation in the primary burial microenvironment; burials in at least some tomb chambers may have come from different primary burial sites.

Tomb A87, Chamber Southeast

The MNI recovered from this chamber is five. Two of the five are adults. One adult is a male 20–25 years of age; the other is a female skeleton of indeterminate age. There are three subadults. One is in the 15–18-year range and may be male. The others consist of the skeletons of a 14–15-year adolescent and a 6–10-year-old child.

Burial 1 is the female adult skeleton of indeterminate age. There is no evidence of significant skeletal disease in the bone fragments available for analysis. Burial 2 is a subadult 14–15 years of age at death. The left distal humerus has reactive changes in the three fossae that may have been the result of hyperflexion and hyperextension. The right humerus is not available for comparison. These lesions are probably associated with strenuous physical activity. However, this is an unusual manifestation in someone this young. Burial 50 consists of fragments of a 15–18 year old who most likely was a male. There is no evidence of skeletal pathology in the skeletal elements available for study.

Burial 51 is the fragmentary skeleton of a 6–10-year-old child. The body of a lumbar vertebra exhibits hypervascularity with some evidence of reactive woven bone formation. This may be within the normal range associated with the growth and development of the vertebrae. However, the presence of woven bone suggests the possibility of an early stage of an infectious disease such as tuberculosis. Burial 52 consists of the fragmentary skeletal remains of an adult male between 20 and 25 years of age. In the right proximal femoral shaft there is a circumscribed, depressed lesion with

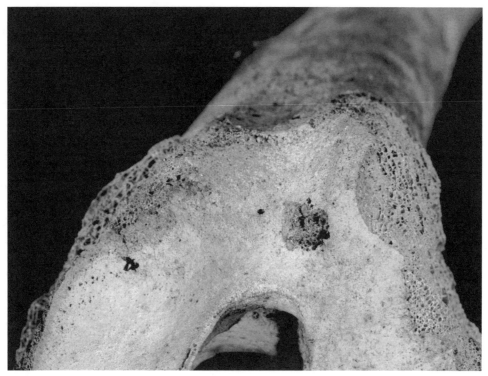

Figure 6.38. Partial destruction of subchondral bone of the left lateral condyle in Burial 3, A86SW. This abnormality also involves the overlying cartilage and is a manifestation of osteochondritis dissecans.

a sclerotic margin known as a fibrous cortical defect. This is an asymptomatic condition that today is usually an incidental radiographic finding.

TOMB A88

This tomb is poorly defined relative to a shaft. A shaft may have existed, but its location could not be determined with certainty during excavation. The tomb consists of a single chamber that contains two stratified levels, upper and lower (Figure 6.39). Major silting occurred in the chamber, making excavation and recovery of chamber contents particularly challenging. The human remains are fragmentary and skeletal elements for each burial include less than 20%, with the exception of Burial 1, which is associated with the upper level. Similar to what occurs in a few other EB IA chambers, the bone elements vary relative to preservation of cortical bone. In some burials the outer layers of bone are cracked. There is also evidence that some of the broken edges of the skull fragments were abraded before excavation. This probably occurred during transfer from the primary to the secondary burial site and interment in the chamber. Another feature of interest in the bone fragments is the presence of many bore holes (Figure 6.40) caused by insect (probably dermestid beetles) larvae (Ortner 1981). This is the only tomb chamber where this condition is present in skeletal remains. It is not clear if the holes were created at the primary or secondary burial site. Dermestid beetles normally do not penetrate soil for more than a few (ca. 15) centimeters. Either the larvae or beetles could have been carried into the chamber during the interment in the secondary burial site. The lack of evidence of this insect activity in other chambers excavated at the site argues against this possibility. If the boring occurred at the primary burial site, the temporary grave would have had to be on the surface or very close to it. The MNI buried in both levels of this tomb is twelve. These consist of nine adults, two subadults, and one burial in which the age category could not be determined.

Tomb A88, Upper Level

Burial 1, the only burial associated with this level, was an articulated female skeleton (Figure 6.41) placed behind a stone barrier. The estimated age is 45 years. Antemortem tooth loss is apparent in the mandible. The maxilla is missing. There is widespread evidence of joint disease, including osteophyte formation at the margins of the vertebral bodies as well as some porous degeneration also in margins of the vertebral end plates. Enthesopathies are apparent at muscle attachment sites, particularly in the innominates.

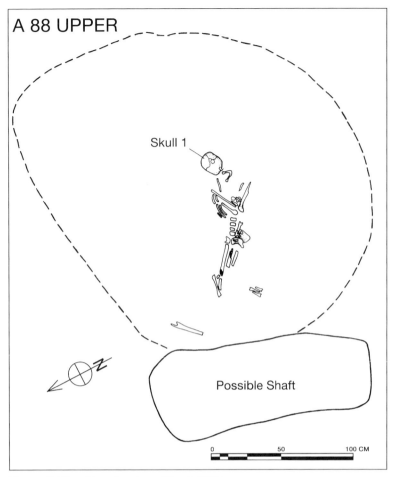

Figure 6.39. Plan drawing of Tomb A88U.

The bones of this skeleton are unusually light relative to other skeletons of the same sex and similar size. Cortical thickness of the long bones is diminished. Both of these abnormalities are indicative of osteoporosis. This condition is seen in other skeletons recovered from the cemetery. In a female 45 or more years of age this is probably due to hormonal changes associated with menopause, but it is important to remember that other metabolic problems, including malnutrition, can also cause osteoporosis.

Tomb A88, Lower Level
Most of the burials and tomb gifts were located on this level of the tomb chamber. They were not arranged in the typical pattern apparent in most of the other EB IA tomb chambers at Bâb edh-Dhrâ' (Figure 6.42). There were two large bone piles, but there was a separate pile for a small skeleton, and two of the skulls (Burials 2 and 3) were separated from the other skeletal assemblages. Because of the highly fragmented nature of the burials, the association between skeletal elements, particularly the skull and postcranial bones, is problematic. Burial 2 is a fragmented female skeleton that was found in a separate bone pile. The estimated age is 35 or more years. There is no significant evidence of skeletal disease. Burial 2a was located near Burial 2 but has skeletal elements that could not be part of Burial 2, even though the burial was that of an adult female, as was Burial 1. There is no evidence of skeletal disease in the bone fragments available for study.

Burial 50 is a fragmentary skeleton of a 12-year-old child. There is a depressed lesion in the outer table of the right parietal bone. These have been associated with depressed fractures caused by a blow to the skull. However, they also occur as the result of a pressure erosion caused by a cyst. The latter condition is entirely asymptomatic and an incidental finding in modern clinical cases. Burial 51 is the fragmentary skeleton of a child about four years of age. There is a lesion reflecting increased vascularity on the right scapula. A lesion at this site is associated with scurvy but, lacking evidence of that disease in other skeletal elements, this can only be considered a possibility; there is more convincing evidence of scurvy in other juvenile skeletons from the EB IA phase of the cemetery.

Burial 52 consists of the fragmentary bones of a robust adult male. Possible evidence of strenuous physical activity is apparent in the juxta-articular erosion of cortical bone in the right proximal humerus. The left humerus is not available for study. Burial 53 is the skeleton of an adult female. There is no significant evidence of skeletal disease in the bone fragments available for study. Burial 54 is also the fragmentary skeleton of an adult female. In the available fragments there is no evidence of skeletal pathology. Burial 55 is highly fragmentary and probably the skeleton of a robust young adult male. In the skull fragments there are three lesions that were the result of skull trauma (Figure 6.43). All appear to be the result of a cut into the skull elements by a sharp weapon such as an axe or sword. All of the lesions show evidence of healing but two more so than the third, in which woven reactive bone is still present, indicating that the most recent trauma took place a few weeks before death. None of the injuries would have compromised the dura, so the injury could have caused life-threatening concussion,

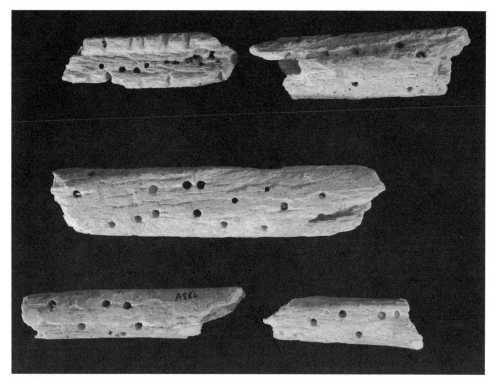

Figure 6.40. Holes in bone from A88 probably caused by dermestid larvae.

Figure 6.41. Burial 1 from the upper level of A88. Bones are in approximate anatomical relationship. Burial is from an adult female about 45 years of age.

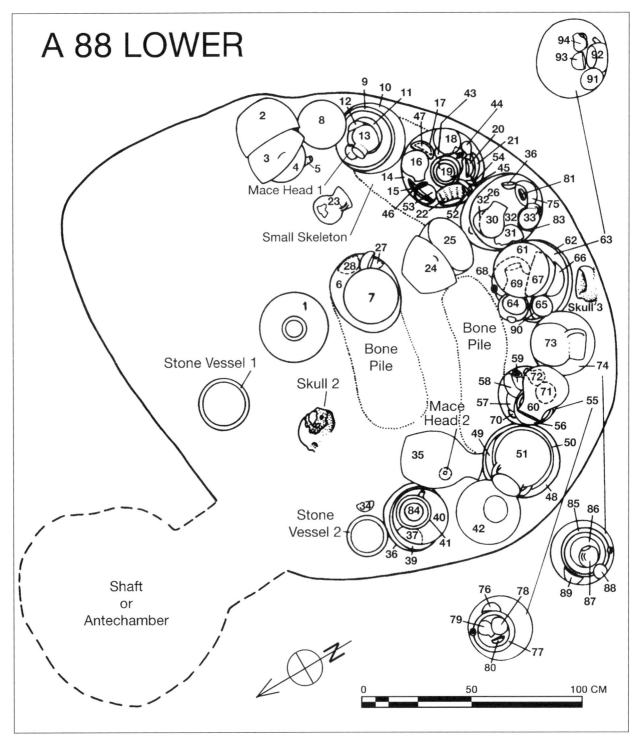

Figure 6.42. Plan drawing of Tomb A88L.

but the even more serious complication of meningitis would have been avoided. Evidence of skull trauma associated with interpersonal violence occurs but is rare in the EB IA skeletal remains.

Burial 56 is also highly fragmented and is the skeleton of an adult female. The diaphysis of the right humerus is abnormally angled, suggesting poor alignment of a fracture resulting in deformation. If a fracture occurred, it did so many years before death, since the diaphysis is otherwise normal. The left humerus is unavailable for study. Burial 57 is the highly fragmented skeleton of an adult female. There is no evidence of skeletal disorder apparent in the elements available for study. Burial 58 is poorly preserved. The bones are riddled with holes

Figure 6.43. Fragments of the skull from Burial 55, A88L. There are multiple depressed lesions in the fragments. The most apparent of these lesions is seen in the second fragment from the right on the bottom row.

approximately two millimeters in diameter of very uniform size and perfectly round (See Figure 6.40). The bone is too fragmentary to determine age or sex with any certainty. However, the skeleton is gracile, and if it is an adult it is more likely to be female.

There is a pathological frontal fragment that cannot be associated with a specific burial. The defect is a porous lesion in the external surface of the outer table. The diploë is unaffected. Porosity occurring in otherwise normal cortical bone is the result of hypervascularity that can be caused by a variety of pathological conditions, including scalp infection, but trauma or scurvy are also associated with this type of disorder.

TOMB A89

Unfortunately, no photographs of this tomb were taken and the field notes are less than ideal. However, a plan drawing of the tomb complex (Figure 6.44) indicates the presence of three chambers. All of the chambers were badly silted, making recovery of chamber contents difficult. The human burials are highly fragmentary and incomplete, with most burials represented by less than 20% of the total skeleton. The fragmentary nature of the burials combined with the secondary burial tradition made association of some bones to a specific burial difficult in all cases and impossible in some. In some of the burials, the taphonomic processes varied within the burial, indicating varying microenvironmental factors affecting preservation within the burial.

Tomb A89, Chamber Northwest

This chamber was completely filled with silt. The chamber also contained an unusually large assemblage of tomb gifts. There appear to be three levels of matting with a base layer that had some bone elements on it and some bone elements and fragments between the upper two levels and on top of the highest level. The significance of the levels remains unknown. It is clear that elements of more than one burial were associated with a single layer, so it seems unlikely that the layers represent different burials. The MNI associated with this chamber is five, including three adults and two subadults. Preservation was poor and no burial had more than about 20% of the elements available for analysis. Because of poor preservation, no burial numbers were assigned during excavation. It seems likely that the different layers are the result of multiple interment events with silting occurring between the events. These events may have been separated by several months, depending on how many silting episodes occurred. There is additional

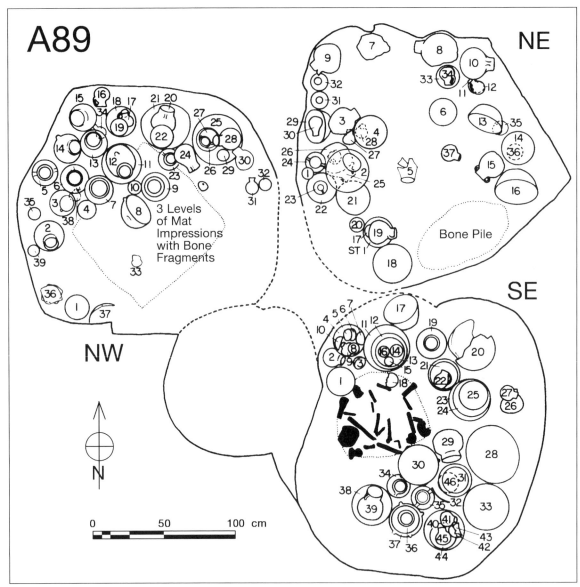

Figure 6.44. Plan drawing of Tomb A89.

evidence supporting silt flow through an open shaft and chamber entrance as the source of at least some of the siltation in the tomb chambers.

Burial 50 is the incomplete skeleton of an infant between two and four years of age. Fragments of both the skull and postcranial skeleton are present, and there was no evidence of skeletal disease. Burial 51 consists of a few very gracile and fragmentary postcranial elements of an adult female. A more precise age estimate is not possible. The postcranial long bone fragments have unusually thin cortices and narrow diameters. This may be a reflection of the overall gracility of the individual, but osteoporosis needs to be considered as well.

Burial 52 is the incomplete skeleton of a robust adult male about 25 years of age. There is evidence of enthesopathy suggestive of vigorous physical activity. Destructive, cystic lesions are present on the margins of the right humeral head. The lesions have a sclerotic margin indicating that remodeling had occurred before death. The left humerus was not recovered during excavation. Marginal cystic lesions are most commonly associated with erosive joint disease, but this diagnostic option cannot be made with certainty because of the incomplete and fragmentary nature of the burial.

Burial 53 is the fragmentary skeleton of an adult male about 30 years of age. There is no evidence of skeletal disease in the fragments available for study. Burial 54 consists of cranial and dental fragments of a child seven to eight years of age. There is no evidence of skeletal disease.

Tomb A89, Chamber Northeast

This tomb chamber was completely silted, and removal of chamber contents, including the human remains, was difficult. This was a factor in the very incomplete and poorly preserved condition of most of the burials. Within burials, there was considerable variation in preservation. The poor burial conditions and preservation made the association of skeletal elements to a specific burial uncertain in most cases. The MNI interred in the chamber was five, including one adult female, two adult males, and three subadults ranging from a late term fetus to about four years of age.

Burial 50 is the skeleton of a very gracile adult female. There was no significant evidence of disease in the very incomplete skeletal remains available for study. Burial 51 is the one burial in this chamber that was relatively complete, with more than half of the skeletal elements represented, although preservation of the bones is rather poor. The burial consists of the skeleton of a robust adult male 45 or more years of age. A fragment of the frontal sinus exhibits porous reactive bone on the internal surface that probably was caused by a chronic sinus infection. The lesions are undergoing remodeling, indicative that the condition was not active at the time of death. Additional bone lesions are suggestive of osteoarthritis and vigorous physical activity.

Burial 52 is the skeleton of a child about four years of age. There is no evidence of skeletal disease in the very incomplete skeletal remains available for study. Burial 53 is the skeleton of a late-term fetus or neonate that is very incomplete. The bone fragments available for analysis show no evidence of disease. Burial 54 consists of incomplete cranial elements along with a few postcranial fragments of a child between one and three years of age. There is no convincing evidence of skeletal disease in this burial.

Tomb A89, Chamber Southeast

The MNI associated with this tomb chamber is thirteen. These consist of seven adults, five of which were male skeletons, two female. The six subadult burials range in age from late-term fetal to 10–12 years of age. Sex was not determined for any of the subadult skeletons. None of the burials in this chamber was identified as a specific burial during excavation. Because of this, numbers were assigned to burials during laboratory analysis beginning with Burial 50.

Burial 50 is the skeleton of an adult male with approximately 30% of the skeletal elements represented. The age at death was about 19 years. An interesting and potentially important pathological feature of this burial is the presence of fine, porous, woven bone formation on the subchondral surfaces of several joints (Figure 6.45). The layer of abnormal bone is less than

Figure 6.45. Fragment of the left acetabulum from Burial 50 with subchondral bone porosity. The burial is a male about 19 years of age from tomb chamber A89SE.

a 0.5 millimeter thick and is a condition unlikely to be seen in modern clinical radiographs or CT scans. The cause of this abnormality cannot be established with certainty, but diagnostic options include one of the erosive arthropathies. The widespread and bilateral bone formation apparent in this case makes rheumatoid arthritis unlikely. The bilateral involvement tends to rule out most of the seronegative erosive arthropathies, but early stage psoriatic arthritis is possible. However, specific diagnosis is unwise, particularly in view of the early stage in the disease process, which is unlikely to be well known or understood in a modern clinical context.

There is an additional abnormality of both femora, the shafts of which are flattened in the mediolateral axis. There is also abnormal bowing both in the anterio-posterior and mediolateral axis. The most likely cause is the vitamin D deficiency disease, rickets. In the vertebral fragments, there is some evidence of a destructive process affecting the lower lumbar vertebrae, which could have been caused by tuberculosis. Other cases of probable tuberculosis have been identified in other tombs, so it is likely that this disease was endemic in these people. Although this individual was male, the skeleton is not robust, which may reflect debilitating illness beginning in early childhood and affecting the development of the individual.

Burial 51 is the skeleton of a female 17–19 years of age. This burial is one of the examples where the taphonomic conditions vary within a single burial. The bones of the upper body are preserved in relatively good condition, but the preservation progressively deteriorates toward the bones of the feet. Both the acetabular portions of the hip joint exhibit subchondral bone destruction that may reflect hard physical activity in a young adult.

Burial 52 is the skeleton of an adult male. A more specific estimate of age is not possible because of the poor preservation of skeletal elements. There is evidence of strenuous physical activity resulting in subchondral bone breakdown, osteophyte and enthesophyte formation. An unusual abnormality is apparent in a left 9th or 10th rib. In this rib, there is bone remodeling and hypertrophy of the posterior end and the bone articulating with the vertebra is absent. The most likely cause of this abnormality is fracture and non-union of the posterior rib.

Both Burials 53 and 54 consist of the very incomplete and poorly preserved skeletons of very robust, adult males. Aside from some minor evidence of degenerative joint disease, there is no significant evidence of skeletal disease in the fragmentary bones available for study. Burial 55 is the incomplete skeleton of a child probably between six and eight years of age. There is no evidence of skeletal disease. The partial skeleton of another child comprises Burial 56, which includes cranial and postcranial elements. The estimated age is 10–12 years and there is no evidence of disease in the fragmentary and incomplete skeleton.

Burial 57 consists of cranial and postcranial fragments of a child probably between one and two years of age. The cranial fragments exhibit extensive evidence of porosity of the outer table and porous reactive bone formation on the inner table. This pattern of porosity and porous reactive bone formation can be caused by more than one disorder, including scurvy, but reactive bone formation on the inner table is more commonly seen in cases of infection resulting in meningitis (Schultz 2003). Evidence of probable tuberculosis in other Bâb edh-Dhrâ' EB IA burials supports the conclusion that this disease was endemic in the human society at this time. Tuberculous meningitis, of all the potential causes of meningitis, is most likely to provide the time between onset of meningitis and death to permit some reactive bone formation.

Burial 58 consists of a fragment of the distal humerus and diaphysis of the femora. Little can be said about these fragments other than they are probably from an infant between birth and one year of age. There is no evidence of skeletal disease. Burial 59 contains only postcranial elements, and there remains some question regarding the association of all the elements with this burial. The bone fragments available for study suggest that they were from an adult male. There is no evidence of skeletal disease.

Only postcranial skeletal elements are associated with Burial 60. As is apparent in several other burials in this and other tombs, there is considerable variation in bone preservation within a single skeleton. The bones present are from a gracile adult female. A specific age estimate is not possible for this burial. Osteoarthritis affecting both the right and left acetabulum suggests an age past early adulthood. There is no other evidence of significant skeletal disease.

Burial 61 is represented by the fragmentary cranial and postcranial elements of a late-term fetus. Burial 62 only includes the left petrous portion of the temporal bone, also from a late-term fetus. There is no evidence of skeletal disease in either burial. The late stage of development of these fetal remains suggests the possibility of aborted twins or the death of the mother late in pregnancy.

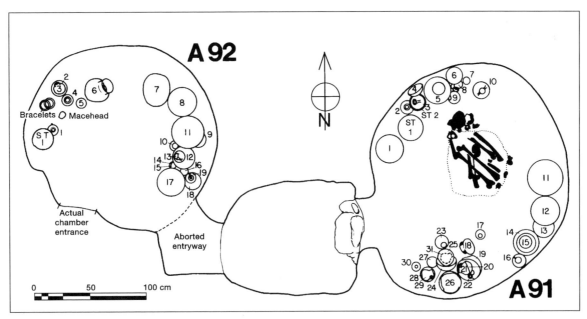

Figure 6.46. Plan drawing of Tombs A91 and A92.

There are several bone fragments from this tomb chamber that exhibit evidence of skeletal disease but that cannot be associated with a specific burial. These skeletal disorders include a lytic lesion of an iliac fragment that could have been caused by an infectious disease such as tuberculosis, endosteal reactive bone formation of the skull possibly linked to infection, porous lesions of the outer table of cranial fragments suggestive of scalp infection, and enamel hypoplasia and caries of several teeth.

TOMB A91

The relationship between tombs A91 and A92 is unclear (Figure 6.46). The excavation and proximity of the two chambers suggests a linkage, but A92 appears to have an independent entryway. For this reason, the two chambers will be treated as separate tombs, but the possibility remains that they were part of a single tomb.

Within the chamber, there was a slight amount of silting that particularly affected the smaller bones lying on the original surface of the chamber floor. There also was some ceiling fall. Element completeness of the human burials generally was poor, although Burial 60 was represented by several bones and was more complete than the other burials in the chamber. Preservation also varied between burials. Some bone elements were in good condition, with little evidence of cracking of cortical bone. However, the cortical bone of most elements was badly cracked and brittle. The MNI buried in this chamber is 13, which includes three adults, two probable adults, and eight subadults. Burial numbers begin with 60.

Burial 60 in Tomb A91 is better preserved and more complete than other burials in this chamber. It consists of the remains of a 14-year-old who was probably female. The angle of the right femoral neck relative to the shaft is smaller than normal, indicating a developmental defect that would have affected the gait of the individual. This condition is known as coxa vara and would have been associated with abnormal biomechanical loading of the knee. The diaphysis of both tibiae exhibits formation of striated porous woven bone that was remodeling into compact bone at the time of death. The pathology was much more pronounced on the right than the left. This is suggestive of a systemic infectious condition, although other conditions stimulating bone formation such as trauma cannot be ruled out.

The very fragmentary remains associated with Burial 61 appear to be those of a young adult female about 19 years of age. A right proximal fragment of the radius that probably is associated with this burial has an abnormal angulation suggestive of a green-stick fracture in childhood with less than ideal alignment on healing. Burial 62 is the very fragmentary and incomplete remains of a young adult that is probably female. No evidence of skeletal disorder is present in the few elements available for analysis.

Burials 63 and 64 are incomplete and fragmentary remains of two children aged eight to ten and six to eight years respectively. There is no evidence of skeletal disorders in either of the burials. Burial 65 is the incomplete postcranial skeleton of a child between one and two years of age. There is reactive porous bone formation on the diaphysis of the tibia suggestive of trauma or infection.

The incomplete postcranial skeletal elements of an adult male constitute Burial 66. The bones are robust, with evidence of bone formation in the olecranon fossa of the left humerus. The cause of this bone abnormality is not certain, but chronic hyperextension of the elbow is a plausible possibility. Hyperextension can be the result of several biomechanical actions, such as frequent spear throwing.

Burial 67 consists of cranial fragments of an infant about one year of age. The right orbital fragment has evidence of cribra orbitalia. The left orbital roof is not present. Burial 68 consists of cranial elements of a child between three and four years of age. Porous reactive bone formation is apparent on several elements, including the alveolus of the left maxilla, the floor of the nasal aperture, and the infraorbital foramen. The right maxilla is not present. Similar lesions also occur on the right zygomatic bone, the endocranial surfaces of several cranial fragments with greatest severity associated with the occipital bone. One of the parietal fragments exhibits abnormal thickness of the diploë. A proximal left rib fragment has a woven bone lesion that is probably early stage callus resulting from fracture.

Burial 69 is represented by a single bone of the cranial base (pars basilaris) of a late fetal to perinatal infant. There is no skeletal disease associated with this element. Burial 70 consists of postcranial fragments of a 12–14-year adolescent. There is no significant evidence of bone disorder. Burial 71 includes only a right calcaneus, and Burial 72 consists of a left calcaneus that is different in size from that of Burial 71. Both calcanei are normal.

There are skeletal elements that cannot be associated with a specific burial. Cranial elements, including the frontal, the right parietal, and a fragment of the occipital, are associated with a cranium assigned the number B-1a. There is a healed lesion of reactive bone on the internal vault. Both trauma and infection should be considered in differential diagnosis. Cranial elements assigned to B-1b include right and left parietal fragments and a partial occipital bone. There is no significant evidence of disease in these cranial elements.

Additional unassociated elements of interest include a fragment of a left mandibular ramus of unknown age or sex with porous reactive bone lesions. The lesion is most apparent on the coronoid neck and process where the temporalis muscle inserts. A specific diagnosis is not possible without additional evidence, but this is a site and a type of lesion associated with scurvy in subadults. There is also a right upper canine that has marked dental hypoplasia indicative of a significant episode of disease stress at the time this tooth was developing. A fragment of the distal radius, side undetermined, has a woven bone lesion that is probably early stage callus resulting from a fracture.

TOMB A91(92)

This tomb was located just to the west of Tomb A91, and as with A91 only a single chamber was identified. It remains possible that it is linked with A91. There was little silting in the chamber of Tomb A91(92), but there was considerable ceiling fall. Some of the bones in direct contact with the floor of the chamber were affected by the minimal silt, but most of the bones were unaffected except by the damage that may have been done by the ceiling fall. Both preservation and skeletal elements recovered are poor, and this limits the observations that can be made. There was a minimum number of five individuals buried in this chamber, including three adults and two subadults. Burial numbers in this chamber begin with 63.

Burial 63 consists of a male between 20 and 25 years of age. No skeletal disorders were found in the fragmentary remains available for analysis. Similarly, no evidence of skeletal disease was found in Burial 64. This burial included the fragmentary skeletal elements of a young to middle-aged female.

Burial 65 is the very incomplete remains of a male in his early 20s. The right acetabulum is abnormal, with subchondral erosion and reactive bone deposition on associated surfaces. The left acetabulum is not present. Trauma, infection, and arthritis are all options in differential diagnosis, although bone deposition is less likely in trauma-induced damage to the joint.

Burial 66 is the fragmentary remains of a child between six and eight years of age. There is porous reactive bone formation associated with the proximal left femur. The abnormality is not present on the right femur fragment. Trauma, infection, and cancer can all stimulate this type of lesion, although infection is the most common cause. Burial 67 consists of a single tibial fragment of a late-term fetus. There is no evidence of significant skeletal disease associated with this fragment.

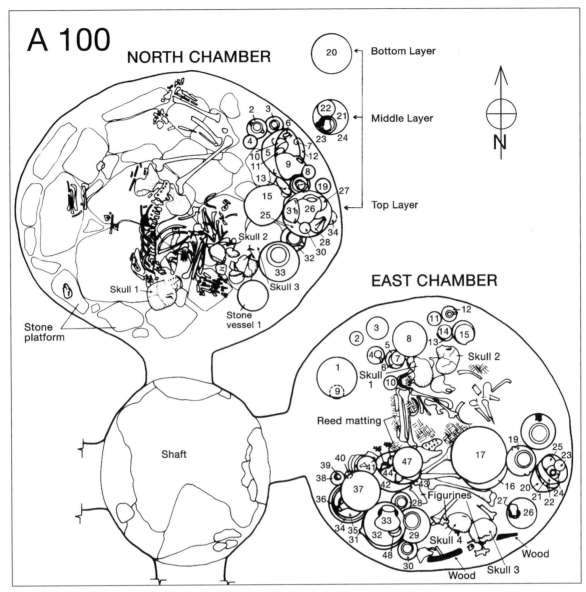

Figure 6.47. Plan drawing of Tombs A100N and A100E.

TOMB A100

This tomb consists of a central shaft and four chambers radiating from the base of the shaft (Figures 6.47 and 6.48). The basic architecture of the tomb is typical of the EB IA burial tradition. However, the north chamber is unlike the other three chambers in that it was cleared of its earlier burial contents and reused in EB IB. Also, the floor of the chamber was covered with carefully fitted flat stones (Figures 6.49 and 6.50). Between two of these stones a fragment of the iliac crest of an adult innominate (os coxae) was recovered. There was only one other adult burial in the tomb, and this fragment is not from this burial, so it represents a remnant of the previous use of the chamber. It is possible that the stone floor was present in the earlier inter-

ment and the fragment was not detected and removed because of its location between two stones. The fact that no other small bones or bone fragments from the previous interment were recovered during excavation argues against this option. More likely it was simply a fragment left behind when the chamber was cleared for EB IB use and, either by chance or by action of the people reusing the chamber, became lodged between the stones used for the floor. The pottery in this chamber is all dated to the EB IB period, indicating that the original tomb gifts had been removed as well. Unlike the burials in all the other shaft tomb chambers, the burials present in A100N are all primary, with the possible exception of the infant burial, and reflect at least two episodes of interment. Preservation of some of the

Shaft Tombs Excavated in 1977 89

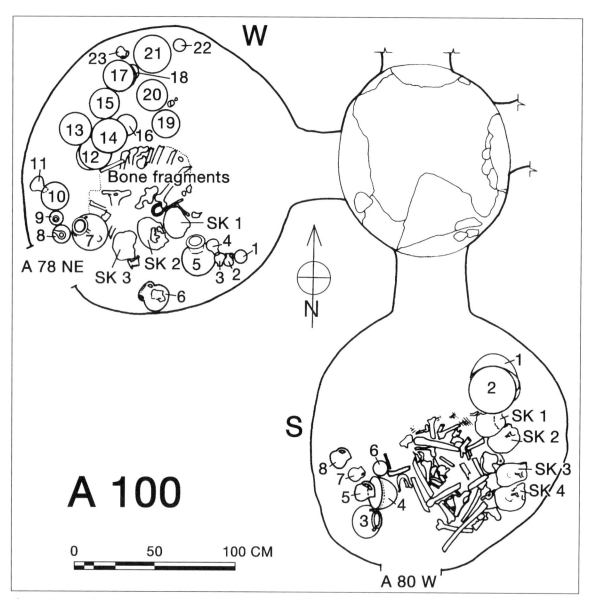

Figure 6.48. Plan drawing of Tombs A100S and A100W.

burial artifacts is remarkable, with dried grapes found in one juglet (Figure 6.51) and dried protein residues found in a sealed jar (Figures 6.52a and 6.52b).

The time difference between the initial interment in A100N, probably in EB IA like the other chambers in this tomb, and the subsequent clearing and reuse imply that the location of the tomb was well known for an extended period of time. This could have resulted from the shaft being kept open for many years or from some method of identifying the location of the shaft.

We can only speculate on what happened to the burials disinterred when the EB IB burials were placed in the north chamber. We have been unable to match the iliac fragment found in the chamber with any of the burials recovered from the three other chambers that are all associated with the EB IA period. This suggests that the EB IA skeletal remains and cultural artifacts from the north chamber were simply discarded. However, the unusually large number of individuals (24) interred in A100E and the two distinct bone piles make this chamber the possible place where the EB IA burials from A100N were moved (Schaub 1981b).

Tomb A100, Chamber North
The north chamber contains the remains of four individuals (Figure 6.49). There was a slight amount of ceiling fall in the back portion of the chamber, but this had no significant effect on the preservation of the

90   CHAPTER 6

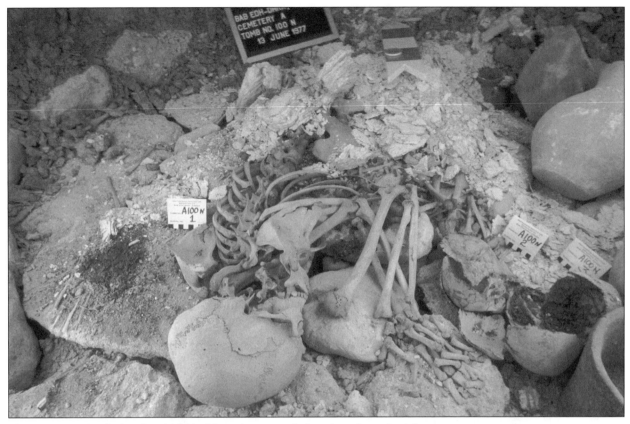

Figure 6.49. Tomb chamber A100N with paving stones below the primary burials.

Figure 6.50. Paving stones on the floor with lining stones on the far wall of tomb chamber A100N after burials and cultural artifacts in the chamber had been removed.

Shaft Tombs Excavated in 1977

Figure 6.51. Small juglet from A100N containing dried grapes.

burials. Two of the burials were children apparently interred as primary burials in the first EB IB burial episode. Burial 2 is the complete skeleton of a child about four years of age and Burial 3 the complete skeleton of a child about seven years. An adult male, primary burial (Burial 1) about 40 years of age was interred subsequently with sufficient elapsed time between the burial episodes so that some of the bones of the children were easily displaced. Burial 4 is the complete skeleton of an infant. The completeness associated with this burial argues for it being a primary burial, but the bones had been somewhat disturbed, probably at the time of interment of a

Figure 6.52a. Large sealed jar from A100N containing dried food residues.

subsequent primary burial. However, the disturbed distribution of the bones creates the possibility that the infant remains had been buried elsewhere initially, although this seems unlikely. Burial 1 has two healed depressed lesions on the cranial vault (Figure 6.53). A healed lesion associated with an infectious focus is possible, but a more likely explanation for both lesions is a depressed fracture resulting from a blow to the head. Significant dental plaque is associated with the buccal surfaces of most teeth (Figure 6.54a). Severe dental caries resulting in exposure of the pulp cavity and root canal is apparent in the upper left second molar (Figure 6.54b). Burial 2 has a circumscribed destructive lesion of the cranial base (Figures 6.55a and 6.55b). Tuberculosis of the skull does produce lesions of this type in children, so TB is a diagnostic option, but other conditions, including a fibrous cyst, are possible. Burials 3 and 4 have no skeletal evidence of disease.

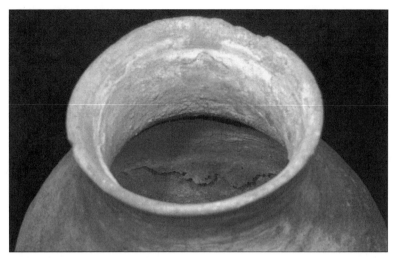

Figure 6.52b. Interior view of sealed jar from A100N with dried food residues adherent to jar wall.

Tomb A100, Chamber East

There was very little ceiling fall and no silting present in the east chamber of A100 (Figure 6.56). The slight ceiling fall occurred in the right rear quadrant of the chamber over one of the piles of long bones. Reed matting was present under the skulls. Soil constituents were adherent on some of the skeletal elements

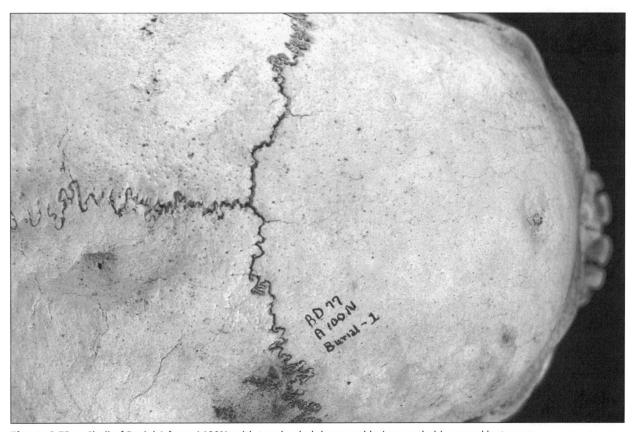

Figure 6.53. Skull of Burial 1 from A100N, with two healed depressed lesions probably caused by trauma.

(Figure 6.57) that were either from the primary burial site or had been applied to the elements during the transition from the primary to the secondary burial site. There were two clusters of skeletal elements in the chamber (Figure 6.58), which may reflect separate burial episodes. Cultural artifacts placed in the chamber included an unusual number of pottery vessels (Figure 6.59), three unfired clay figurines, and a wooden stick. This chamber is particularly notable for the number of unusual cases of skeletal disease. These include hip dysplasia (Figures 6.60a and 6.60b), in which the acetabulum is too shallow to provide biomechanical stability to the joint. This results in chronic partial dislocation, which is eventually complicated by osteoarthritis. Another skeletal disease is a skin ulcer associated with osteomyelitis (Figure 6.61a) in an older woman who also had osteoporosis (Figure 6.61b). Perhaps the most important example of skeletal disease is a probable case of tuberculosis in a young male skeleton. All of these cases of skeletal disease will be discussed in greater detail in the descriptions of the individual burials. Another notable aspect of this chamber is the unusually large number of burials included. The MNI interred in the chamber was 24. This very large number is additional evidence supporting at least two episodes of interment in the chamber. As noted in the introduction to the description of this tomb, one hypothesis is that this is the chamber where the EB IA remains removed from the north chamber were placed when the north chamber was cleared for reuse during EB IB. There was one fragment of an adult innominate (os coxae) that had been left behind in A100N. This fragment cannot be associated with any of the adult burials in A100E, so the hypothesis must remain one possible explanation for the large number of burials in this chamber. The 24 burials included eight adult males and two adult female skeletons. There were fourteen subadult burials ranging from late-term fetal remains to adolescent. In general, preservation of all the skeletal remains was better than average.

Burial 1 is the robust skeleton of an adult male that includes cranial and postcranial elements. Estimated age is 50+ years. Widespread moderate osteophyte and enthesophyte development are indicative of osteoarthritis and a lifetime of strenuous physical activity. This is remarkable since this burial has dysplasia of the right hip that would certainly have had an adverse ef-

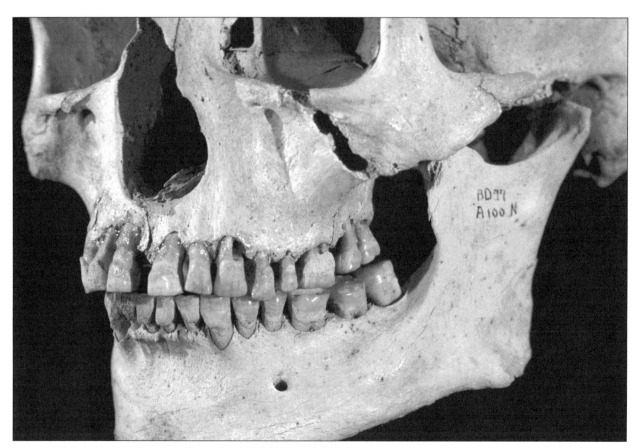

Figure 6.54a. Dental plaque on the teeth of Burial 1 from A100N.

fect on locomotion. This condition was almost certainly present from birth, so this individual would have needed to adjust to this disorder from very early in life. The robusticity of the skeleton and the general evidence of a physically active life testify to the potential of human beings to compensate for physical disability.

Burial 2 is the skeleton of a female probably in excess of 50 years of age at the time she died. This is the skeleton highlighted in the introductory comments for this tomb that had evidence of chronic infection associated with a skin ulcer overlying the left tibia. This disorder resulted in a generalized periostosis of both the left tibia and fibula. The abnormal bone generated by this condition resulted in almost complete fusion of the two bones (Figure 6.61a). This individual also had severe osteoporosis, with a bone weight at least 35% less than the average of other similar skeletons from this site. Today, in the Western world, the most common cause of osteoporosis is the hormonal change in women associated with menopause. Burial 2 may be an example of this aging phenomenon (Figure 6.61b). However, other problems can cause osteoporosis as well, including severe malnutrition such as occurs with starvation. Elsewhere in the skeleton the subchondral bone of the sacroiliac joint is covered with fine woven reactive bone. The cause of this disorder is unknown but one of the erosive arthropathies is possible.

Burial 3 includes a remarkably well-preserved skull and postcranial bones with little evidence of postmortem damage. This burial is a male between 35 and 40 years of age at the time of death. There is widespread evidence of dental hypoplasia (Figure 6.62) indicative of one or more physiological stress events during childhood. There is evidence of moderate to severe tooth wear indicative of a diet with a high concentration of abrasive materials. Dental caries present in some teeth is sufficiently severe to permit access of mouth bacteria to the root canal and the alveolar bone surrounding the roots. This has resulted in two fairly minor dental abscesses. Evidence of moderate to severe osteoarthritis is present in several of the major joints. This combined with enthesopathy associated with several muscle/tendon attachment sites implies vigorous physical activity during life.

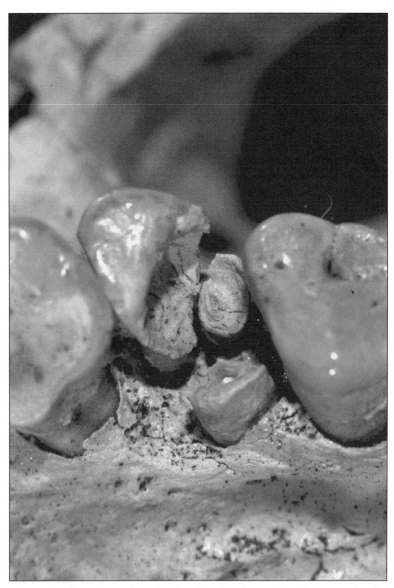

Figure 6.54b. Severe dental caries affecting the maxillary left second molar of Burial 1 from A100N.

Burial 4 is the poorly preserved remains of a robust adult male who was unusually tall. Enthesophyte development is associated with several long bones. In the distal right humerus there is an abnormal depression that was being filled with porous reactive bone. An infectious focus is the most probable explanation, but the defect was in the healing stage at the time of death. There is also evidence of osteophyte development at the margins of the knees. The patellar groove is shallow with an abnormal groove that has a porous surface. The cause of this abnormality is not known.

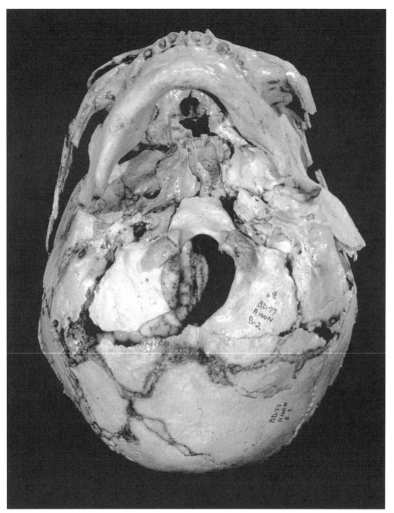

Figure 6.55a. Skull base of Burial 2 from tomb chamber A100N with a destructive lesion of the sphenoid that could have been caused by tuberculosis.

Unlike most of the burials in A100E, Burial 5 consists of the poorly preserved postcranial remains of a subadult. The poor preservation prevents a more specific estimate of age. There is no corresponding subadult skull. No evidence of skeletal disease was apparent on the postcranial bones available for analysis.

Burial 50 consists only of the incomplete right innominate of a male about 18 years of age. There is no evidence of skeletal disease associated with this element. Burial 51 includes two tibiae, two femora, and an incomplete right innominate. There is no evidence of significant skeletal disease in these skeletal elements.

Burial 52 is the skull of an infant about one year of age. There is no evidence of skeletal disease on the skull elements. Burial 53 includes postcranial skeletal elements of a child about six years of age. There is no evidence of skeletal disease in the elements available for analysis. The number 54 was not assigned to any of the burials in this chamber.

Burial 55 consists of the fragmentary postcranial remains of an adolescent about 13 years of age. There is no evidence of skeletal disease. Burial 56 consists of a single radius probably of a female about 18 years of age. The bone is normal.

Burials 57, 58, 59, 60, 61, and 62 are the remains of late-term fetuses or neonates. All the bones except for the left tibia of Burial 60 are normal. This bone has a lesion consisting of woven bone on the medial anterior crest. The right tibia is not available for study. Burial 63 includes both cranial and postcranial skeletal elements of an infant about 6 months of age. There is no evidence of skeletal disease in this burial.

Burial 64 consists of postcranial skeletal elements of a child about eight years of age. There is no evidence of disease in the bones available for analysis. Burial numbers between 65 and 70 were not used. Burial 71 consists of the incomplete postcranial bones of a child about five years of age. There is no evidence of skeletal disease. Burial 72 includes some of the major long bones of a child about ten years of age. The bone fragments are normal.

Burial 73 consists of postcranial skeletal elements of a male skeleton about 19 years of age. Two of the lower lumbar vertebrae have lytic lesions with some reactive bone formation. The lytic lesions occur in the vertebral bodies and are associated with extensive porosity affecting the arches of the affected vertebrae. Given the presence of probable cases of tuberculosis in other EB I burials at Bâb edh-Dhrâ', this disease is a plausible diagnostic option for this burial, although vertebral arch involvement would favor another infectious disease such as brucellosis.

A left tibia and two humeri are the only bones associated with Burial 74. These bones indicate that the individual was a robust adult male. Extensive enthesopathies in these bones indicate vigorous physical activity and probably an age beyond young adulthood. Burial 75 consists of a right femur of an adult male with an estimated age in the early 20s. As with Burial 74 there is evidence of moderate enthesopathy; size and robusticity preclude this femur from being associated with the femur in Burial 74. One of the enthesopathies is a depressed lesion of the posterior distal metaphysis superior to the condyles. This is where the plantaris and the two heads of the gastrocnemius muscles originate, and tears at the site are well known and fairly common. In addition, there is some marginal osteophyte formation, particularly on the condyles. This argues for vigorous physical activity during life.

There are several skeletal elements that cannot be associated with a specific burial that exhibit evidence of skeletal disorders. One is a left patella with marginal osteophyte formation and bone deposition on the subchondral surface that may have been caused by chronic subluxation of that bone. A similar condition is apparent in another right patella that is not associated with the left patella described above. Lesions apparent on the other skeletal elements are associated with osteoarthritis of varying severity.

Tomb A100, Chamber South
The south chamber of Tomb A100 had negligible ceiling fall and silting. The chamber contents were arranged in the way typical of most EB IA chambers, with crania aligned to the left of the chamber entryway and postcranial bones and sometimes mandibles in a central bone pile (Figure 6.63). Bone preservation as well as skeletal element completeness was above average, with most individuals having more than 50% of the bones present. The wall in the rear of the chamber broke through into Tomb chamber A80W (Figure 6.64). The human remains in A100S give clear evidence of adherent soil from the primary burial site. Components of this soil were retained in anatomical features such as the zygomatic arch and eye orbits (Figures 6.65 and 6.66). Some of this retained soil is suggestive of deposits from silting, which may indicate that the primary burial site was an underground burial chamber subjected to periodic siltation during the rainy season.

The MNI associated with this chamber is 11, including four adult and seven subadult burials. Burial 1 includes cranial and postcranial elements of a child

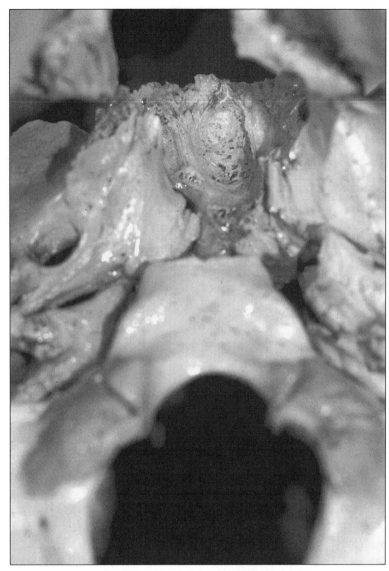

Figure 6.55b. Detail of a sphenoid lesion, Burial 2 from A100N.

about eight years of age. In the cranium, there is bilateral cribra orbitalia. There is a bilateral abnormality of both femora, with reactive bone formation suggestive of callus following fracture. The alignment of the femora remains normal, so if fracture is the cause of the lesions it is most likely a green-stick (partial) fracture and not a transverse fracture entirely through the cortex. Greenstick fractures usually occur in subadults, and this is the likely cause of the bilateral lesions apparent in the femora of this child.

Burial 2 is the partial skeleton of an adult male at least 40 years of age. Marginal reactive bone formation is apparent in the bones of the right knee. The lesions include periarticular bone formation, cystic excavation of cancellous bone below the joint, and subchondral bone destruction. Comparable bones for the left side were not recovered.

Figure 6.56. Tomb chamber A100E.

Figure 6.57. Skulls in tomb chamber A100E with soil probably from the primary burial site still adherent to the bone surface.

Figure 6.58. Detail of two clusters of skeletal elements in A100E.

Figure 6.59. Unusual arrangement of pottery and skeletal elements in A100E.

Shaft Tombs Excavated in 1977

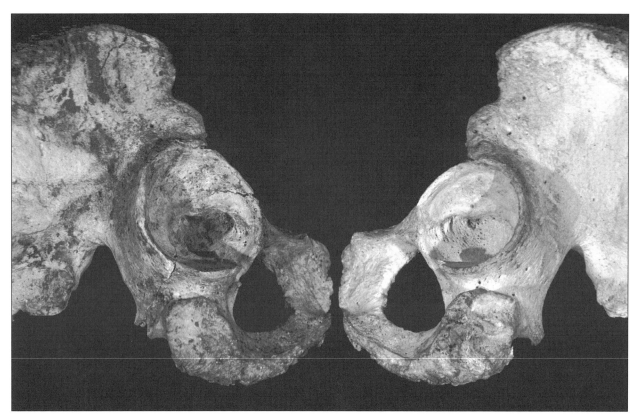

Figure 6.60a. Chronic subluxation of the right hip resulting from an abnormally shallow acetabulum from Tomb A100E. The right hip is compared with the normal left hip.

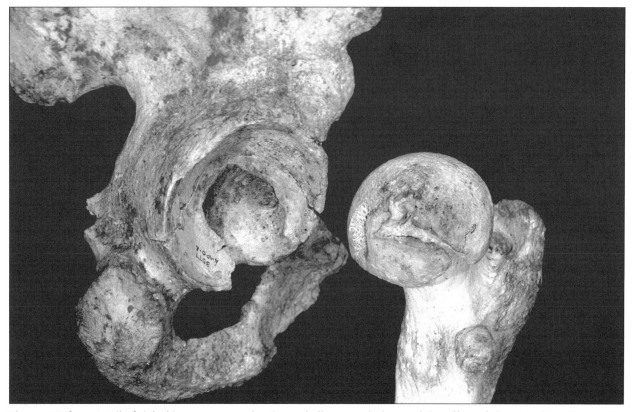

Figure 6.60b. Detail of right hip components showing a shallow acetabulum and the effect of chronic subluxation on the subchondral bone of the femoral head of Burial 1 from Tomb A100E.

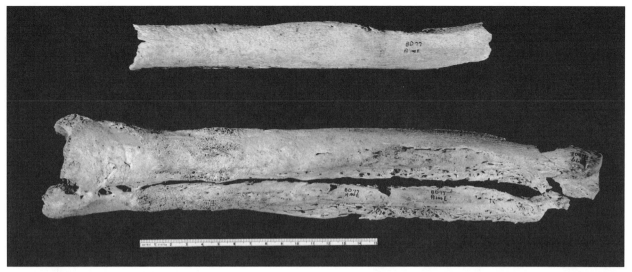

Figure 6.61a. Abnormal left tibia and fibula compared with a relatively normal right tibia from Burial 2, a woman 50+ years of age from Tomb A100E. The abnormal bone formation apparent in the left tibia and fibula is the result of chronic infection associated with an ulcer.

Burial 3 is the partial skeleton of a female between 17 and 20 years of age. There is a bilateral abnormality of the femoral condyles suggestive of chronic subluxation of the knee. Unfortunately, sites of attachment for the cruciate ligaments were damaged so that supporting evidence for subluxation of the knees is not observable.

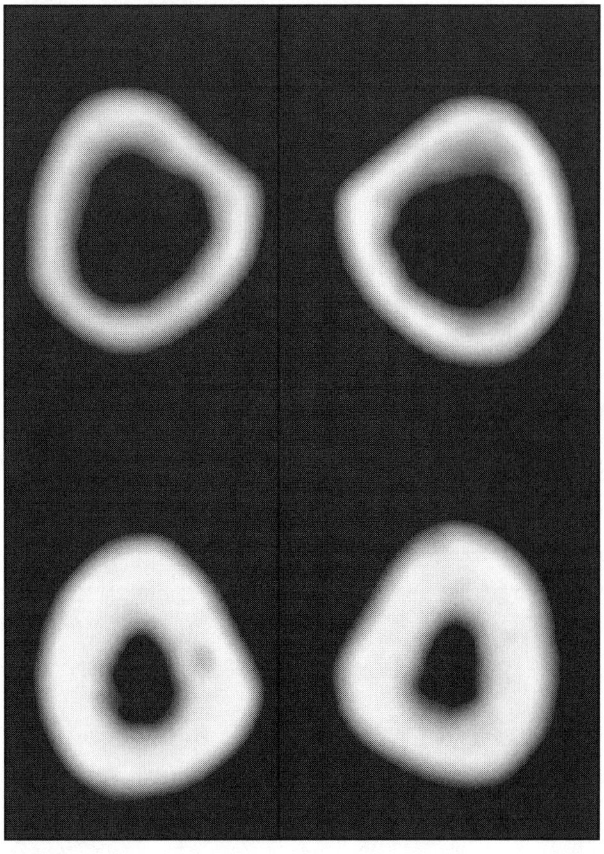

Burial 4 included both cranial and postcranial elements of an adult female whose age is in excess of 25 years. Anatomical features needed for a more precise estimate of age are not present. In the skull, there is considerable antemortem tooth loss. There is also slight cribra orbitalia in the right orbit, and the right maxillary and frontal sinuses have reactive bone formation that is probably the result of chronic infection. This chronic condition may be associated with the lytic lesion that occurs in the region of the right zygomaticomaxillary suture. There is little evidence of disorder in the postcranial skeleton. An exception to this is reactive bone formation associated with the left elbow that probably was the result of chronic hyperextension of the joint. This may reflect significant biomechanical stress. Also, a postmortem break through the right femoral head revealed a cavity with a well-defined sclerotic margin. A fibrous or, more likely, a benign cartilaginous tumor seems a likely cause for this lesion.

Burial 50 is the partial skeleton of an adolescent between 12 and 14 years of age. The cranium is fragmentary and incomplete. The inner table of the cranial vault has multiple depressions indicative of abnormal intracranial pressure. Several conditions can cause this, including hydrocephalus and pituitary tumors. There are bilateral lesions of the acetabulum characterized by porous woven bone on the subchondral surfaces of the joint. This lesion does not occur on any other subchondral bone surface in this skeleton. There is also a

Figure 6.61b. CT image through the midshaft of the humerus in Burial 2, A100E (upper) compared with a normal female (lower). The major loss of cortical bone was probably caused by postmenopausal osteoporosis.

Shaft Tombs Excavated in 1977 101

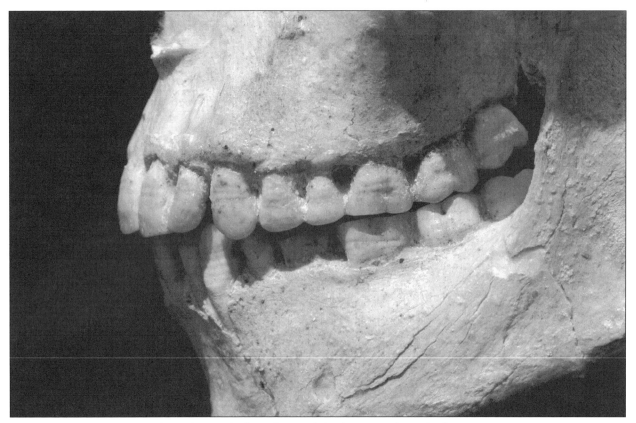

Figure 6.62. Linear enamel hypoplasia in the lateral view of the dentition of Burial 3 from A100E.

Figure 6.63. Tomb chamber A100S.

Figure 6.64. Defect in the wall of tomb chamber A100S. This defect probably occurred after the tomb was sealed, since no effort was made to repair the damage.

Figure 6.65. Soil from the primary burial site is still adherent in the zygomatic arch of the skull from A100S, seen on the right margin of the photo.

Shaft Tombs Excavated in 1977

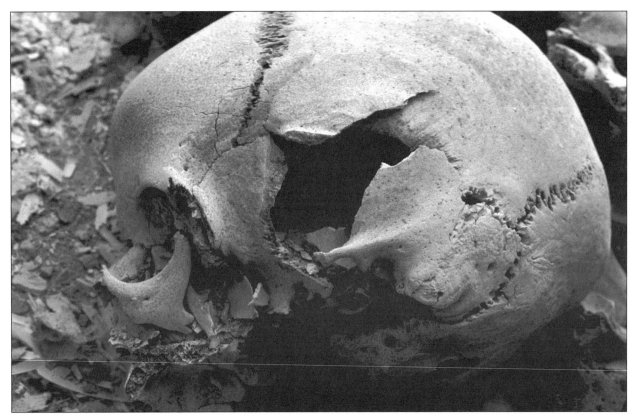

Figure 6.66. Soil from the primary burial site in the eye orbit of a skull from A100S.

bilateral, diaphyseal periostosis of the tibia. The tibiae are also abnormally bent in the mediolateral axis, suggestive of rickets, perhaps at an earlier age, which this individual survived.

Burials 60, 62, and 64 are the skeletal remains of infants ranging from late-term to less than one year of age. There is no evidence of skeletal disease in any of the burials. Burial 61 is the partial skeleton of a three-year-old child. Abnormal bowing is apparent in the clavicles and the tibiae that is probably the result of rickets.

Burial 63 is an incomplete skeleton of an adolescent between 12 and 15 years of age. There is no evidence of skeletal disease in the fragmentary bones available for analysis. Burial 65 includes only a few skeletal elements of what is probably an adult male. There is bilateral evidence of osteoarthritic lipping of the margins of the acetabulum.

Tomb A100, Chamber West
The contents of the west chamber of A100 are arranged like most of the other EB IA tomb chambers (Figures 6.67 and 6.68). There was minimal silting and only slight and scattered collapse of the ceiling. There is a deformity in the marl forming the rear wall of the chamber (Figure 6.69) that was probably caused by ground shifting associated with an earthquake subsequent to the creation of the tomb chamber. The location of the cemetery site in the north end of the Great Rift system makes earthquakes a likely factor in the ceiling fall seen in so many of the tomb chambers. In the rear of the chamber wall there is a break that provides a small opening into tomb chamber A78NE. The MNI interred in this chamber is eleven, including four adults and seven subadults.

Burial 1 is the skeleton of an adult female between 18 and 25 years of age and includes both cranial and postcranial elements. The posterior portion of the skull has a circular depressed lesion in the left lambdoid suture (Figure 6.70). Both the inner an outer tables were affected, and there is little doubt that the lesion was the result of a blow to the back of the head. The extensive remodeling associated with the lesion indicates long-term survival after the trauma. There is also a degree of marginal lipping of the occipital condyles that is unusual in a person of this age. A plausible cause of this condition might be using the head to carry heavy loads.

Burial 2 consists of cranial and postcranial elements of an adult female between the ages of 25 and 29. Abnormalities of the skull include enamel hypoplasia indicative of at least one significant stress event during

Figure 6.67. Tomb chamber A100W.

Figure 6.68. Arrangement of skeletal elements in tomb chamber A100W.

development of the teeth. Reactive bone is present in the maxillary sinuses. Bilateral cribra orbitalia is present, as is a healed reactive bone lesion associated with the left infraorbital foramen, which may be indicative of an episode of scurvy, probably sometime during the developmental period.

Figure 6.69. Deformity in the rear wall of A100W, probably associated with an earthquake after the tomb chamber was created.

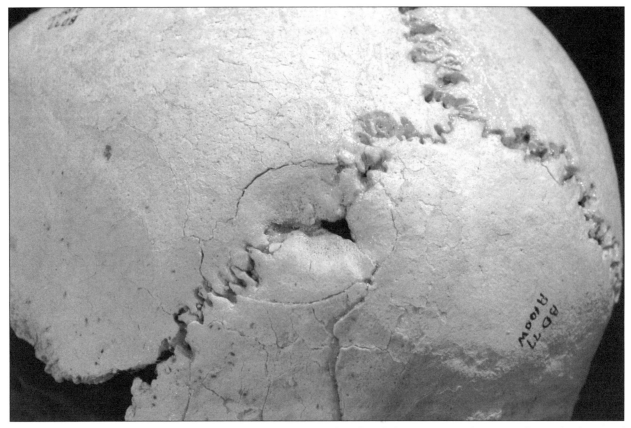

Figure 6.70. Depressed lesion in Burial 1 from A100W. This burial is a female between 18 and 25 years of age. The depressed lesion had healed well before death. The cracks in the bone surrounding the lesion are probably postmortem damage.

Burial 3 is the incomplete skeleton of an adult male about 25 years of age and includes both cranial and postcranial elements. There is a focal lesion at the midline of the frontal bone that is characterized by a remodeled central destructive focus surrounded by a zone of reactive bone formation, creating a crater-like lesion. This type of lesion is most commonly associated with an infectious focus that is lytic centrally but blastic peripherally. There is also evidence of perimortem trauma in the right temporal squama that may be associated with trauma at the time of death. However, postmortem damage cannot be ruled out.

Burial 65 includes only scapular and clavicular fragments associated with what was probably an adult male. The subchondral bone of the glenoid fossae has evidence of a destructive process that also resulted in reactive bone formation, primarily near the margins of the joint. This is suggestive of an erosive arthropathy of some type, but a more specific diagnosis is not possible on the basis of the available evidence.

Burials 60, 61, 63, 64, 66, and 67 are the skeletal elements of late fetal or infant remains. No evidence of skeletal disease was present in any of the bones. This is not the case with Burial 62, which is also the skeleton of a perinatal infant. The tibiae of this burial are abnormally bowed in the A-P axis, and both also have evidence of reactive periostosis (Figure 6.71). Periostosis is also apparent on the right proximal femur. Bowing can occur in infection such as congenital syphilis but is more commonly seen in rickets. The periosteal lesions are also compatible with a diagnosis of rickets, and this seems the likely cause, although rickets in a newborn is unusual.

There are a few fragments found in this chamber that show evidence of skeletal disease but cannot be attributed to a specific burial. One of these fragments is from the frontal bone of an infant or young child that has reactive woven bone on the cortical surface. Scurvy is certainly possible, but infectious conditions and tumor are as well. There is a fragment of a lumbar vertebra, possibly the fourth, that has a localized hyperostosis of the right margin of the body. This is a typical location for hyperostosis associated with DISH, but the lack of evidence for additional lesions of this type in any other vertebral elements recovered from the chamber prevents certainty about the diagnosis. There is also an unsided fibular fragment that has reactive woven bone formation that is remodeling into compact bone. This could be the result of trauma or infection.

TOMB A101

Tomb 101 consists of a central shaft with four chambers radiating, with openings to the chambers at the base of the shaft (Figure 6.72). The entryway of the southwest chamber contained the incomplete skeleton of a gazelle (Figure 6.73). This was almost certainly a sacrifice associated with the funerary ceremony, since a right foreleg and left hind leg as well as some other parts of the body had been removed prior to interment (Hesse and Wapnish 1981). Preservation of the chamber contents is poor, with extensive ceiling fall a major factor in the fragmentary nature of the human burials. However, the burials interred in the chambers are poorly represented by the number of skeletal elements originally placed in the chamber. There were no blocking stones sealing off the chamber entryways, so it is remarkable that silting was not a significant factor in any of the chambers. No information is recorded regarding the contents of the southwest chamber of this tomb.

Tomb A101, Chamber North
Very substantial ceiling collapse occurred in this chamber, and this is most likely the major factor in breakage apparent both in the human remains and in the pottery that had been placed in the chamber. The skeletal remains include both cranial and postcranial

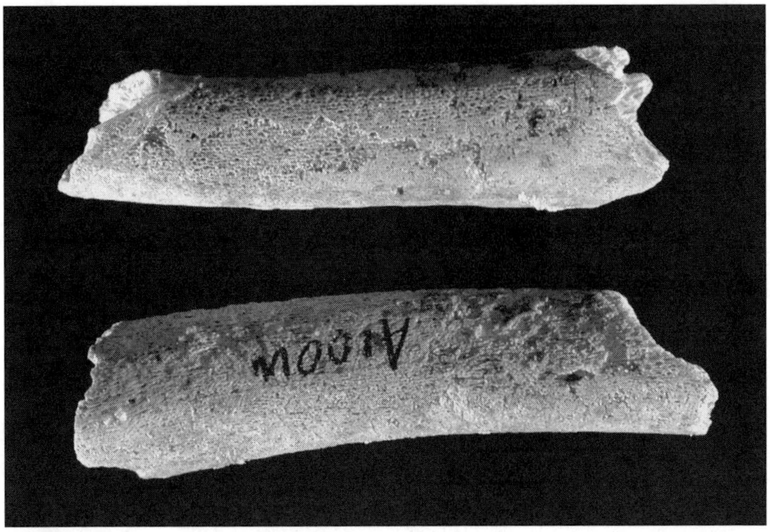

Figure 6.71. Abnormal tibiae of a perinatal infant (Burial 62, A100W) with abnormal curvature of the shaft and reactive periosteal bone formation on the cortical surface.

elements, but are very fragmentary and incomplete. The MNI in the chamber is three, including single adult male and female skeletons and a young child one to two years of age.

Burial 50 is the very fragmentary and incomplete skeleton of a 20–25-year-old female. The dentition associated with this burial has multiple hypoplastic lines in the teeth, indicative of several stress events during the development of the teeth in early childhood. Burial 51 consists of an isolated right ulna of an adult that probably was male in view of the size and robust nature. There is no evidence of skeletal disease associated with this bone. Burial 52 is the fragmentary and incomplete skeleton of a child between one and two years of age. There is no evidence of disease apparent in the remains available for study.

Tomb A101, Chamber East

This chamber was affected by some ceiling fall, but much less so than other chambers in this tomb. Nevertheless, the human remains are fragmented and incomplete. The MNI associated with this chamber is two, including an adult male and an older adult female.

Burial 50 includes both cranial and postcranial skeletal elements of an older adult female. Cortical bone associated with this burial is thinner than normal, and exposed cancellous bone trabeculae are sparse and thin. Both these conditions occur with normal aging but are, most commonly, particularly severe in postmenopausal women and are manifestations of osteoporosis. Burial 51 consists of fragmentary cranial and postcranial elements of a robust adult male. There is a lytic lesion associated with the alveolar bone containing the upper right lateral incisor that is probably the result of an abscess. The cranial fragments have an abnormally enlarged diploë that may be associated with an abnormal increase in the space needed for red blood cell formation that is most often caused by anemia.

In vertebral fragments that cannot be attributed to a specific burial there is evidence of osteoarthritis of the apophyseal facets.

Tomb A101, Chamber South

Preservation was adversely affected by substantial ceiling collapse. The skeletal remains include both cranial

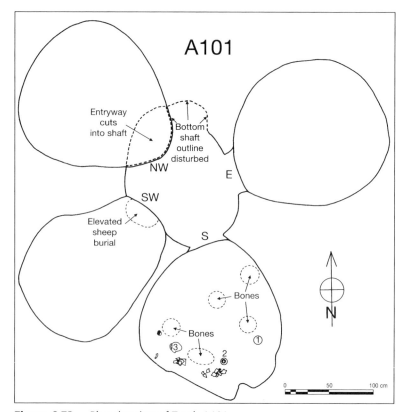

Figure 6.72. Plan drawing of Tomb A101.

and postcranial elements and are badly fragmented and incomplete, with less than 10% of the skeletal elements available for analysis. The MNI interred in this chamber is seven, including two adults and five subadults.

Burial 50 is the skeleton of a robust male adult. There is insufficient data for a more precise estimate of age. The left talus has a destructive lesion that affects both the medial subchondral bone and the underlying cancellous bone. Margins of the lesion show compact bone remodeling indicative of healing. An erosive arthropathy is possible, but septic arthritis is also a possible diagnostic option. The right talus is unavailable for analysis. The left patella has some osteoarthritic bone features, suggesting that the age of this individual was past the early adult stage. The right patella was not recovered during excavation. Burial 51 is the very incomplete skeleton of an adult female. A more precise age estimate is not possible. There is no evidence of skeletal disease in the bone fragments available for analysis.

Burials 52 (age estimate three to four years), 54 (infant), 55 (infant), and 56 (perinatal) show no evidence of skeletal disease in the very fragmentary remains available for analysis. Burial 53 is the skeleton of a two year old. The tibiae of this burial both show evidence of periostosis of the diaphysis. This condition

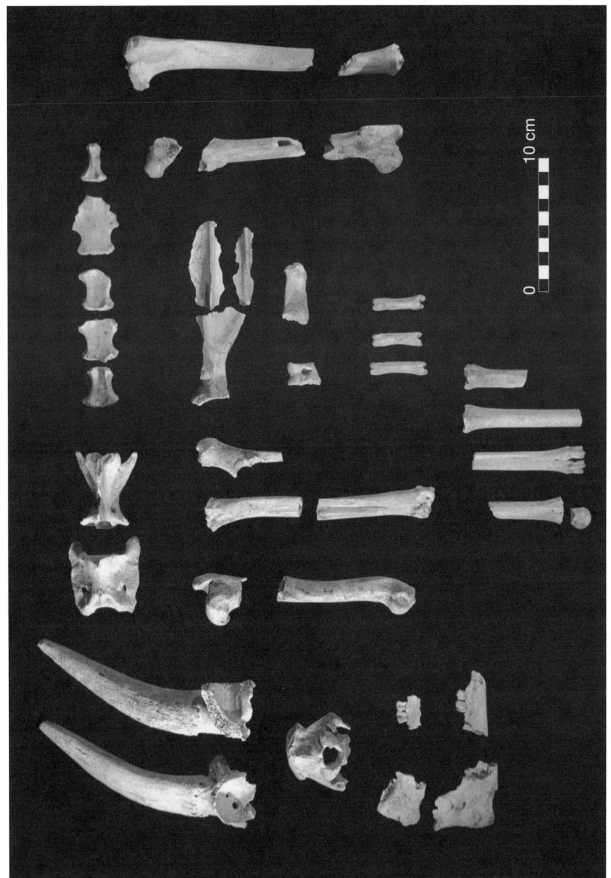

Figure 6.73. Skeletal elements of a gazelle probably sacrificed as part of the funerary ceremony associated with Tomb A101.

is also present on both proximal ulnae. These lesions are probably associated with a systemic, fairly chronic infection, although other disorders can produce this type of lesion.

TOMB A102

This tomb included a central shaft and four chambers. There was no evidence of blocking stones and silting, and ceiling fall filled the chambers. There is very little evidence of cultural artifacts or burials in the chambers. The south chamber had four pots. None of the chambers had human remains other than small fragments of bone, none of which provided any data. Indeed bone preservation was so poor that no skeletal remains were salvaged from the chambers. The absence or scarcity of cultural artifacts in the chambers raises the possibility that at least most of the chambers had not been used for interment or, more likely, the tomb had been robbed in antiquity.

TOMB A120 AND A121

Tombs A120 and A121 designate two tomb chambers connected by a poorly defined shaft (Figure 6.74). The reason for this designation is that there may have been a second shaft associated with tomb chamber A121 that cut into the shaft of A120. The poor definition of the entryways for each chamber leaves the possibility that A121 may have been a separate tomb with a single chamber. If so, this is the only example in the 1977 excavations where a mistake appears to have been made in the placement of a tomb. Ceiling fall and silting made recovery of skeletal remains and cultural artifacts very difficult in both chambers. Tomb chamber contents were encased in the chamber fill, and this greatly limited the recovery of both human remains and cultural artifacts.

Tomb A120, Chamber Description

The skeletal remains in this chamber were embedded in the silt (Figures 6.75a and 6.75b). Skeletal elements were largely fragmentary, poorly preserved, and incomplete. The arrangement of the skeletal elements

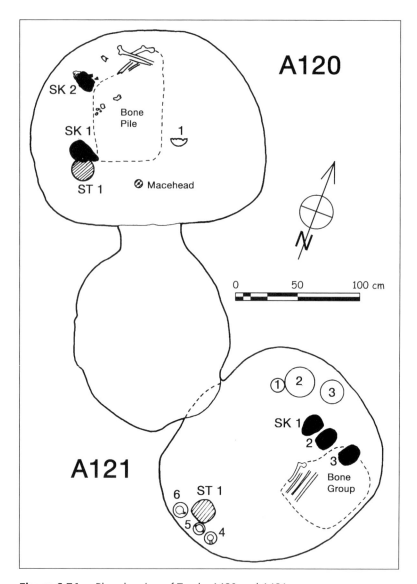

Figure 6.74. Plan drawing of Tombs A120 and A121.

was typical of remains in the other EB IA tomb chambers, with the skulls arranged separately from the postcranial bone pile. The MNI included in the chamber is five, including an adult male, an adult female, two adolescents, and one child.

Burial 1 includes fragments of both cranial and postcranial bones and is the remains of a female about 30 years of age at death. The maxillary molars were all lost antemortem, with considerable alveolar resorption that may be indicative of a somewhat older age. There is no evidence of skeletal disorders in the fragments available for study. Burial 2 is the skeleton of a late adolescent female about 16 years of age. The parietal bones of the skull have a bilateral area with abnormal thinning. This has been considered by some observers as a senile condition, but Barnes (1994) has

Figure 6.75a. Tomb chamber A120S.

Figure 6.75b. Bone pile embedded in silt in tomb chamber A120S.

clearly demonstrated that it is a congenital defect in the formation of the skull. Burial 3 also includes cranial and postcranial skeletal elements of a female, about 18 years of age. There is no evidence of skeletal disease apparent in the elements available for analysis.

Burial 50 consists of the mandible and postcranial remains of a robust 40+ male. There is widespread enthesopathy in the skeletal elements that is a fairly common feature of male skeletons in the Bâb edh-Dhrâʻ EB I skeletal sample. There are multiple factors that contribute to this condition, but the major factor is sustained and hard physical activity. This cause is further supported by the presence of marginal porosity of both patellae that is probably associated with osteoarthritis. Burial 51 includes three postcranial elements of a child about eight years of age. There is no evidence of skeletal disorder in the limited bone fragments available for analysis.

One of the skeletal elements that cannot be linked to a specific burial is an upper lumbar vertebra from a subadult. This element has a groove in the posterior, superior end plate that is indicative of a posterior herniation of the intervertebral disk. This probably was caused by trauma.

Tomb A121, Chamber Description

This chamber is in close proximity to the shaft of A120. Its relationship with A120 remains unclear, but it could have been a chamber as part of the A120 complex. Because of the ambiguity in its association, we have designated it as a separate tomb complex. There was no evidence of ceiling fall, but some silting occurred (Figure 6.76) and embedded the skeletal elements. The MNI associated with this chamber is nine, including four adults, one adolescent, one child, one infant, and one perinatal skeleton. Preservation is variable and there is evidence of damage to the skeletal elements in antiquity, presumably during the transfer from the primary burial site to the tomb.

Specific burials were not identified during field excavations, so the burial numbers begin with 50. Burial 50 has only postcranial elements and is the skeleton of a female about 18 years of age. None of the skeletal elements show evidence of disease. Burial 51 includes only postcranial remains. The bones are those of a robust male between 23 and 30 years of age. There is a subchondral depression in the medial condyle of the right femur. Damage to the left femur prevents determination of the presence of this abnormality on this

Figure 6.76. Tomb chamber A121.

bone. However, the fact that there is a similar bilateral defect on both patellae suggests that the condylar abnormality was bilateral. The cause of this condition is unknown, but it probably reflects a problem in the overlying joint cartilage, perhaps resulting from trauma.

Burial 52 is represented only by postcranial bones and is both incomplete (less than 10%) and fragmented. The skeleton is from a robust adult male; a more precise age estimate is not possible. There is no significant evidence of disease present in the bones available for analysis. Burial 53 consists only of postcranial remains (less than 10%) from a robust adult male about 35+ years. There is no evidence of skeletal disease in the elements present.

Burial 54 is the skeleton of a female about 30 years of age. The cortical bone is abnormally thin. It is not clear whether this condition is because of a problem during growth or occurred in adulthood as the result of an adult disorder. Metabolic problems such as severe malnutrition can cause cortical thinning. The estimated age is too young for postmenopausal osteoporosis, although an early onset variant of this disease is possible. However, other options are more likely.

Burial 55 includes less than 10% of the skeletal elements of a robust adult male. A more precise age is not possible. There is no evidence of disease in elements present. Burial 56 is represented by a right temporal fragment and the distal femur. Both elements indicate a probable neonatal age, and there is no evidence of skeletal disease. Burial 57 includes only the distal femora of an infant between six months and one year of age. The bone fragments are normal. Burial 58 consists only of a single left deciduous first molar. Its development provides an estimated age of six years.

In postcranial elements that cannot be linked to a specific burial, there is evidence of moderate to severe osteoarthritis in two associated lower thoracic vertebrae. An upper lumbar vertebra, probably from another burial, has a defect in the posterior inferior margin of the vertebral body that continues through the posterior margin. The most likely cause of this defect is a traumatic herniation of the vertebral disk.

TOMB C9

This tomb consists of a single chamber containing a minimum number of eleven burials. Unfortunately, field documentation of this tomb is less than ideal, but the difficulty in excavation is revealed by the very incomplete and poorly preserved skeletal elements. Although skeletal preservation and element recovery of bone was generally poor, the bones of the hands and feet are remarkably well represented, indicating considerable care in the transition from the primary burial site to final interment in the tomb chamber.

Of the eleven burials identified in the chamber, there are three adults, including two males and one female. There are an additional two probable adults, one of which is likely a female, the other of undetermined sex. There is one adolescent between 12 and 15 years of age, four children ranging in age from four to nine years and a single infant about one year of age. Five skulls were given burial numbers 1 through 4. The skull of Burial 4 is from an adult who was probably female but also included cranial fragments of a subadult designated as Burial 4a. The remaining burials are designated with numbering beginning with 60.

Burial 1 is the partially reconstructed but incomplete cranium of a robust adult male that probably is associated with the postcranial remains designated as Burial 63. The bones of the cranium are normal. In the postcranial skeleton, marginal osteophytes and cysts in the long bones available for study are caused by osteoarthritis. There is no other evidence of significant skeletal disease.

Burial 2 consists of incomplete cranial elements that could be from a gracile adult female or a subadult of either sex. There is no evidence of skeletal disease in the cranial elements. Because of the ambiguity regarding both the age and sex of this cranium, we have not associated it with a specific postcranial burial, but it is likely associated with one of the female or subadult postcranial burials.

Burial 3 is the partially reconstructed cranium of a child five years of age or slightly older. Slight cribra orbitalia is present in the right orbital roof. The left orbit is unavailable for comparison. It is likely that this cranium is associated with a postcranial skeleton of one of the children's burials, but a specific association cannot be made.

Burial 4 is the incomplete cranium of an adult female. Although it is likely that this cranium is linked to one of the postcranial burials, a specific association cannot be made. There is no evidence of significant skeletal disease in this burial. Burial 4a consists of cranial elements of a subadult at least five years of age. This burial contains no evidence of skeletal disease and cannot be associated with specific postcranial skeletal elements.

Burial 60 is the very incomplete skeleton of a child about seven years of age. This burial includes fragmentary skeletal elements of the cranial and postcranial skeleton. There is a unilateral lesion (periostosis) apparent in the left proximal femur. The cause of the abnormality is not known, although infection is the most common disorder causing this type of skeletal lesion. The teeth associated with this burial have dental hypoplasia indicative of at least one major stress event during development.

Burial 61 includes cranial and postcranial skeletal elements. The estimated age is between nine and ten years. Preservation of the skeletal elements present is good. The mandibular dentition has multiple hypoplastic lines associated with separate stress events occurring during development of the teeth. One of these lines is particularly well-defined, indicating a severe stress event.

Burial 62 consists of the very incomplete cranial and postcranial elements of a child about six years of age. The right humerus has a reactive periosteal lesion on the posterior distal diaphysis. This is a nonspecific lesion, but may be indicative of infection. Unfortunately, the left humerus is not present, so we cannot determine if the lesion was bilateral.

Burial 64 is a very incomplete skeleton of a mature adult female. The diaphyseal cortex is abnormally thin, suggestive of metabolic disease. Given an older age estimate, postmenopausal endocrine disorders may be the cause, but malnutrition may also have been a factor.

Burial 65 is the fragmentary and incomplete skeleton of a four- to five-year-old child. There is no evidence of disorder in the bones available for analysis. Burial 66 includes incomplete fragments of both humeri and a fragment of the left clavicle. The remains are probably those of an adult female. There is no evidence of lesions in the fragments available for study. Burial 67 is also very incomplete, with only the left distal humerus and both clavicles available for analysis. The estimated age is young adult and the bones are very robust, indicating male sex. There is no evidence of skeletal disease. Burial 68 consists only of a left clavicle fragment that cannot be associated with any other burial. The fragment is most likely from an adult, but a more specific age cannot be determined. There is no evidence of skeletal disease.

Burial 69 includes only teeth and the left clavicle. There may be some cranial fragments that might be associated with this burial, which are the skeletal elements of a 12–15-year-old adolescent. The teeth show multiple hypoplastic lines resulting from multiple stress events during the formation of the teeth. Burial 70 consists of cranial fragments, some teeth, and a fragment of the C1 vertebra of a one-year-old infant. There is no evidence of skeletal disease.

There is some evidence of skeletal disorders apparent in skeletal elements that cannot be associated with a specific burial. These disorders include orbital roofs of two subadults showing evidence of cribra orbitalia. Another case of skeletal disorder is apparent in the contiguous L2–L5 lumbar vertebrae. The vertebral bodies are biconcave rather than flat. The most common cause of this condition is a metabolic problem such as osteoporosis, although trauma may cause this abnormality. These vertebrae appear to be robust and are likely from a male skeleton, so the cause of this osteoporosis would probably be malnutrition. Another unassociated vertebral body has evidence of a herniated disk with deformity at the anterior inferior margin associated with osteophytes. Trauma is a likely cause of this abnormality.

TOMB C10

Tomb C10 consists of a poorly defined central shaft and two equally poorly defined chambers (Figure 6.77). Within the chambers, there was considerable ceiling fall and silting, and preservation of chamber contents was very poor. The plan drawing (Figure 6.77) indicates that human remains were identified in both the east and west chambers. The human remains in the west chamber were very poorly preserved and no observations on them are possible. Skeletal remains in the east chamber were both fragmentary and very incomplete (Figure 6.78), making observations on the remains problematic at best.

Tomb C10, Chamber East

The human remains in this chamber are represented typically by 10% or less of the skeletal elements. The bones that are present are also fragmentary, making statements about age, sex, and the presence of skeletal diseases no more than tentative. The best estimate regarding the MNI interred in the chamber is 14, including 7 adults. Three of the adults are probably female, two are male, and sex could not be determined in the remaining two adult burials. There are seven subadults ranging in age between neonatal and 12 years.

Burial 1 consists of cranial elements that cannot with certainty be associated with postcranial remains. The estimated age of the burial is about 25 years and the sex is female. There is a depressed lesion on the left side of the frontal bone. This could be a traumatic depressed fracture from a blow to the head. There is no involvement of the inner table. The lesion also

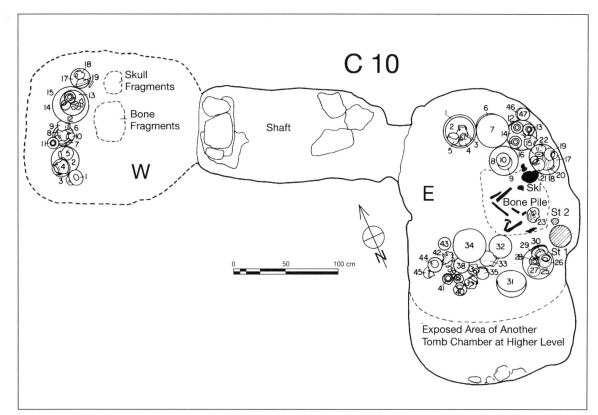

Figure 6.77. Plan drawing of Tomb C10.

Figure 6.78. Tomb chamber C10.

Shaft Tombs Excavated in 1977

could be caused by pressure erosion from an overlying sebaceous cyst.

Burial 2 includes cranial elements of an adult male. The right parietal bone has a rounded compact bone growth about a centimeter in diameter on the ectocranial surface, but is otherwise unremarkable. The bone lesion is probably a benign tumor. Burial 3 consists of cranial elements of a gracile adult about 25 years of age. Sex cannot be determined with certainty, but is slightly more likely to be male despite the gracility. There are two depressed lesions on the frontal bone that may be the result of a blow to the head, but could also be due to remodeling in reaction to a cyst. The ectocranial surface of the frontal, sphenoid, and parietals have reactive areas of striated bone deposits. The cause of these deposits is unknown, but options include infection, scurvy, and trauma. Burial 4 consists of badly damaged and incomplete fragments of the cranium from an adult of unknown sex, but who was slightly more likely to have been a small male. There is no evidence of skeletal disease.

Burial 50 is represented only by a fragment of an unfused pars basilaris, a mandibular second molar, and a proximal, left femoral fragment. The individual was a child about five to seven years of age, and there is no evidence of skeletal disease. Burial 51 is represented by a normal fragment of the pars basilaris of a child about two years of age. Burial 52 includes the pars petrosa and an unidentified long bone fragment from a neonate with no evidence of skeletal disease. Burial 53 consists of cranial, dental, and postcranial elements of a child about eight years of age. The elements available for study show no evidence of disease. Burial 54 also includes a few cranial, dental, and postcranial elements of a five-year-old child with no evidence of skeletal disease. Burial 55 consists of a fragment of the frontal bone and some fragmentary postcranial elements of a very robust adult male. There is no evidence of significant skeletal disease. Burial 56 is represented by a distal femoral fragment of a late-term fetus with no skeletal disorders. A fragment of the right posterior mandible is the only bone associated with Burial 57. The gracility of the fragment suggests that it was from a female. The third molar is in the early stage of eruption, indicating an age between 15 and 18 years. Burial 58 is represented only by a normal sacrum, which is wide relative to the individual's overall size, indicating that it was from a female. The estimated age is between 20 and 25 years. Burial 59 includes only diaphyseal fragments of the femur and tibia. They are from a child 10–12 years of age. The fragments are normal. Burial 60 is represented only by a fragment of the left petrous portion of the skull. The fragment is from an adult, but the sex cannot be determined.

Some of the skeletal and dental elements could not be linked to a specific burial. Several unassociated teeth showed evidence of dental hypoplasia, an infant frontal bone has a depressed lesion that may have been caused by trauma, and three fragments of the orbit have cribra orbitalia that could have been caused by infection, scurvy, or anemia.

TOMB G2

Tomb G2 consists of a shaft and a single chamber entered from the base of the shaft (Figure 6.79). The chamber walls are marl and partially consolidated layers of sand and gravel (Figure 6.80). The gravel layers were particularly unstable, which led to considerable ceiling fall that filled the lower half of the chamber. Not surprisingly, bone preservation is very poor, with most of the bones broken and skeletal elements for each of the burials very incomplete. Nevertheless, we were able to identify a minimum number of five individuals associated with this chamber, including three adults and two subadults. No specific burial was identified during excavation, so burial numbers have been assigned in the laboratory, beginning with the number 50. None of the skeletal elements for which burial numbers are provided show any evidence of skeletal disease. Among the cultural elements recovered from the tomb chamber were three unfired clay female figurines (Figure 6.81).

Burial 50 consists of the poorly preserved and very incomplete remains of a young adult female. Burials 51 and 52 are adult males; a more precise age cannot be estimated because the skeletal elements were incomplete and poorly preserved. Burial 54 is the normal remains of an infant about six months of age. Burial 55 is a subadult whose specific age cannot be estimated.

There is a fragment of a lumbar vertebral body that has a destructive lesion of the end plate with some reactive bone formation. The cause of this abnormality cannot be determined, but possible causes include infection, erosive arthropathy, and trauma. There is also a cranial fragment, most likely a parietal bone, that has abnormal porosity of the inner table. The porosity appears to be in the healing stage with evidence of remodeling and repair. The individual appears to have survived whatever disorder caused the abnormality. The prognosis for meningitis is poor, so an infection seems unlikely. Scurvy or some other disorder that

Figure 6.79. Shaft for Tomb G2, with entry for chamber at the base of the shaft.

Figure 6.80. Tomb chamber G2.

produces increased vascularity and reactive bone formation may have been the cause.

TOMB G4

Like Tomb G2, Tomb G4 consists of a shaft and a single chamber (Figures 6.82 and 6.83). The shaft is partially lined with a stone retaining wall that may have been used to stabilize marl/gravel sediments that formed the shaft and the chamber wall as well as the ceiling of the chamber. At the time of excavation, the chamber was completely filled with ceiling fall and silt. Exposure of the chamber floor revealed the presence of six crania and postcranial skeletal elements (Figure 6.84) of an unknown number of burials. Also recovered from the chamber were the cultural materials, including a wooden artifact of unknown utility that had been placed on top of the bone pile and pottery. Preservation of bone in this tomb was much better than was the case in tomb G2. The MNI associated with tomb G4 is 13, including five adults, two adolescents, four children, and two infants. Some of the burials were identified during excavation. However, some of these burials subsequently needed to be divided, as skeletal elements from more than one individual were identified during laboratory analysis in some burials. We have, as much as possible, retained the field burial numbers but have assigned additional burial numbers starting with 50 for the burials identified during subsequent analysis.

Burial 1 is represented by cranial and dental remains. Initially postcranial bones were attributed to this burial, but current research indicates that the cranial and dental elements are from a mature adult female 40 or more years of age. The postcranial bones are from a younger individual and are now designated as Burial 51. The dentition of Burial 1 is severely worn, with significant antemortem tooth loss and remodeling of alveolar bone. In some molars, tooth wear has almost obliterated the crown of the tooth. There is a large apical abscess associated with the right upper second molar. The abscess affected the alveolar bone but also penetrated the right maxillary sinus and would have led to chronic sinus infection. There is evidence of reactive bone formation on the interior surface of the sinus. The cranial orbits have bilateral cribra orbitalia that is probably the result of chronic infection, but other causes of this condition cannot be ruled out.

Burial 2 consists of both cranial and postcranial remains of an adult female forty or more years of age. About 30% of the skeletal elements are present, and all indicate a fairly gracile person, in contrast with the evidence of hard physical activity apparent in the severity of the enthesopathies found particularly in the bones of the upper extremity. The teeth are heavily worn, although there is no evidence of antemortem tooth loss. Reactive bone, probably the result of chronic infection, is apparent in the right maxillary sinus. However, this bone was remodeled, indicating recovery and healing from the infection. A somewhat unusual manifestation of osteoarthritis is apparent in the subchondral bone and margins of the right hip. In the superior acetabulum, the subchondral bone is eburnated, indicating that the articular cartilage had been destroyed and the subchondral bone of both the superior acetabulum and superior femoral head had been in direct contact. There is also subchondral bone deposition in the superior acetabulum adjacent to the eburnated area associated with a corresponding groove in the femoral head. All of the hip abnormalities are indicative of significant osteoarthritis in this joint. Unfortunately, the elements of the left hip are not present, so the presence of bilateral osteoarthritis of this joint cannot be determined. There is additional evidence of possible osteoarthritis apparent in the left lateral clavicle associated with the acromioclavicular joint. This evidence suggests widespread osteoarthritis in this burial.

Figure 6.81. Clay figurine in chamber G2.

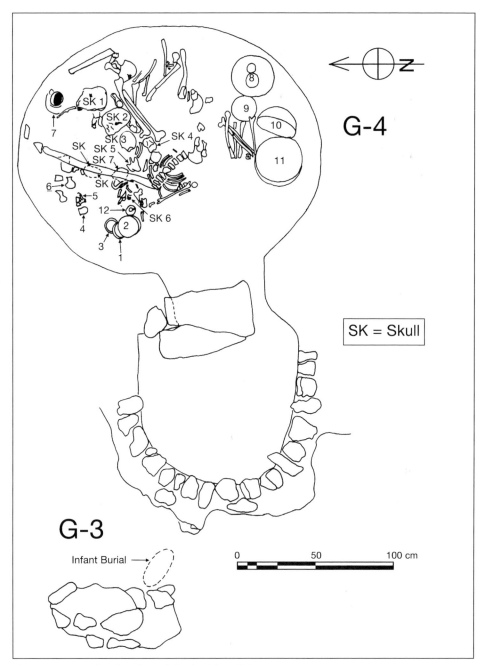

Figure 6.82. Plan drawing of Tombs G3 and G4.

Burial 3 includes both cranial and postcranial elements of a young female around 17 years of age. Skeletal disorders include at least two stress events resulting in enamel hypoplasia. Cribra orbitalia is also present bilaterally on the orbital roofs. Burial 4 contains the cranial and postcranial bones of an adult male between 25 and 30 years of age. The skull contains areas of abnormal porosity of the outer table in the posterior frontal bone and possibly in the parietal bones as well. There is no evidence of marrow hyperplasia, so it seems likely that the abnormality is the result of scalp infection and not anemia. Scurvy is possible, but there is no supporting evidence of this disease in other parts of the skull. Significant interdental caries occurs in the left maxilla between the first and second molars. This resulted in a dental abscess that affects the alveolar process with the formation of porous reactive bone surrounding the abscess but also drained into the maxillary sinus, stimulating chronic infection and the presence of reactive bone. The dentition contains five well-defined hypoplastic lines indicative of five significant stress events during dental development.

Shaft Tombs Excavated in 1977

Figure 6.83. Floor of tomb chamber G4, with contents exposed.

Figure 6.84. Skulls embedded in silt in the G4 chamber.

Burial 50 consists only of postcranial elements of an adolescent about 12 years of age. The sex cannot be determined with confidence. There is no evidence of skeletal disease in the bones available for study. Burial 51 includes only postcranial bones of an adult female between 25 and 29 years of age. There is no evidence of pathology in the elements present. A right innominate (os coxa) is the only bone associated with Burial 52. The estimated age is 25 years, but the sex cannot be determined with certainty, partly because the bone is incomplete, but also because the existing evidence regarding sex is ambivalent.

Burial 60 includes only postcranial remains; completeness is 10% or less. The estimated age is between 1.5 and 3 years. The sex is unknown. The two tibiae exhibit abnormal periostosis. The bilateral nature of these lesions argues for a systemic problem, most likely infection, but scurvy is also possible. Burial 61 is also an infant, between one and two years of age. There probably are cranial elements associated with this burial, but the relationship with postcranial remains cannot be established with certainty because remains of several individuals of a similar age were recovered from the tomb chamber. Like Burial 60, this burial has bilateral periostosis on both tibial shafts, although the disorder is more severe on the left. Again, the bilateral distribution implies a systemic condition and most likely infection. Burial 62 is an infant between one and two years of age.

There is no evidence of skeletal disease in the few postcranial skeletal elements available for study. Burial 63 also contains no cranial elements and only two postcranial bones. The estimated age is between six months and one year. There is no evidence of skeletal disease present. Burial 64 comprises femoral and ulnar fragments of a perinatal infant. There is no evidence of skeletal disease in the fragments. Burial 65 is represented by the teeth of a six-year-old child that could not be attributed to any other burial. The teeth were normal.

Burials 30 and 31 are subadult cranial elements that could not be associated with a specific postcranial skeleton. Burial 30 consists of the cranial remains of an infant one to two years of age that could be associated with Burials 60, 61, or 62. The bone is normal. Burial 31 includes only occipital fragments of an infant/early child between one and three years of age. The fragments are normal.

There are a few abnormal skeletal elements that could not be associated with any specific burial. Four rib fragments, at least two of which are from the same individual, have callus formation indicative of rib fracture. Other evidence of possible trauma includes lumbar vertebrae with evidence of disk herniation and a thoracic vertebra with a depression in the vertebral body that could be caused by compression fracture. Other skeletal elements have evidence of osteoarthritis.

CHAPTER 7

The EB IB Charnel House (G1): Excavated in 1977

DONALD J. ORTNER, LIESE MEIER, AND BRUNO FROHLICH

The G1 charnel house was excavated during the 1977 field season in an effort to salvage as much as possible from an EB IB exposed burial structure composed primarily of mudbrick. A brief and preliminary report on the G1 burials was included in a general paper on cemetery excavations conducted during the 1977 field season (Ortner 1981); a specific, although still tentative, report on the human remains from the G1 Charnel House was published a year later (Ortner 1982). A short description of the archaeology of the tomb was included in a review of the burial traditions associated with the Bâb edh-Dhrâ' cemetery (Schaub 1981b). Unlike the square or rectangular charnel houses more typical of the EB II and III periods at Bâb edh-Dhrâ', this charnel house was round with a maximum internal diameter of 3.84 meters (Figure 7.1). Remnants of the mudbrick wall remained around most of the perimeter (Figure 7.2). At the time of excavation, the stone orthostats and threshold of the entryway were in place, as was the large blocking stone (Figure 7.3). Silt and windblown sand had covered much of the remaining cultural artifacts and human remains (Figure 7.4).

Although disintegration and natural erosion of the original structure undoubtedly began in antiquity, this process was exacerbated by significant recent looting,

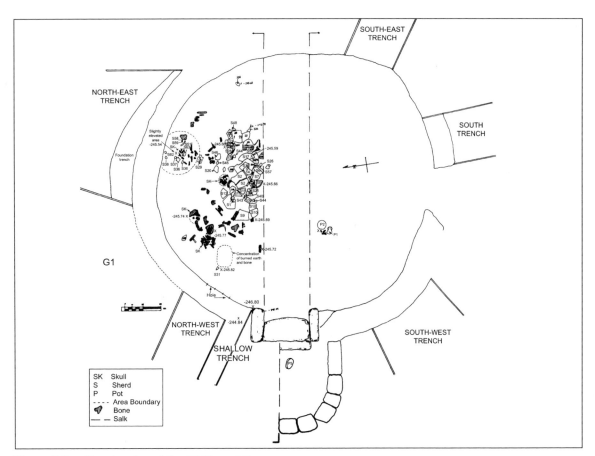

Figure 7.1. Plan drawing of the G1 Charnel House.

Figure 7.2. Early stage in excavation of G1. Note the stone orthostats and the large blocking stone still covering the entryway.

Figure 7.3. Early exposure of G1 contents with some recovery of skeletal elements and potsherds.

Figure 7.4. Careful removal of dirt matrix surrounding a skull, mandible, and long bone.

presumably in response to the value artifacts have had in recent times in the antiquities market. The cemetery site is also located near the border with Israel, and military construction had affected some structures within the cemetery. However, there is convincing evidence both in the human remains and cultural artifacts that significant damage to the charnel house and its contents occurred in ancient times. Virtually all the pottery and human remains were broken, but more convincing is the evidence of burning that is apparent in the human remains and on the pottery sherds. In some cases, the heat was sufficiently intense to badly deform the bone and convert it into something with physical properties similar to ceramic artifacts (see Figure 3.9). This deformation can only occur if the burials were recently interred at the time burning occurred and had retained substantial amounts of soft tissue.

In the section profile of the charnel house and the fill that covered its contents in the years following its use, there is an ash lens near the top of the profile and another on the floor (Schaub 1981b). The ash located on the floor seems likely to have been from the burning event associated with the destruction of the charnel house in antiquity. This lower ash layer is probably linked with the evidence of fire that affected the pottery and human remains. The timing significance of the ash lens near the top of the profile remains problematic, but it is unlikely that the fire associated with this ash layer is associated with the evidence of the initial burning event that affected the cultural artifacts and bones.

One of the charnel houses excavated in 1977 (A56) and dated to the EB II period at Bâb edh-Dhrâ' was small and round with a conical mudbrick structure that probably continued to the peak of the structure (Schaub 1981b). The slight inward curve of the G1 Charnel House walls continues this basic structural feature, but the size of the tomb would have made extending the mudbrick to form a complete ceiling/roof problematic at best. Almost certainly, the larger charnel houses, including G1, had roofs probably made of wooden poles extending across the top of the walls and providing support for reed matting that was covered with mud (Schaub 1981b). What is probably a dried mud remnant of the ceiling/roof retains the woven pattern of the reed matting (see Figures 7.5, 7.6). The poles as well as the reed matting would have provided

Figure 7.5. First author inspecting reed impressions in dried mud presumed to be part of the roof that collapsed when the charnel house was burned in antiquity.

Figure 7.6. Detail of the imprint of woven reed in the mud roof of the charnel house.

fuel for the fire. What ignited the fire in the charnel house is unknown. Possibilities include lightning, accidental ignition associated with a funerary ceremony, and intentional vandalism by raiding bands attacking the settlement.

Like the cultural artifacts found in the charnel house, the human remains are badly fragmented and very incomplete (Figure 7.7). Charnel houses dated to the EB II–III period are thought to contain predominantly primary burials. Initially we assumed that the G1 Charnel House would prove to be associated with a similar burial tradition. Indeed we do have direct evidence in the north chamber of Tomb A100 that primary burial did occur in the EB IB period. However, as we will emphasize later in this chapter, the data on skeletal elements present in the charnel house do not support this conclusion, and it seems likely that many, if not most or all, burials in the G1 Charnel House continued the secondary burial tradition apparent in the EB IA shaft tombs. However, unlike burials in most of the shaft tomb chambers, there were multiple interment events in the charnel house in which one or more burials were interred. As new bodies were placed in the charnel house, earlier burials were disturbed and both the cultural artifacts and human remains were scattered and probably broken.

This conclusion is supported by the extensive breakage apparent in the human remains but particularly in the cultural artifacts placed in the charnel house. This damage appears to be most severe near the base of the walls, where the earlier interments would have been moved as new burials were added to the charnel house. Some breakage and scattering before the burning event is apparent in some of the pottery (Figure 7.8). This early destruction is further complicated by more recent damage to the charnel house that adds to the challenge of interpreting the significance of the contents. Distinguishing between destruction in antiquity versus recent times was usually not possible. Nevertheless, at least a few bones were in correct anatomical relationship (Figure 7.9), suggesting that, for some burials at least, relatively short time periods elapsed between primary and secondary interment. This observation is also supported by the fact that several of the bones were badly deformed by the heat, which can only happen if the bones are fresh, with at least some overlying soft tissue.

The combination of vandalism in antiquity and activities in recent times that caused further dam-

Figure 7.7. Exposure of cultural artifacts and human remains showing both the fragmentary nature of the contents and the scattered relationship between the bones and artifacts.

Figure 7.8. Reconstructed large bowl from G1. The discontinuity in the dark line caused by burning indicates that the bowl had been broken before the burning event took place. This may reflect damage to cultural artifacts that occurred when additional burials and funerary gifts were added to earlier burials within the charnel house.

Figure 7.9. Partial articulation of upper extremity in a burial from the G1 Charnel House. It is likely that soft tissue was still present when these bones were placed in the tomb.

age greatly limits the data that can be extracted from the human remains and the conclusions that can be made. Identifying more than one bone associated with a single burial is a very rare circumstance. This means that much less can be stated about the skeletal biology, including stature and demography of the people. Also, the limited observations on the presence of skeletal diseases that might provide evidence regarding the health of the EB IB people are likely to represent only a partial amount of the evidence of skeletal disease associated with the people represented by the G1 skeletons.

GENERAL OSTEOLOGY

One of the interesting findings is that there are very few fetal or infant (birth to one year of age) bones in the remains that were recovered (Table 7.1). This is in marked contrast with the shaft tombs where fetal and infant remains constitute about 28% of the total sample. There were only three burials in the infant age range recovered from the G1 Charnel House. In the shaft tombs, only about 40% of the burials are from adults. This difference could be the result of taphonomic processes and the difficulty associated with recovering the human remains from the G1 Charnel House. However, the fact that some subadult remains were found argues against this explanation, as does the relatively good preservation of subadults in the shaft tomb chambers. The more plausible alternative explanation is that infants and children usually were not buried in the G1 Charnel House. They may have had a special structure or place for burial, or infants and very young children may simply have been discarded without formal burial rituals. The fact that very young individuals were not commonly interred in the G1 Charnel House contrasts with the EB IA shaft tomb burial tradition, where infants and children were carefully interred in the tomb chambers along with adults.

Our most recent analysis indicates that there was a minimum number of 74 adults 18 years of age and older (Figures 7.10a, b) and 41 subadults interred in the G1 Charnel House (Figures 7.11a, b; 7.12a, b; 7.13a, b; 7.14a, b; 7.15). These estimates are based on the maximum number associated with the most frequent specific skeletal element recovered. In the case of adult burials, the minimum number is based on the total number of the left petrous portion of the temporal bone. There are four fewer right petrous fragments of the temporal bone in the adult sample.

A cautionary note is in order regarding this frequency. The temporal bone grows rapidly, as do all the bones of the cranial vault. It reaches maximum size in childhood, and distinguishing an adult fragmentary petrous portion of the temporal bone from that of a subadult in late childhood or adolescence is not possible without additional information, which is not available in the G1 remains. This means that some of the petrous elements included in the adult subsample may be subadults and that the converse is also possible. In the subadult sample between the ages of birth and 15 years, the MNI is based on the proximal femur fragments. In the >15 subadult remains, the minimum number of burials is based on the mandibular body.

Although there is a high degree of congruence between the number of left and right skeletal elements, the numbers are not exactly the same. Furthermore, the very fragmentary and often distorted bone elements make it virtually impossible to match pairs of bones from a single burial. For both of these reasons it remains possible if not probable that some of the right side bone elements from a burial are not associated with a left element. This means that in the case of both adults and subadults the actual number of individuals interred in the charnel house is likely to have been at least slightly greater than the estimated minimum number. Unfortunately, the fragmentary nature of the human remains makes a more precise estimate of the total individuals interred in the charnel house impossible.

One of the factors that has important implications for the analysis of the G1 human remains is the almost universal burning that has affected the bones. The effects range from bone that has shrunk in size and become very distorted in shape to bone that has been discolored by the heat but minimally changed in size or shape. Much of what we know about the effect of burning on bone is based on forensic research in which bone associated with various amounts of soft tissue has been subjected to different temperature conditions in a controlled experiment. This research has demonstrated that burning of bone produces a variety of effects depending on the condition of the bone relative to the presence of soft tissue and the hydration of the bone at the time of incineration. Another

Table 7.1. Distribution of Estimated Ages in Subadult Postcranial Remains from the G1 EB IB Charnel House

Age category	N
Neonatal to 4.9 years*	7
5–9.9 years	14
10–14.9 years	8
15–18 years	12
Total	41

*In this age category, three of the individuals were birth to one year of age.

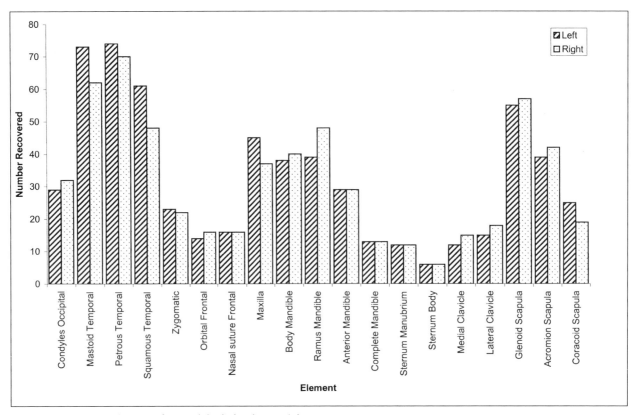

Figure 7.10a. Distribution of G-1 Adult Skeletal Material.

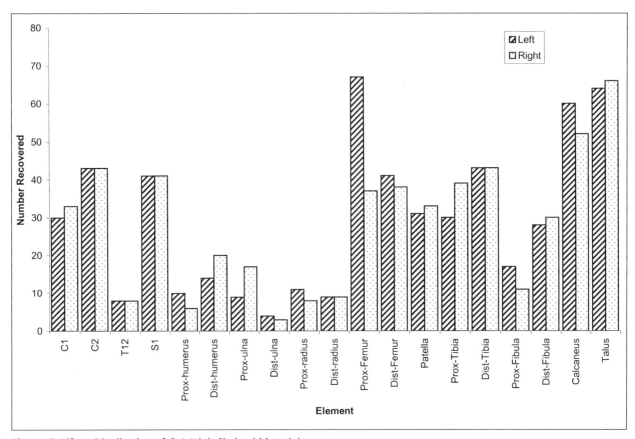

Figure 7.10b. Distribution of G-1 Adult Skeletal Material.

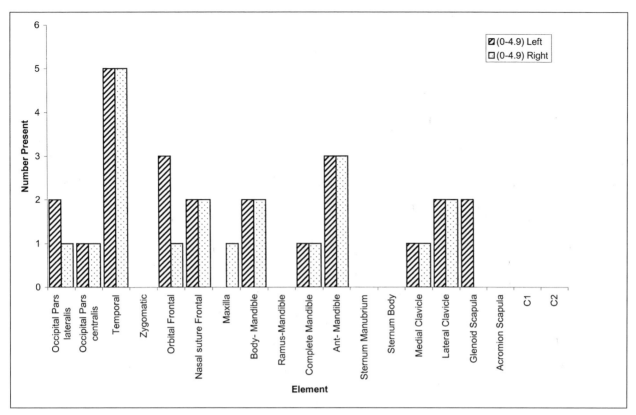

Figure 7.11a. Distribution of G-1 Subadult Skeletal Material (0–4.9 years).

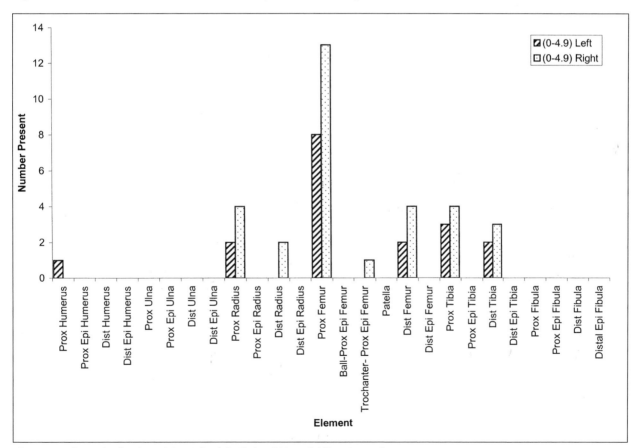

Figure 7.11b. Distribution of G-1 Subadult Skeletal Material (0–4.9 years).

EB IB Charnel House (G1): Excavated in 1977

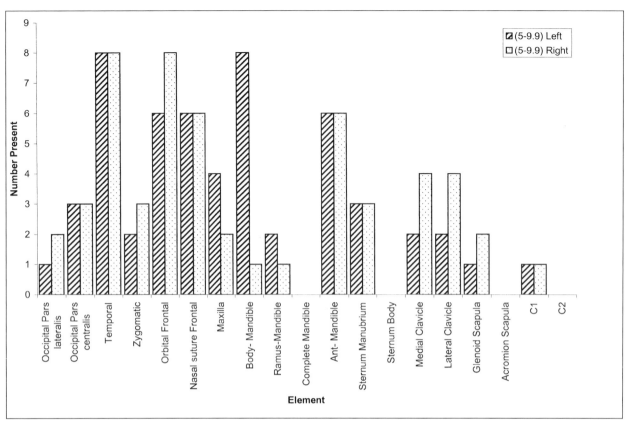

Figure 7.12a. Distribution of G-1 Subadult Skeletal Material (5–9.9 years).

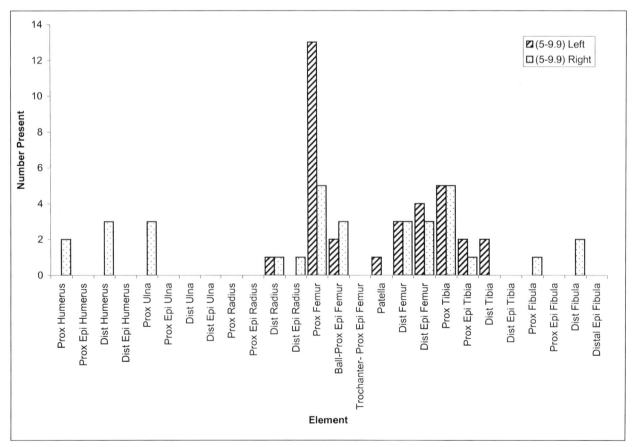

Figure 7.12b. Distribution of G-1 Subadult Skeletal Material (5–9.9 years).

132   CHAPTER 7

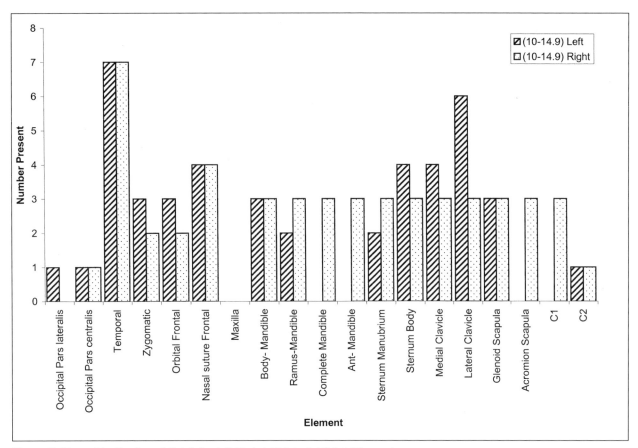

Figure 7.13a. Distribution of G-1 Subadult Skeletal Material (10–14.9 years).

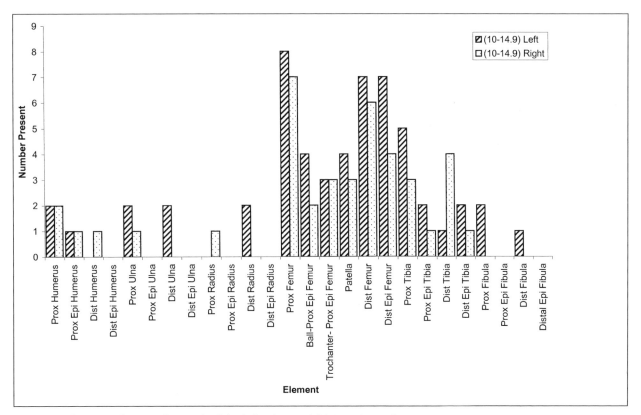

Figure 7.13b. Distribution of G-1 Subadult Skeletal Material (10–14.9 years).

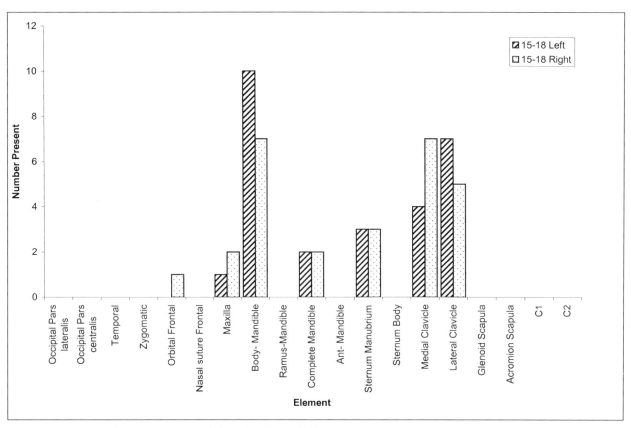

Figure 7.14a. Distribution of G-1 Subadult Skeletal Material (15–18 years).

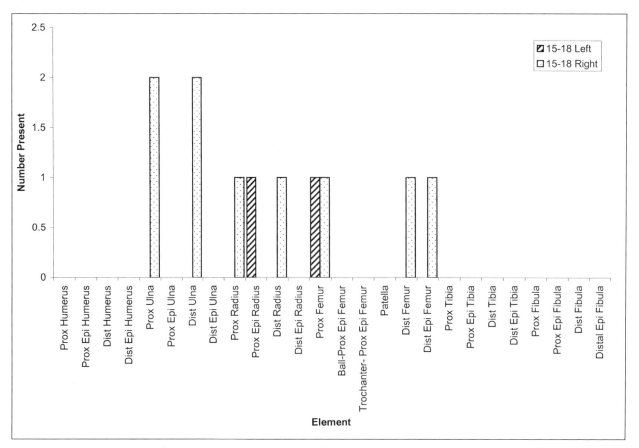

Figure 7.14b. Distribution of G-1 Subadult Skeletal Material (15–18 years).

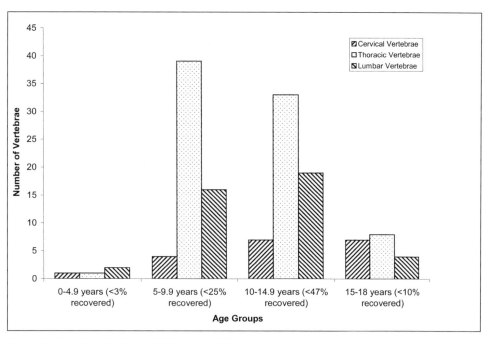

Figure 7.15. Distribution of G-1 Subadult Vertebrae.

variable of obvious importance is the temperature to which bones are exposed.

There are three general categories of bone relative to its condition at the time burning takes place. The first category includes bone still surrounded by substantial soft tissue mass. A second category is bone that has been defleshed but still contains tissue in the marrow space and significant water in the mineralized tissue. The third category includes bone that does not have any associated soft tissue and has also lost much of its original water. Differentiating the effect of burning between dry bone and the other two categories is relatively easy, since the size and shape of dry bone usually remains intact through the burning process. Burning in the two other categories has different effects depending on the temperature as well as the amount of soft tissue associated with the bone and the amount of original water still present at the time of incineration.

Most of the following observations regarding the effect of burning on bone are based on the research of Baby (1954), which relies on a skeletal sample from an archaeological site where the bones had been burned at the time of burial. Lower temperatures result in incomplete incineration and in "smoking" of bone. The other factor is the amount of combustible material associated with bone. The mineral phase of bone, which constitutes about 70% of the bone mass, contains little combustible material; most of the burning of bone depends on its protein component and the associated soft tissue if the latter is present.

Dry bones, presumably with reduced protein and water content, may crack minimally, but there is no uniform color change. Fleshed bones do not burn uniformly either; some parts become calcined before other parts have reached the initial "smoke" stage. In this case, the differences in tissue depth and composition affect the amount of change, including deformation that takes place. The presence of fresh tissue also tends to result in noticeable warping and cracking of the shaft during incineration. Fresh bones undergo the most uniform color change overall, although they still exhibit some warping and cracking from the burning process.

Calcination is a more severe result of higher temperatures than those associated with smoking and includes color transformation of the bone to white, gray, or pale blue. Dry bones crack more at higher temperatures than at lower temperatures, but very much less than the cracking seen in fresh and fleshed bones under similar circumstances. Dry bones take on a tan color instead of white, gray, or blue or the color change apparent in calcination. In dry bone there is a gradient of color change from the tan external surface, which becomes darker immediately below the periosteal surface and changes to a white near the endosteal surface.

The fresh and fleshed bones have more pronounced longitudinal cracking, while the dry bones exhibit superficial transverse cracks. There is evidence that, when burned at high temperatures, fleshed bones produce a pattern of concentric fractures on the popliteal region of the femur because the increased amount of organic material present there allows for greater intensity in burning.

The temperatures required for smoking and calcination to take place are somewhat variable. There is general agreement that smoking occurs at temperatures up to 800°C, while temperatures higher than this produce calcination. However, Buikstra and Swegle (1989) have produced calcined bones at significantly lower temperatures (400–600°C) with a more tightly controlled experiment in which they measured the actual temperature of the bone by placing thermocouples on the bone being incinerated and evaluating the result after cooling. The difference in the estimates of the temperature needed for calcination probably reflect different analytical methods and the difference between oven temperature and the temperature of the bone itself. However, this research does not change the fundamental principle that calcination occurs at higher temperatures than smoking.

The shrinkage that occurs in bone also affects the microscopic structure of bone. This has implications for forensic contexts where bone microstructure is used to estimate age at death in forensic cases. However, it also may be important in situations where similar methods are used to estimate age at death in human archaeological skeletal samples. In an earlier study, Bradtmiller et al. (1984) found no alteration in bone microstructure in burned bone that was sufficient to interfere with microscopic aging techniques. In more recent research, Nelson (1992) found that there is significant reduction in the size of the microstructural elements that can affect age estimates.

The disparity between opinions of the two researchers may be related to the more widespread disagreement about the overall degree of shrinkage that bone undergoes as a result of burning. Studies have shown between 10% and 25% shrinkage, but the results are not consistent. It does seem that the specific temperature plays a role in the amount of shrinkage, as does the condition of the bones when burned. Bones with soft tissue still adherent seem to show more shrinkage. Ubelaker (1989) suggests that researchers allow for the maximum amount of shrinkage, that is, 25%.

The effect of burning that took place on the human remains in the G1 Charnel House ranges from smoky discoloration of the bone surface to calcination with severe shrinkage and deformation. Virtually all the bones show evidence of heating. Although reduced size and abnormal shape of many of the skeletal components is clearly demonstrated, one cannot be certain that some reduction in size did not occur in bone elements that appear normal. Since much of what we are able to reconstruct about the skeletal biology of archaeological human skeletal samples depends on having complete skeletal elements in which the size of bones is important, there are real limits on the confidence one can have in making generalizations about biological variables in the G1 sample. This will become very apparent in the discussion below about the sex distribution based on the size of the femoral head.

DISTRIBUTION OF SKELETAL ELEMENTS

If all the bones of an individual had been interred at the time of burial, as would be the case if the body were placed in the charnel house at the time of death, the skeletal elements recovered during excavation should be present in the same relative frequency as the bones of a living person. This, of course, assumes that taphonomic processes have not caused the destruction of some bones and all the bones present in the tomb were excavated with appropriate care. If the distribution of skeletal elements does not reflect the bones present in a living individual, several factors could have contributed to this condition. Taphonomy could be a factor, with local conditions of the charnel house contributing to some skeletal elements disintegrating faster than others. Indeed, as Ubelaker emphasizes for forensic situations, microenvironmental variability can exist within an assemblage of burials that will result in variation in the preservation of bones (Ubelaker 1997). Taphonomy could also be a factor in poor representation of bones from fetal and infant skeletons, since smaller, less mineralized bone disintegrates more rapidly than larger, more highly mineralized bone. A major factor, and one that we argue is the primary cause of the uneven distribution of skeletal elements from the G1 Charnel House, is the probability that burials are mostly secondary and that recovery from the primary burial site was incomplete, as was certainly the case with the shaft tomb burials.

Subadult Distribution of Skeletal Elements

As we have seen earlier (Figures 7.11 through 7.15), very few bones of the cranium from subadult remains were recovered from the G1 Charnel House. The cranial elements present include a few maxillary and temporal bone fragments. Subadult mandibles are more common. If we take the maximum number of mandibular fragments for each age category (Figure 7.16), we establish a minimum number of 41 subadults represented by mandibles. In the postcranial subadult

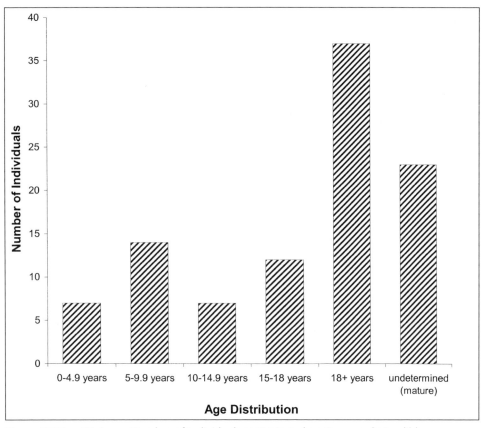

Figure 7.16. Minimum Number of Individuals at G-1 Based on Recovered Mandibles

subsample, there is some variation (Figures 7.17, 7.18) in the bone components recovered between the age categories (birth to 4.9 years, 5–9.9 years, 10–14.9 years, and 15–18 years). In the age categories between birth and 14.9 years, by far the most frequent bone element recovered was the proximal femur. However, in the oldest subadult category the most frequent skeletal component recovered was the body fragment of the mandible. Relatively few subadult vertebrae were recovered (Figure 7.15). Given the minimum number of subadults identified in the G1 Charnel House, the expected number of subadult vertebrae is about three times greater than what was recovered, and very few of the elements were from the youngest age category.

Adult Distribution of Skeletal Elements
In the adult subsample (Figures 7.10a, 7.10b), the temporal bone of the skull is the most common bone recovered from the charnel house. Elements of the proximal left femur were next in frequency. These were followed in frequency by the two largest bones of the feet, the calcaneus and talus. Fragments of the scapula, the right femur, and both right and left tibia were all recovered with about the same frequency. Adult vertebrae were counted by anatomical category, that is, cervical, thoracic, and lumbar. Relative to the expected amount based on the minimum number of adults recovered from the charnel house, the best represented category of vertebrae is the lumbar, with slightly less than 40% of the expected vertebrae recovered (Figure 7.19).

In other studies of archaeological human bone assemblages with a secondary burial tradition, the highest skeletal element prevalence varied somewhat. In Ubelaker's study (1974) of two Native American ossuaries excavated in Maryland, USA, the most common elements in the adult subsample were the right tibia and right mandible. In the subadult subsample, the most common elements were the left temporal and the left femur. In his analysis of the Native American ossuary sample from Fairty, Anderson (1964) found the talus and temporal bones most common among the adult subsample, and among the subadult remains the humerus and femur were most common. In earlier research Stewart (1940) argued that the temporal bone was the best element for estimating the minimum number of individuals from an archaeological human bone assemblage. Possible reasons for this variability

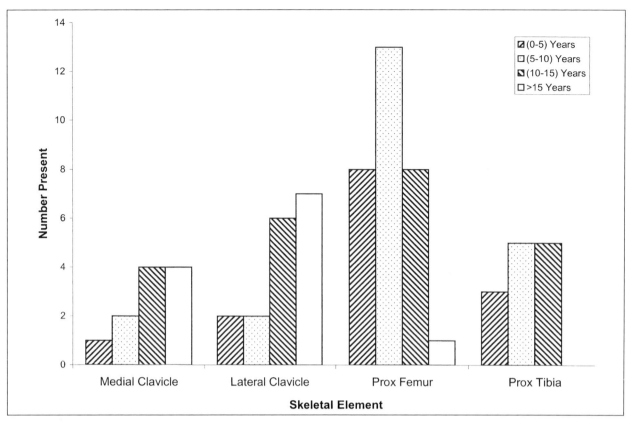

Figure 7.17. Subadult Distribution of Select Elements (Left Side).

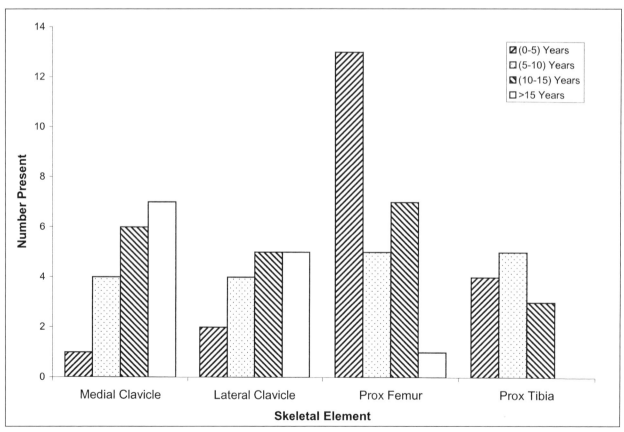

Figure 7.18. Subadult Distribution of Select Elements (Right Side).

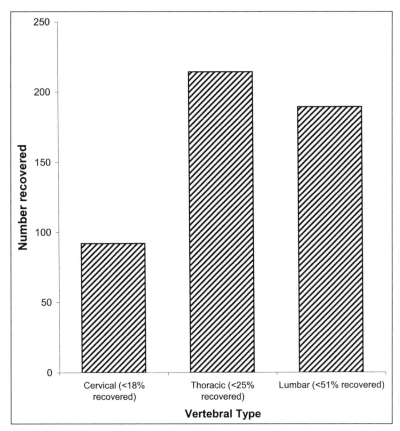

Figure 7.19. Distribution of G-1 Adult Vertebrae.

in the most prevalent skeletal elements are reviewed in the concluding section of this chapter. Certainly the size and density of the bone is important, since both large bones and those with a greater proportion of dense compact bone are more likely to survive the burial environment and be noticed during the recovery and transfer process from the primary to the secondary burial site.

It is possible that the relatively high frequency of temporal components in the adult burials from the G1 Charnel House is a function of preservation in the primary and secondary site, perhaps combined with a cultural emphasis on the recovery of bones of the skull from the primary burial site for reinterment in the charnel house. This does seem to be the case in the shaft tomb chambers of the EB IA period excavated in the 1977 and 1979 field seasons. In these chambers, the preservation and reinterment of the skull/mandible bones was more complete than for the postcranial bones.

There is no obvious explanation for the paucity of other bones of the skull in the G1 Charnel House, particularly if there was a cultural emphasis on recovery of the skull. The postcranial bones that are more frequent probably reflect a combination of being relatively large in size and easily noticed when recovery from the primary site occurred. However, there are curious inconsistencies in the long bones that were recovered. All of the bones of the upper extremity are very low in prevalence, despite the fact that at least the humerus is a large bone and should easily have been seen during recovery from the primary burial site.

It is tempting to argue for a cultural component in the selection of skeletal elements to be recovered and interred in the charnel house. It is at least possible that more cultural value was attributed to some bones than others or, as Ubelaker (1974) suggests in the context of his analysis of an ossuary, that some bones may have been retained, unburied, for ceremonial purposes. This would imply knowledge of skeletal anatomy and the ability to identify the more important skeletal elements needed for final burial. Arguing for culturally determined selection of skeletal elements to be reburied on the basis of the prevalence of skeletal elements recovered from the G1 human remains is obviously speculative. The petrous portion of the temporal bone, the glenoid portion of the scapula, the talus, and the calcaneus are the bone elements best represented in the adult remains. These are all elements that tend to be well preserved because of their structure and/or the relative proportion of dense compact bone. The implication of this is that taphonomic processes may have been a factor in the bones transferred from the primary to the secondary burial site. Nevertheless, the very irregular prevalence of major long bones in the G1 remains is not easily explained by taphonomic factors. As additional charnel house contents are analyzed, additional insight into the factors contributing to this discrepancy may emerge. It is also important to emphasize that there are very few complete bones of any type in the G1 sample. This adds to the challenge of reconstructing many aspects of the skeletal biology of the people represented by the G1 bones.

The care exhibited in antiquity in recovering bone from the primary burial site is substantially greater in the shaft tomb burials. Typically, most of the major long bones were found in the EB IA tomb chambers. As we have seen, this is not the case with the G1 remains.

This suggests that there may have been a decline in the social significance of the funerary tradition in the EB IB period or, at the very least, a change in the tradition regarding what is recovered from the primary burial site and reburied in the charnel house.

Sex Distribution

Very little data are available on the sex ratios for the G1 remains. The only skeletal element recovered that permits reasonable determination of sex is the femoral head. Unfortunately, most of the femoral heads have been affected by burning of the charnel house. For a few of the heads, the effect of heating appears at least minimal. Figure 7.20 shows the distribution of maximum diameters for those adult left femoral heads (N = 18) where we were fairly confident that the dimension had not been seriously affected by heating. The femoral heads of the left side occurred most frequently. Combining right and left does provide a somewhat better distribution (Figure 7.21) but possibly includes femoral heads of the same individual, which would give undue weight in situations where both the right and left from the same person were included.

With most measurements of the skeleton that are influenced by sex, the distribution is bimodal. There is a hint of this in the distribution of the 18 maximum head diameters apparent in Figure 7.20. However, the sample size is too small to be certain about the midpoint between the two subsamples.

In the EB IA shaft tomb burials, where we have a much larger sample of femoral heads from individuals in which sex can be determined with a high probability of being correct, the statistical point of division between the male and female distributions is 43.5 millimeters; that is, a femoral head 44 millimeters or above in diameter is usually male and one 43 or below is usually female. There is, of course, some overlap in the male and female distribution. However, research on modern documented skeletal samples indicates that the accuracy of determining sex on the basis of the maximum diameter of the femoral head is about 90% (Stewart 1979; Ubelaker 1989). In human populations, such as the people of the EB IA phase, where sexual dimorphism is substantial, the accuracy may be even greater. While we cannot be certain that the distribution of femoral head sizes in the G1 Charnel House remains is the same as we found in the EB IA remains, this assumption seems likely.

If we assume that the female range for femoral head sized in the G1 sample is 43 millimeters or less and the male range 44 millimeters or higher, the frequency of femoral heads that can be attributed to the female sex varies somewhat between the two graphs. If we take only the 18 left femoral heads to ensure that two measurements from the same individual are not included, there are 11 in the female range and 7 in the male range. In Figure 7.21, combining measurable right and left femoral heads, there are 19 females and 10 males. Clearly, the male/female ratios in both distributions favor the female sex. It may be that the G1 Charnel House contained more adult females than males. However, the small sample size leaves uncertainty about this conclusion. Another possible explanation for the imbalance between male and female femoral head sizes is slight shrinking of the head caused by the burning. This would reduce the size of the head and increase the proportion of femoral head diameters in the female range.

Even less can be said about the robusticity of the G1 people. We do see evidence of very muscular and robust male skeletal elements and some very gracile female skeletal elements. This resembles the findings of the more adequate sample of EB IA skeletal remains. Given this, it seems likely that the EB IB people were rather sexually dimorphic and similar in robusticity to the EB IA people.

EVIDENCE OF DISEASE IN THE G1 REMAINS

The greatly reduced prevalence of subadult remains, particularly with reference to the EB IA burials, does affect the prevalence of skeletal disease, since subadults have a higher risk, particularly for infectious disease, and are more likely to show skeletal manifestations of disease in noninfectious disorders such as scurvy. Nevertheless, it is surprising, given the presence of plausible evidence of both tuberculosis and brucellosis in the EB IA sample, that there is no evidence of skeletal infectious disease in any of the G1 skeletal fragments, including the vertebrae.

The fragmentary nature of the G1 skeletal sample makes interpretation of the relative health status of the EB IB people hazardous. Nevertheless, the lack of evidence of infectious disease does suggest that the G1 people may have been healthier than those associated with the EB IA phase. However, a cautionary note is necessary for at least two reasons. The first of these is that the G1 remains are certainly not representative of the living population of the EB IB phase at Bâb edh-Dhrâʿ. Subadults are poorly represented. The presence of some remains from young children argues against taphonomy being a significant factor affecting the frequency of subadult remains recovered. It seems likely

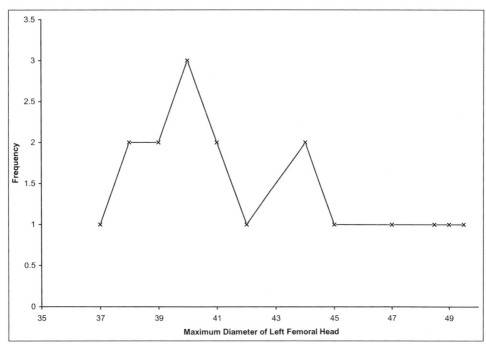

Figure 7.20. Distribution of Femoral Head Size (Left).

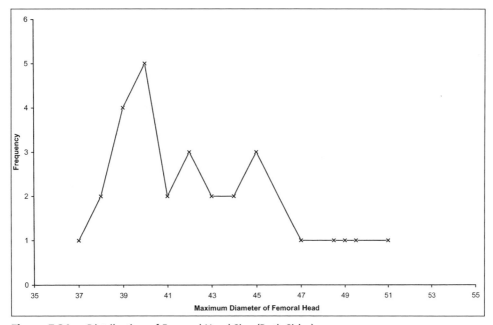

Figure 7.21. Distribution of Femoral Head Size (Both Sides).

that very young subadults were not regularly buried in the charnel house, although it is at least possible that most of the infants and young children were interred in another burial structure or location. Again, this seems unlikely, given the presence of at least a few infants' and young children's skeletal elements. In addition to the subadult sampling problem, the adult remains are very fragmentary and incomplete. It may be that some evidence of skeletal disease was present in the living population represented by the G1 remains but was lost because of breakage and careless recovery at the primary burial site and poor handling in the transfer to the charnel house.

The second factor is that skeletal evidence of disease is usually associated with relatively chronic conditions where the individual survives for a considerable amount of time with the disease. This is particularly the case for infectious disease. Acute diseases, with rapid death, rarely affect the skeleton and will not often be seen in archaeological skeletal samples. The lack of skeletal disease could result from a situation where the people were dying of acute diseases. The other option, of course, is that the people were very healthy and responded well to exposure to pathogens so the skeleton was rarely affected.

One can approach these possible explanations for the lack of skeletal evidence of disease through careful demographic analysis. For example, if life expectancy is significantly reduced and associated with reduced evidence of skeletal disorders, then an increased prevalence of death from acute disease would be implied. Unfortunately, careful demographic reconstruction of life expectancy for the G1 sample is not possible because of the very fragmentary condition of the remains and the evidence that the very young who died typically were not interred in the G1 Charnel House. Nevertheless, the overall appearance of robusticity of the people, combined with the lack of skeletal disease, is most plausibly explained as an indication of improved health relative to the EB IA people. This is unexpected, and should be treated as tentative, since an increasingly sedentary economy should increase exposure to infectious pathogens and could be expected to have increased evidence of skeletal disease (Ortner 2003). However, it is at least possible that the EB IB represents a time of greater food resource adequacy and stability. Evidence of nutritional deficiency is abundant in the EB IA burials. The lack of evidence of this deficiency in the G1 bones does at least suggest that nutrition was more adequate. Because of the significant link between adequate nutrition and the ability to resist infection and other diseases, one possible explanation for the lack of evidence of skeletal disorders other than trauma in the G1 remains may be a more abundant and nutritionally adequate food supply.

In both the vertebrae and the joints of the long bone fragments that are present, there is some evidence of osteoarthritis. We have not been able to quantify this evidence. However, a reasonable generalization is that neither the number of cases nor the severity is more than moderate.

There is evidence of some interpersonal violence probably associated with warfare. One of the fragmentary skulls has a healed wound of the skull vault that was caused by a blow from an axe (Ortner 1982). The blow came at an angle to the outer table of the right parietal near the sagittal suture and penetrated both tables (Figures 7.22a, 7.22b). The axe had a curved cutting edge, which accounts for the deeper penetration through the inner table at the center of the wound and the incomplete penetration at either end of the lesion. The fragment of the bone created by the blow was broken off at the sagittal suture and presumably lost at the time of trauma. The margins of the wound have completely remodeled, with compact bone filling in the exposed diploë. This indicates complete recovery and survival for at least many months, if not years, following trauma. The injury did not penetrate deeply beyond the inner table, and almost certainly the dura remained intact, so the probability of meningitis or encephalitis was greatly reduced. Furthermore, there is no evidence of infection adjacent to the injury site of the outer table, so there may not have been significant secondary infection there either. A very similar injury was seen in a fragment of the skull from another individual (Figure 7.23) in which the injury also penetrated both tables. Remodeling at the margins of the lesion indicates long-term survival following the trauma.

Not all axe wounds of the skull penetrated both tables. In one case, a glancing blow, presumably from an axe, created an oblong depression in the outer table (Figure 7.24). The greatest danger in this type of injury was the possibility of concussion with reactive buildup of cerebrospinal fluid that can cause serious damage to the brain and associated tissues. The evidence of healing and long-term survival seen in all cases of the skull trauma apparent in the G1 remains indicates that secondary complications such as concussion were not a significant problem in the cases of skull trauma recovered from the charnel house.

SUMMARY AND CONCLUSIONS

To the extent that the G1 Charnel House is representative of the burial tradition of the EB IB phase at Bâb edh-Dhrâ', there are significant differences in the burial tradition as compared with the EB IA tombs, other than the emphasis on above-ground charnel houses rather than shaft tombs. The secondary burial tradition was universal in the EB IA phase of the Bâb edh-Dhrâ' people, although there was some variation in the time between primary and secondary interment. This practice of secondary burial was continued in the inhumations of the G1 EB IB Charnel House, but this is one of the few similarities that exist. A major difference is the near absence of neonatal through early childhood burials. Another difference is that there almost certainly were multiple burial events, as recent bodies were added to the earlier charnel house contents. The evidence from the EB IA shaft tomb burials excavated in 1977 and 1979 generally supports

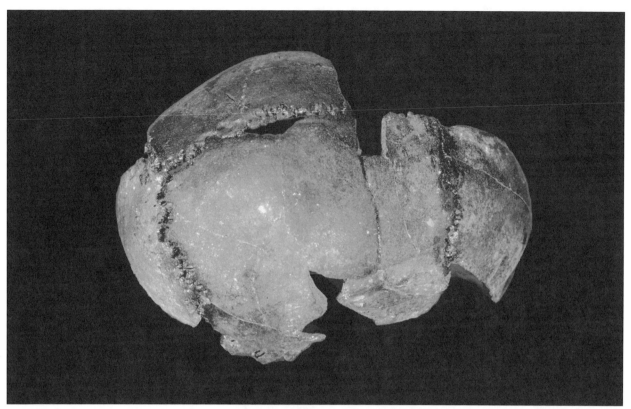

Figure 7.22a. Partially reconstructed skull vault from the EB IB G1 Charnel House with a healed axe wound of the right parietal near the sagittal suture. The wound penetrates both tables.

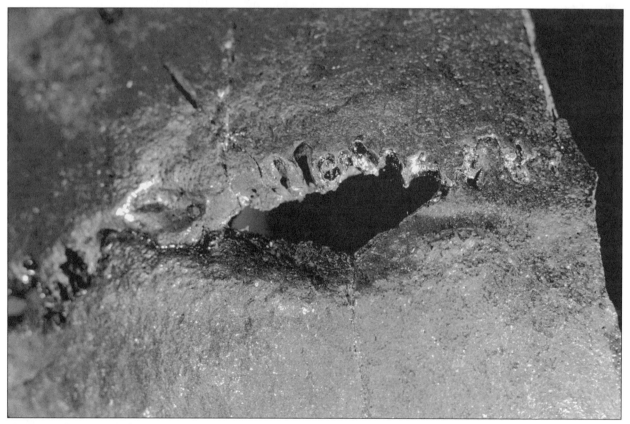

Figure 7.22b. Detail of axe wound in 7.22a.

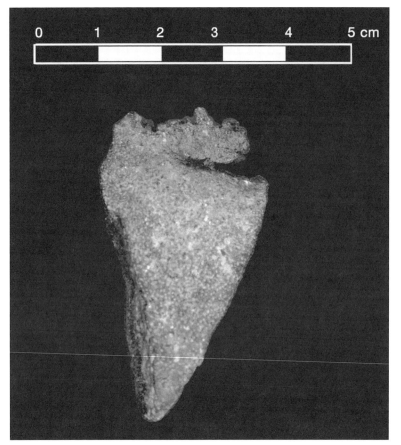

Figure 7.23. Skull fragment with a healed axe wound. The wound penetrates both tables.

When the adult male body was placed in the north chamber of Tomb A100, the bones of two subadult skeletons were disturbed, indicating the probability of at least two burial events. The undisturbed subadult bones were in anatomical relationship, and it seems likely that the two subadults had been placed in the chamber at the time of death or very shortly thereafter and are most likely primary burials as well. The significance of this burial practice in a chamber dated to EB IB on the basis of associated pottery is problematic. This chamber is one of a very few shaft tomb chambers dated to the EB IB phase. Other shaft tomb chambers associated with EB IB include A88 and G2. Tomb A88 includes an upper layer with a single articulated burial and a lower level with several disarticulated, poorly preserved burials. The dating of the lower layer is certainly EB IB, based on the distinctive pottery. There is only a single pot associated with the upper level burial, and this artifact is not sufficiently distinctive to allow certainty about the date (Schaub 1981a). However, the position of this burial above the EB IB level argues for an EB IB or later date. Tomb G2 burials are disarticulated and undoubtedly secondary.

Another distinctive feature of Tomb A100 is that the north chamber is the only one dated to EB IB. The other three chambers associated with this tomb have artifacts linking them to the EB IA phase, and all the burials were secondary. Furthermore, it is certain that the north chamber had previously been used for burials, presumably in the EB IA phase, and had been cleared to make room for the EB IB burials and artifacts. With the possible exception of the articulated skeleton associated with the upper level of Tomb A88, all the other EB IB shaft tomb chambers contained secondary burials. The very unusual circumstances apparent in A100N are probably unique and need not do more than raise a cautionary note relative to the secondary burials that appear to characterize the G1 interments. The variable distribution of skeletal elements in the G1 Charnel House is a strong argument favoring a continuation of the secondary burial tradition of EB IA. However, the paucity of infants' and

a single burial event for at least each tomb chamber, although it remains likely that different chambers of a single tomb were utilized at different times, at least in several of the tombs. A somewhat different picture emerges from the excavations of EB IA tombs in 1981, which took place in a different part of the cemetery (see Chapter 9) and may represent a burial tradition toward the end of the EB IA period. In some chambers, the evidence indicates that one or more earlier inhumations were disturbed as later burials were placed in the chamber.

Even the continuity in the secondary burial aspect of the EB IB funerary tradition is somewhat problematic. There is some evidence of primary burial associated with the EB IB phase, as seen in the north chamber of Tomb A100 discussed in Chapter 6. In this shaft tomb chamber, the one adult male burial was almost certainly placed in the chamber at the time of death. At the time of excavation, all bones were in correct anatomical position and all were present. This would be highly unlikely unless this individual had been interred in the chamber within a very short time period following death.

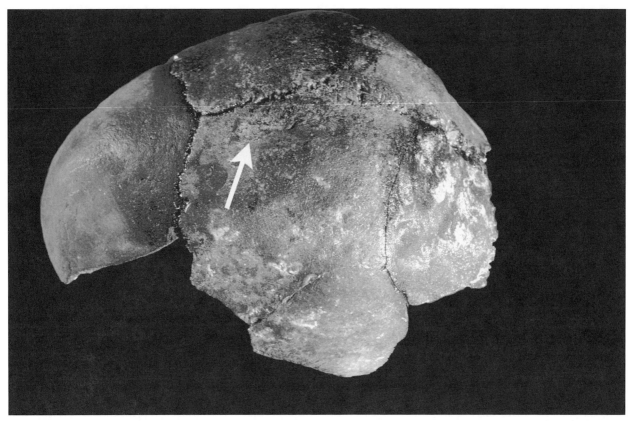

Figure 7.24. Partially reconstructed skull vault from EB with a healed axe wound (arrow). In this case only the outer table was affected by the trauma.

young children's skeletons is in marked contrast with the prevalence of burials in this age range in the shaft tomb chambers.

There are several potential reasons for the skeletal element variation in frequency. In his analysis of two adjacent Native American ossuaries in Maryland, USA, that were secondary burial sites, Ubelaker (1974) listed five possibilities for the variability apparent in these ossuaries: (1) problems in the transfer of human remains between the primary and secondary burial sites, (2) cultural selection of skeletal elements, (3) taphonomic factors affecting the preservation of some skeletal elements more than others, (4) loss of skeletal elements, particularly smaller bones, at the time of archaeological excavation, and (5) loss of bone subsequent to excavation during curation and analysis. Ubelaker highlights his experience in excavating one of the ossuaries in which bones of infants were poorly preserved and disintegrated during excavation.

The very unusual distribution of G1 skeletal elements apparent in Figures 7.11 through 7.14 can best be explained as a result of a secondary burial practice in which not all bones of a specific burial were recovered for interment at the secondary burial site. This, of course, does not mean that all interments in the G1 Charnel House were secondary, but it seems likely that at least most burials were. What remains unknown is an explanation for the variation between the frequencies of the skeletal elements. The bones interred in the G1 Charnel House do tend to be the larger and more easily recognized and recovered from a primary burial site. However, there are unusual gaps in the prevalence of skeletal elements. Very few bones of the face were recovered. These bones tend to be easily broken, and damage resulting in their loss is possible during recovery from the primary site and/or transportation back to the secondary burial structure. However, it is important to note that facial bones were often intact in the EB IA burials that were also secondary. Because of the obvious relationship between the skull and mandible with the appearance of the person before death, one could argue that these bones would have particular cultural emphasis in the funerary tradition with the transfer from the primary to the secondary burial site. One thinks of the remarkable restoration of soft tissue features that is present in some of the PPNB burials from Jericho (Kenyon 1960) that highlight the appearance of the living individual.

The great care manifested in the recovery and reinterment of the EB IA burials included very small bones from late-term fetal and neonatal remains. The preservation and recovery of these bones in the EB IA tomb chambers implies that the taphonomic conditions at both the primary and secondary burial site were relatively good. Assuming that the burial environment for the people represented by the G1 remains was similar if not the same, the effect of taphonomic factors in producing both the different age and skeletal element distribution is minimal.

It is clear that some breakage of bones occurred in the EB IA remains during the recovery from the primary burial site and the transfer to the secondary tomb chambers. This fact coupled with the disproportionate distribution of skeletal elements suggests that cultural factors were important in the element prevalence found in the G1 burials. At least two cultural factors may have affected this prevalence. The first of these is a cultural emphasis on certain bones of the skeleton in the transfer of burials to the charnel house. The second factor is a deemphasis of the importance of the skeleton in the burial tradition, leading to a careless recovery from the primary site and a lack of care in the transportation of the bones to the charnel house. The evidence regarding the degree of care in recovery and reinterment of the human remains in the G1 Charnel House indicates a much more haphazard recovery, and this reflects a substantial difference from the EB IA burials.

Very little can be concluded about the health of the G1 people. There is minimal evidence of infectious disease. Keep in mind that every bone and bone fragment recovered from the charnel house was examined for evidence of abnormality. Although the bones are usually incomplete, it remains surprising that there is minimal evidence of reactive bone porosity or abnormal bone formation on any of the fragments. In addition, there is no evidence of destructive lesions of any bone. This is particularly significant relative to the vertebrae, which are present in sufficient numbers so that tuberculosis, if present in the society, should have affected some of them. The few examples of abnormal bone formation are associated with callus formation following fracture. As noted above, there is some evidence of trauma from intentional violence, as evidenced from the healed axe wounds in three of the skulls (Figures 7.22 through 7.24). This finding corresponds well with the additional evidence of intentional skull trauma in shaft tomb chambers dated to the EB IB phase. There is proportionally less evidence of interpersonal skull violence in the EB IA material, which suggests that this type of violence increased in the transition between EB IA and EB IB. There is also some evidence of accidental fracture, but this is very limited, to some compression fractures of vertebral bodies and a few fractures of ribs. Evidence of joint disease does occur in some of the vertebral bodies and some of the subchondral bone surfaces but, again, there is relatively little evidence of this disorder.

If, in fact, the EB IB people were healthier than the people of EB IA, there are some troublesome questions. Certainly an increase in sedentary permanent occupation of a site with increased population density is associated with an elevated exposure to infectious pathogens in most human societies. Given this, one would expect to see more evidence of disease in the G1 remains compared to the EB IA burials. There is no evidence that this is the case, which raises issues about the relationship between the EB IA and EB IB people.

We do see evidence of malnutrition in the EB IA remains along with probable evidence of infectious diseases, including osteomyelitis, tuberculosis, and possibly brucellosis. Malnourishment does increase the probability of morbidity among people exposed to infectious pathogens, although rapid death with no skeletal involvement is more likely as well. However, the diet of more nomadic people tends to be varied and protein sources adequate relative to their more sedentary agriculturist cousins (Cohen and Armelagos 1984). Given this, we would expect that nomadic pastoralists would be relatively healthy when compared with more sedentary agriculturists living in larger, more densely populated habitation sites. The high proportion of infant burials in the EB IA tomb chambers can be interpreted as indicative of high birth rates. This is associated with a population that was increasing in size (Bocquet-Appel 2002; Bocquet-Appel and Naji 2006) and thriving. The obvious bias in burial practice apparent in the G1 burials precludes a comparison of birth rates, but the cultural development apparent in the large charnel houses and the large fortified town that developed during EB II and III at the site argue for a population that was stable and perhaps growing in size. It is at least plausible that the more stable food resources available from heavier dependence on food crops created a better nutritional environment and improved health despite the potential for increased exposure to infectious agents.

CHAPTER 8

The Tombs and Burials of the EB I People: Tombs Excavated in 1979

Donald J. Ortner, Evan M. Garofalo, and Bruno Frohlich

IN THIS CHAPTER WE PROVIDE a general statement about the human remains found in each of the EB I tombs and tomb chambers excavated during the field season of 1979. The major objective of excavations in the Bâb edh-Dhrâ' cemetery area was to increase the sample size of the human remains excavated in the 1977 field season. As with Chapter 6, the main emphasis in this chapter is to describe the human remains from the tomb chambers to provide a general understanding of the number of individual burials excavated, determine the age distribution of the burials, and describe the evidence of skeletal disease that may be apparent in the burials. A more detailed analysis of the dental disorders is presented in Chapter 13, and data about the metric and nonmetric aspects of the human remains are discussed in Chapter 10. All the burials were secondary, with varying times between interment in the primary site and final burial in the shaft tomb chambers. The reader is referred to the introductory comments in Chapter 6 for additional details on the basic burial tradition and the shaft tomb structure.

TOMB A102

There are four chambers associated with the central shaft of this tomb (Figure 8.1). The chambers were cut predominantly in marl, but the layers closest to the

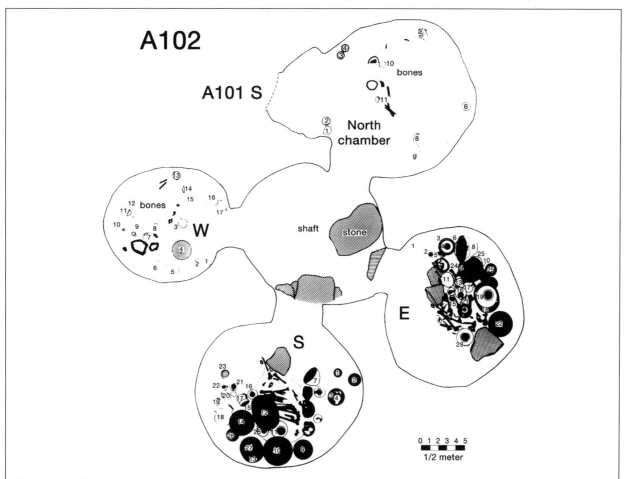

Figure 8.1. Plan drawing of Tomb A102.

chamber floor had a significant component of sand and gravel, which tend to erode onto the chamber floor. The basic arrangement of the tomb chamber contents is typical of all the shaft tomb chambers in the EB IA phase, that is, skulls arranged carefully to the left of the entryway, postcranial bones in a central pile, and burial gifts on the margin of the right area of the floor.

Tomb A102, Chamber Northeast

The absence of blocking stones at the entryway of this chamber provided easy access for extensive silting of the chamber. This was further exacerbated by some ceiling collapse. The combination of silting and ceiling collapse contributed to very poor preservation of the chamber contents (Figure 8.2). A feature seen in a few chambers excavated in 1977 was a break in the wall of this chamber connecting with an adjacent tomb chamber, in this case the southwest chamber of Tomb A101, excavated in 1977.

The MNI in this chamber is ten, including three adults and seven subadults. One of the adult burials is a male 40 years of age or more. There were two adult female burials in this chamber, one of which is about 30 years of age; a specific age cannot be estimated for the other.

Burial 1 includes very fragmentary and incomplete cranial and postcranial skeletal elements. The estimated age of this burial is ten. There is no skeletal evidence of disease in the elements available for analysis. Burial 2 is represented by the fragmentary cranial remains of an adult female about 30 years of age. There are postcranial skeletal elements that could be attributed to this burial, but they are more likely associated with Burial 61 described below. There is no evidence of skeletal disease.

Both cranial and postcranial elements make up Burial 3 in the best-preserved skeleton associated with this chamber. About 25% of the skeletal elements are present in this burial. The remains are from a very robust male 40 or more years of age at the time of death. There is widespread evidence of moderate osteoarthritis associated with many of the joints available for analysis. Burial 4 is the very incomplete cranial and postcranial remains of a child about eight years of age. The teeth probably associated with this burial have at least two hypoplastic lines indicative of two significant stress events earlier in life.

The remaining burials were not identified until the bone assemblage was analyzed in the laboratory. Burial

Figure 8.2. Northeast chamber of A102. Both ceiling fall and silting contributed to the very poor condition of the burials.

61 consists of very incomplete postcranial bones of an adult female. The patellae have joint disorders with partial destruction of the subchondral bone and reactive bone formation on some of the subchondral surfaces. This evidence of arthritis suggests an age past early adulthood. There is an abnormality in the development of some of the bones of the feet, including the tali and the left navicular (the right was not recovered), which has reduced growth in the anterioposterior axis. Unfortunately, the bones of the feet are too incomplete to permit confidence about the cause of this abnormality, but one possibility is club foot deformity.

Burials 62 and 63 are very incomplete skeletons of an infant and a seven-year-old child. None of the skeletal elements associated with these burials has any evidence of disease. Burial 64 consists of both cranial and postcranial elements but is very incomplete. The estimated age of this burial is perinatal. Periostosis is apparent in both the upper and lower extremities. A systemic condition of this type is most likely caused by infection, but other disorders are possible as well. Given the neonatal age, the disorder is most likely congenital. A similar disorder is apparent in Burial 66, which consists of postcranial, mandibular, and dental elements. The estimated age is six months. Periostosis occurs on the anterior crest of both tibiae. Burial 65 includes postcranial and dental elements of a child between three and five years of age. None of the elements shows any evidence of disease.

Tomb A102, Chamber East

Some silting and considerable ceiling fall had occurred in this chamber before excavation (Figure 8.3). The general arrangement of skeletal remains and tomb artifacts is similar to other shaft tombs with the skulls placed to the left of the chamber entrance and postcranial bones in the center of the chamber floor. Some of the bones may have been partially articulated at the time of secondary burial. An example of this is apparent in the proximity of the mandible and skull of Burial 1 (Figure 8.3). This suggests a relatively short time between primary and secondary burial, unlike most of the burials excavated in 1977. Nevertheless, most of the postcranial bones were mixed and haphazardly placed in the center of the chamber (Figure 8.4). Silting had embedded most of the bones and displaced the pots as they floated on the silt. Preservation of the

Figure 8.3. East chamber of A102. Note the large piece of ceiling fall in the foreground.

Figure 8.4. Careless deposit of postcranial bones in the east chamber of A102.

skeletal remains was fair, with completeness ranging from less than 5% to more than 20% of the skeletal elements. The MNI in this chamber is 14, including six adults and eight subadults ranging in age from infancy to early adolescence.

Burial 1 includes cranial, dental, and postcranial material and is remarkably complete. The estimated age is between five and seven years. There is no evidence of skeletal disease in the bones available for analysis. Burial 2 consists of very incomplete postcranial elements, including both tibiae and a complete innominate. The estimated age is about three years. There is no evidence of disease in the few bones available for analysis.

Burial 3 was originally thought to consist of a single individual. Laboratory analysis revealed that the burial included elements of three individuals. Burial 3.1 is the most complete skeleton in this chamber. It contains postcranial bones and the maxilla and mandible with the associated teeth, although several of the teeth were lost antemortem. There is also a fragment of the right mastoid process. The estimated age is between 23 and 28 years and the sex is male. There are multiple, well-defined enamel hypoplasias apparent in the dentition indicative of multiple stress events during dental development. There is also considerable evidence of joint disease, probably osteoarthritis, that is unusual in a young adult. This is probably indicative of hard physical activity. This observation is further supported by the evidence of enthesopathies at several of the muscle origins and insertions.

Burial 3.2 includes skeletal elements of a male in the same age range as Burial 3.1. The dentition has two distinct hypoplastic defects, indicating severe stress events during development of the teeth. Burial 3.3 consists of both cranial and postcranial elements of a female about 16 years of age at the time of death. Evidence of skeletal disease is minimal. However, the presence of bilateral cribra orbitalia does suggest the presence of a chronic disease. Diagnostic options include scurvy, anemia, and chronic infection. With respect to the last option, trachoma is a chronic infectious disease, particularly in many third world countries, in which flies are the vector. The disease does affect the eyes, often resulting in blindness. The antiquity of this infection extends back to at least 3000 BC in Egypt (Aufderheide and Rodríguez-Martín 1998), but may extend back much farther in time (Webb 1990). The chronic nature of the disease and the predilection for affecting the eyes make it a potential cause for skeletal lesions causing a vascular response in the form of increased porosity of bone associated with the site of infection.

Burial 5 consists of the fragmentary skeletal elements of an adult male probably in the 35–40-year range at the time of death. There is evidence of moderate osteoarthritis and associated enthesopathy. Periostosis occurs bilaterally on the femora and tibiae. The cause of this abnormality is not certain, but the bilateral involvement of multiple bones argues for a systemic disorder, the most common cause of which is infection.

Burial 50 was not identified until laboratory analysis of the skeletons in this chamber. It consists of cranial and dental elements of a child five to seven years of age. Both orbital roofs exhibit porosity. Additional evidence of porous periostosis occurs in the infraorbital foramina of the maxilla, the zygomatic bone, and the alveolar process. This distribution of porous lesions is seen in scurvy, and this is the most likely cause of the bone lesions.

Burial 51 includes only the incomplete sacrum and possibly a fragment of the right clavicle of a young adult about 18 years of age. The anatomy of the sacral element is indicative of the male sex. There is no significant evidence of skeletal disease. Burial 52 consists of very incomplete skeletal fragments of an infant about one year of age. There is no evidence of skeletal disorders in the elements available for analysis.

Burial 53 consists of right fragmentary maxillary and mandibular fragments of a child about 11 years of age. Dental enamel of the teeth in these fragments exhibits dental hypoplasia indicative of at least two episodes of severe stress during dental development. Burial 54 also includes right fragmentary maxillary and mandibular fragments of a child somewhat younger than Burial 53. As with Burial 53, the dental enamel has multiple hypoplastic defects associated with stress during development. No other evidence of skeletal disease occurs in the very fragmentary elements available for analysis. Very fragmentary skeletal elements of a child about five years of age comprise Burial 55. There is no evidence of skeletal disease apparent on the elements available for analysis.

Burial 56 consists of incomplete skeletal elements of an adult female probably 50 years of age or older at the time of death. The skeletal elements exhibit multiple fractures with both callus formation and non-union. Many of the vertebral bodies have the biconcave remodeling (codfish vertebrae) most commonly associated with postmenopausal osteoporosis. Other conditions, including cancer and malnutrition, can cause osteoporosis, but given the length of time this individual had the disorder, as evidenced by multiple fracture sites in various states of healing, severe postmenopausal osteoporosis is the most plausible cause. Burial 57 consists of the right and left distal radii and possibly some cranial fragments of a child about two years of age. There is no evidence of skeletal disease in the skeletal elements available for analysis.

Additional evidence of skeletal disorders present in skeletal elements that cannot be associated with a specific burial include osteoarthritis, fracture, and dental hypoplasia. Also apparent in unassociated tarsal bones is the development of lesions that are most likely associated with some type of erosive arthropathy. The cyst-like lesions occur in the distal juxta-articular areas of metatarsals. A common cause of this type of disorder is gout, but other conditions are possible as well. There is also a rib fragment that has reactive bone on the pleural surface that was probably caused by a lung infection. Given the evidence of probable tuberculosis in other skeletons from the EB 1A tombs, this seems a likely cause.

Tomb A102, Chamber South

Despite both substantial ceiling fall and silting (Figure 8.5), bone preservation in this chamber was remarkably good (Figure 8.6) but somewhat uneven. Some of the very small distal phalanges of the feet and other small bones of the hands and feet were recovered, along with several late-term fetal and neonatal skeletal elements, whereas many of the larger skeletal elements were missing. The MNI recovered from this chamber is 17, including three adults and 14 sub-adults. The problem encountered was in associating cranial and postcranial skeletal components. This task is complicated when age and size differences between many of the burials are minimal.

Burial 1 includes both cranial and postcranial bone fragments of a child about five years of age. The only evidence of skeletal disorder is remodeled porosity in the orbital roofs of both orbits. Chronic infection of the eye is possible, but if it were something such as trachoma, healing by the time the child died would have been unlikely. A more probable diagnosis is scurvy from which the child recovered long enough before death that the orbital lesions would have been partially remodeled.

Burial 2 consists of a well-preserved cranium that is most likely associated with postcranial elements

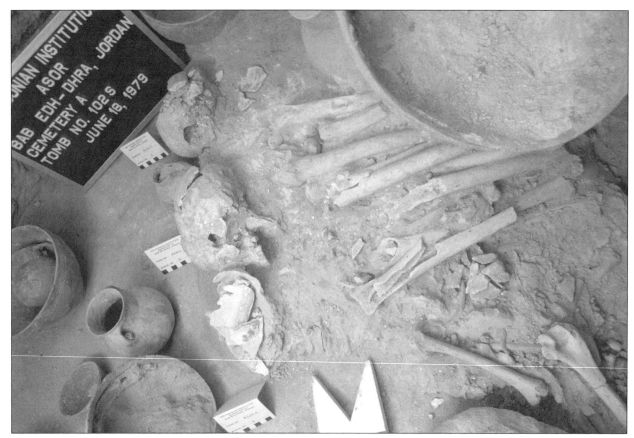

Figure 8.5. South chamber of A102. Despite significant ceiling fall and silting, bone preservation in this chamber was good. Note the more careful placement of the postcranial bones.

that in an earlier stage of the research was designated as Burial 52. Both are now designated as Burial 2. The age at death of the burial is about 30 years and the sex is male. The interesting feature of the skull is a bone-forming tumor in the right parietal (Figure 8.7). The lesion predominantly involves the outer table, but the diploë and inner table are also affected by the growth of the tumor. The tumor process is, in effect, remodeling the bone affected by removing some or all of the existing normal bone and replacing it with tumor bone. The small size of the lesion and the relatively well-organized tumor bone argue for a benign process, although an early stage of a malignant tumor needs to be considered as well. Also apparent in the skull is slight porosity of the orbital roofs. The dentition exhibits dental hypoplasia indicative of a single stress event during development of the teeth. About 15% of the postcranial elements of Burial 2(52) were recovered. No significant bone disorder is apparent in these bones.

Burial 3 consists of the relatively complete cranium that is now linked to postcranial elements formally designated as Burial 50. The individual represented by these bones was a female about 35 years of age at the time of death. There was considerable antemortem tooth loss with extensive remodeling of the alveolar processes. There is some evidence of osteoarthritis in the subchondral joint surfaces, but this is not extensive or severe.

Burial 51 consists of postcranial bones only of an adult male between 40 and 45 years of age. The only evidence of skeletal disorder is bilateral osteoarthritis affecting the hip. Burials 40 and 41 are incomplete skeletons of infants about one year of age. There was no evidence of skeletal disease. Burial 60 is the incomplete skeleton of an infant about six months of age. Burial 61 is based on skeletal elements of an infant about nine months of age. Both the left and right femora have an abnormal degree of bowing in the anterio-posterior axis. This may be an example of normal variability, but a diagnosis of a mild case of rickets is also possible.

Burial 62 includes postcranial and possible cranial elements of an infant about six months of age. There is evidence of abnormal cortical porosity and reactive porous bone formation. The cause of this abnormality is not known, but the most probable diagnostic options

Figure 8.6. South chamber of A102. Pottery surrounds the skulls and postcranial bones, and a large bowl rests on top of the postcranial bone pile.

Figure 8.7. Bone tumor of parietal in Burial 2, A102S.

EB I Tombs Excavated in 1979

include systemic infection and scurvy. Burial 63 consists of postcranial elements, and some cranial elements that may be associated with the postcranial bones. The age is three to six months. There is no evidence of skeletal disease in the bones available for analysis.

Burial 64 is composed of half of the major long bones and some cranial bones that are probably associated with them. Skeletal elements formerly designated as Burials 65 and 71 are now included with this burial because of the distinctive abnormalities seen in the bones. The age is two to three months. All the bones associated with this burial have reactive bone formation. The lesions are bilateral where both sides are available for analysis. The left distal tibia is abnormally deviated medially, which may reflect a pathological fracture resulting from inadequate trabecular bone in the distal metaphysis. The most likely cause of this is scurvy, but rickets or the presence of both disorders should be considered as well.

Burial 70 is the incomplete skeleton of a perinatal infant. There is no evidence of skeletal disease. Burial 72 is represented only by the right and left tibiae of a three- to six-month infant. Both bones exhibit anterio-posterior bowing that is probably due to rickets. A similar condition is apparent in Burial 73, also with an estimated age of three to six months. The tibiae have reactive bone formation in addition to anterio-posterior bowing. Some combination of rickets and infection or scurvy is the probable diagnosis in this case.

Burial 74 consists of incomplete cranial and postcranial elements of a perinatal infant. The left tibia exhibits abnormal porosity of the medial diaphysis; the right tibia was not recovered. Diagnosis is uncertain, but options include infection and scurvy. Burial 75 includes only the incomplete maxilla and some of the associated teeth of a two- to three-year-old child. Both the infraorbital foramina and the alveolar process have abnormal porous reactive bone formation. Scurvy is the likely cause of this abnormality. Burial 76 is represented by only dental elements that could not be associated with any other burial. The teeth are from a child about four years of age. They have no evidence of abnormality.

Abnormalities were also identified in miscellaneous skeletal elements that could not be associated with a specific burial. These include a thoracic vertebra with a central lytic focus that could have been the result of an early stage infectious disease such as tuberculosis; a lower thoracic vertebra with evidence of unhealed posterior disk herniation, probably the result of trauma; bilateral spondylolysis of a lower lumbar vertebra (probably L5); and a right and left greater wing of the sphenoid with porous reactive bone formation that may be associated with scurvy.

Tomb A102, Chamber West
Ceiling fall did not occur in this chamber, but it was completely silted at the time of excavation. This condition made recovery of the human remains difficult and undoubtedly is a factor in damage that is apparent in the bones. An intriguing aspect of the recovery is that the bones of neonates were remarkably well preserved. The conventional wisdom is that infant bones will be more likely to be underrepresented because they are more vulnerable to taphonomic processes that degrade bone. This is certainly not the case in this and other shaft tomb chambers at Bâb edh-Dhrâ'.

Four of the burials identified during laboratory analysis were included as Burial 1, assigned during excavation. The four burials were designated as 1a to 1d during analysis. The MNI associated with this chamber is 15, including three adults, two adolescents, two children, seven neonates or infants, and one late prenatal skeleton.

Burial 1a consists of both cranial and postcranial bones but is very incomplete. The burial is that of an adult male, but a more specific age estimate is not possible. There is no evidence of skeletal disease associated with any of the bones. Burial 1b (including Burial 60) includes cranial and postcranial elements of a 16–18-year male. There is no evidence of skeletal disease. There is also no evidence of skeletal disorders in the cranial and postcranial bones associated with Burial 1c, an adult male skeleton. Burial 1d contains the bones of an adult female between 20 and 25 years of age. None of the bones exhibit evidence of significant disease.

Burials 61 and 64 are incomplete skeletons of children between 2 and 4 years of age and have no evidence of skeletal disease. Burials 62, 65, and 66 are infants between birth and six months of age. None of the burials has any significant evidence of disease, with the exception of Burial 65, in which both tibiae have a mild expression of periostosis that could have been caused by infection or trauma. Burial 63 is a late-term fetal or perinatal skeleton with no evidence of disease. Burials 67, 68, 69, and 70 are all incomplete skeletons of infants who died at the time of or shortly after birth. None shows any evidence of skeletal disease. Burial 71 includes the skull base, some teeth, and the sacrum of an adolescent, probably male, between 11 and 16 years of age. The skeletal elements available for analysis are normal. Of the bones that cannot be associated with a specific burial, bilateral acetabula and a left scapula have reactive bone formation in the subchondral bone of the joint. These may have been from the same individual, and the lesions may have been the result of cartilage breakdown from arthritis. The specific type of arthritis is not known.

TOMB A103

Only one chamber was associated with this tomb, and this was badly damaged by major ceiling fall and silting. The human remains were in very poor condition at the time of excavation, and little can be said about the burials. Careful laboratory analysis reveals a minimum number of four individuals associated with this chamber, including three adults and one subadult.

Burial 1 is represented by cranial, dental, and postcranial bones of a young adult male. The interior surface of the cranium has fine porous reactive bone that is probably the result of meningitis that probably was the cause of death. The surfaces of the alveolar processes are porous, probably as the result of chronic infection of the gum, but other conditions such as scurvy should be considered as well. There is some evidence of early stages of osteoarthritis in the long bones.

Burial 50 includes only mandibular fragments of an adult female. There is no evidence of significant skeletal or dental disease. Burial 51 consists of the mandible of a young adult female and possibly some cranial fragments. There is evidence of considerable antemortem tooth loss in the mandible. Burial 60 is represented by both cranial fragments and some postcranial skeletal fragments of a child about nine years of age. There is no evidence of skeletal disease in the fragments available for analysis.

TOMB A105

The shaft of Tomb A105 was identified as a soil discontinuity about 30 centimeters below the existing soil surface (Figure 8.8). Near the top of the shaft fill was an assemblage of artifacts identified as Tomb A104 (Figure 8.9). No human remains were recovered from

Figure 8.8. Test trench for Tomb A105 with the shaft for the tomb in the early stage of excavation.

this shallow pit, so the relationship of the artifacts to the tomb shaft is unclear.

The tomb chambers are arranged around the base of the shaft in the typical cloverleaf pattern (Figure 8.10). Generally, preservation of human remains in all the chambers was poor. At least two of the chambers had no evidence of blocking stones covering the entryway and ceiling fall as well as silting was substantial in all the chambers. This, of course, made recovery of the fragile human remains difficult, and some additional damage to the remains occurred during excavation.

Tomb A105, Chamber Northwest

Both ceiling fall and particularly silting were severe in this chamber, and preservation of human remains

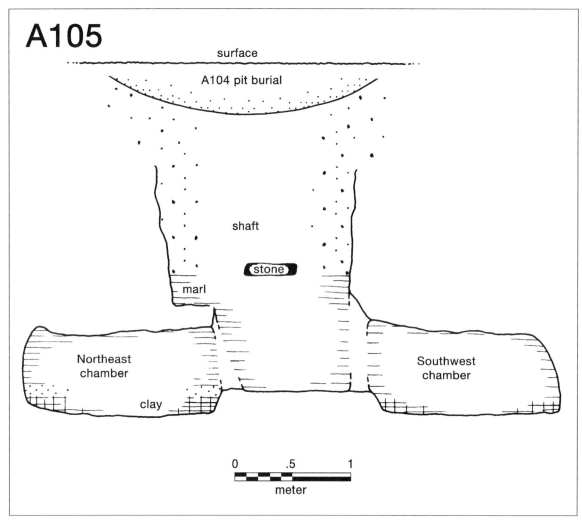

Figure 8.9. Section drawing of tomb chambers A105NE and SW.

is no more than fair (Figure 8.11). The remarkable aspect of the preservation is that subadult bones are better preserved than those of adult skeletons. We have seen this difference in preservation in other tombs at Bâb edh-Dhrâ'. It emphasizes the fact that although burial and taphonomic factors may affect our ability to reconstruct the demography of a human skeletal sample, one cannot assume that subadult, even infant, burials will be underrepresented in all archaeological contexts because of taphonomic processes.

The poor preservation and recovery of human remains make data collection of the human remains from this chamber particularly challenging and raises the specter of either underrepresenting the individuals originally interred in the chamber or overrepresenting the number of individuals because of the difficulty in assigning a bone or bone fragment to a specific burial. Despite these conceptual hazards, we do conclude that the MNI buried in the chamber is 22, which is an unusually large number of burials. These include two third trimester fetuses that may have been aborted or died with the death of the pregnant mother. Five of the seventeen subadults were neonates, and the rest ranged in age from infant to about 12 years. There were five adults, two of which were males.

Bones designated as Burial 1 at the time of excavation have been assigned to another burial. Burial 2 includes both cranial and postcranial fragments of an adult male about 25 years of age. There was no evidence of significant skeletal disease in the elements available for analysis. Burial 3 consists only of cranial fragments of a female about 24 years of age. The orbital roofs had porous lesions that could be the result of either chronic infection or scurvy. Burial 4 is represented by normal cranial and postcranial bones of a child about six years of age.

Burial 5 consists of cranial bones and a fragment of the right femoral diaphysis of a child about three

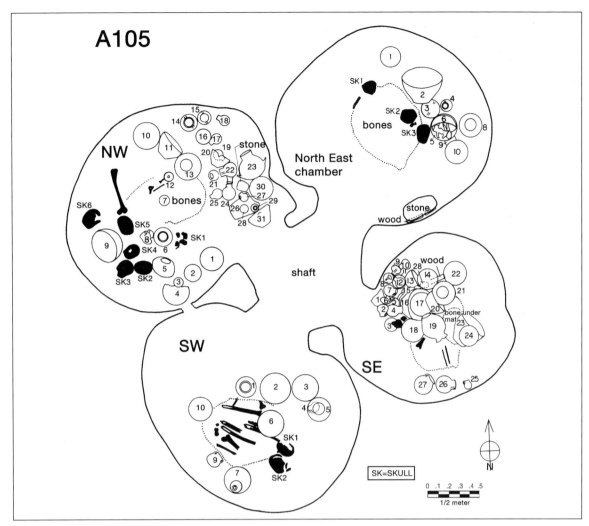

Figure 8.10. Plan drawing of Tomb A105.

years of age. The right orbital plate is porous, which could be caused by infection or scurvy. The left orbital plate is not present. Burial 6 contains both cranial and postcranial skeletal elements of a child about 11 years of age. Dental hypoplasia occurs in the dental enamel, indicative of at least two stress episodes during dental development. Burial 60 consists of femoral fragments of the right diaphysis of a child between two and three years of age. There was no evidence of significant disease. Burial 61 has a right femoral fragment and other postcranial skeletal elements of a child about two years of age. Periostosis is apparent on a left femoral fragment and the distal left humerus fragment. This distribution of abnormal lesions suggests a systemic problem in this burial, the most likely cause of which is infection, but other diseases including metabolic disease and cancer are possible as well.

Burials 62 and 63 are composed of postcranial neonatal skeleton elements. Both burials show evidence of abnormal porosity that cannot be attributed to any specific disease. Burial 64 consists of a fragment of the right femur from a third trimester fetus and is the second example of fetal remains excavated from this chamber. There is no evidence of skeletal disease. Burial 65 is represented by postcranial bones of a child about nine years of age. The subchondral surfaces of the epiphyses are irregular with large pores. The cause of this disorder is not known. Burials 66 and 67 are the incomplete and fragmentary postcranial bones of children also about nine years of age. All of the bones available for analysis in this burial are normal.

Burial 69 consists only of postcranial skeletal elements of an adult female. A more precise age estimate is not possible, but there is evidence of subnormal bone mass usually associated with postmenopausal osteoporosis, so the individual may have been over 50 years of age. The right humerus and right and incomplete left petrous portions of the temporal bone

Figure 8.11. Ceiling fall and heavy silting associated with poor preservation of human burial in A105NW.

make up Burial 73, which is from a child about four years of age. There is abnormal porosity in both the temporal bone fragments that could have been caused by chronic infection.

Burial 74 includes only postcranial skeletal elements of a female about 20 years of age. There is some type of arthropathy apparent in several of the bones characterized by abnormal porosity of the subchondral bone. A more specific diagnosis is not possible. Burial 75 consists of both cranial and postcranial elements of an adult male. The only evidence of disorder is the presence of enamel hypoplasia of the teeth that was caused by a stress event during dental development. Burial 76 comprises two normal radii of an infant corpse. Burial 77 includes fragments of the left ilium and left humerus. There is no evidence of significant disease on either fragment.

Burial 78 is the normal left petrous portion of the temporal bone from a third trimester fetus. Burial 79 consists of the normal right petrous portion of the temporal bone from a neonate. Burial 80 includes both cranial and postcranial skeletal elements of an infant between six and twelve months. There is no evidence of skeletal disorder.

There are some bones and bone fragments that cannot be associated with a specific burial but do show evidence of skeletal disorder. Fractures in various states of healing occur on four elements, including a long bone fragment of a juvenile, the acromion process of an adult male, a midlevel rib, and a juvenile metacarpal. Several unassociated teeth have hypoplastic lines indicating one or more stress events during dental development. Several bones show evidence of reactive porosity with, in some cases, reactive bone formation. The most likely cause is infection, but metabolic disease, trauma, and cancer are possible as well.

Tomb A105, Chamber Northeast

Like other chambers in this shaft tomb, bone preservation was adversely affected by severe ceiling fall and silting (Figure 8.12). All of the designated burials were represented by fewer than 20% of the total number of bones in a skeleton and most by much less. The estimated MNI interred in the chamber is 27, with 8 adults and 19 subadults, including 12 juveniles and 6 late-term to neonatal skeletons.

Burial 1 contains cranial, dental, and postcranial bones of an adult male probably 30 years of age or older. The frontal bone contains three small osteomas, benign tumors of the skull, and two destructive lesions, one of which was active, the other healed at the time of death. The cause of these lesions was probably chronic scalp infection; greater specificity is not possible in this case. The dentition had several teeth lost antemortem with remodeling of the alveolar bone, indicating loss well before death. Skeletal disorders of the postcranial skeleton include defects in the subchondral bone of some joints, probably indicative of trauma to the joint.

Burial 2 consists of cranial, dental, and postcranial remains, although the association of the latter with the

Figure 8.12. Central bone pile in Tomb A105NE. Note the poor condition of the burials and the unusual placement of skulls to the right of the central bone pile.

cranial bones is not certain. The burial is the remains of a male between 23 and 40 years of age. There is a well-remodeled depressed lesion of the frontal bone that is probably a healed depressed fracture, most likely the result of a blow to the head. This lesion is similar to a lesion caused by a sebaceous cyst, and this option remains possible. However, the unequivocal evidence of skull trauma in other skulls from the shaft tombs, although not common, makes this option more likely. The teeth associated with this cranium show evidence of slight dental hypoplasia and considerable buildup of dental calculus. The right canine has slight enamel hypoplasia. There is a fracture of the left proximal radius that was healed but in a misaligned position.

Burial 3 includes cranial, dental, and probably some postcranial bones, although the association of the latter with the cranium is likely but not certain. The skeleton is from a female between 35 and 40 years of age. In the skull, there is porosity of the orbital roof that is probably the result of chronic infection, but scurvy and anemia are possible although unlikely. There is antemortem tooth loss with alveolar remodeling. The teeth that are present have significant tooth wear, in some cases involving most of the dental crown. Caries is also present in the second right upper premolar and the upper right third molar. The postcranial bones available for analysis are normal.

Normal bones of late-term or neonatal skeletons occur in Burials 30, 34, 35, 41, 50, 51, and 52. Normal bones of infants and young children are associated with Burials 42, 80, 81, 82, and 84. Burial 32 includes fragments of the cranium and postcranial elements of an infant about one year of age. There is fine porous reactive bone of the medial left femur fragment. It is likely that other postcranial bones were affected as well, but the remains are too fragmentary to confirm this possibility. Infection is the most likely cause of the abnormal bone, but metabolic diseases and cancer are possible as well.

Burial 69 is the very incomplete skeleton of a male, probably in his 20s. It is represented by a mandibular fragment including some teeth and a fragment of the right innominate; a left femur is also most likely linked to this burial. Most of the teeth associated with the mandibular fragment have evidence of calculus. This abnormality is also probably linked to the evidence of alveolar bone remodeling associated with periodontal disease. The fragment of the right innominate has fine porosity of acetabular subchondral bone indicative of joint disease probably in the early stages.

Burial 72 includes very incomplete postcranial fragments of an adult female. The only evidence of skeletal disorder is a subchondral lesion of the radial head involving eburnation, bone destruction, and marginal reactive bone formation. This is a well-known manifestation of osteoarthritis.

Fragmentary postcranial bone elements of an adult female between 40 and 44 years of age make up Burial 73. In view of the widespread evidence of osteoporosis, this estimate may be low. In addition to the greatly diminished cortex, the right tibia is bowed in the mediolateral axis. The left femur is also bowed, but in both the mediolateral and anterio-posterior axes. This bowing could be the result of rickets in late childhood, with inadequate growth remaining to remodel the bone into a more normal shape, or it could be the result of osteomalacia.

Burial 83 consists of the left and right ileum and the right humerus of an infant about six months of age. There is extensive and abnormal porosity of the skeletal elements indicative of a systemic disorder, probably infection, but metabolic disease or cancer is also possible. Burial 85 includes fragments of the right humerus, right and left innominates, and vertebrae of a young male about 19 years of age. The vertebral bodies of this burial are markedly eccentric, which may be an indication of scoliosis, a spinal deformity. The acetabulum of both the right and left innominate fragments are abnormally shallow and show evidence of abnormal wear affecting the superior acetabular rim. This is almost certainly the result of bilateral hip dysplasia with chronic subluxation of the femoral head.

The right femur and the proximal left humeral epiphysis of a child about eight years of age are the only bones associated with Burial 86. There is no evidence of disease in the bones available for analysis. Burial 87 includes a mandibular fragment and very incomplete postcranial elements of a 25–29-year female. There is some evidence of probable joint disease in the form of abnormal fine porosity on the subchondral bone of some joints. Burial 88 consists of the mandible and right ilium of a child about five years of age. None of the elements available for analysis show evidence of disease. Burial 89 also includes the right ilium as well as the frontal bone of a child four to five years of age. The ilium is normal, but the frontal bone has fine reactive woven bone on the endocranial surface. The outer table has areas of fine porosity through the cortex. Both infection and metabolic disease are possible causes of this disorder. Burial 90 is the very incomplete skeleton of an adult male. The bones available for analysis include cranial fragments and a right distal tibia that cannot be associated with another adult male burial. There is no evidence of skeletal disease. Burial 91 is represented by paired tibia fragments of a child ten or more years of age and cannot be linked with any other burial. There is no evidence of skeletal disease in these bone fragments.

Evidence of skeletal disorders in unassociated bone elements includes additional signs of osteoarthritis, enamel hypoplasia, severe dental wear, and dental calculus. There are three cervical vertebrae that exhibit eccentric spinous processes and an upper thoracic vertebra with an eccentric vertebral body. The abnormalities in these vertebrae are probably due to scoliosis, a deformity of the spine.

Tomb A105, Chamber Southeast
There were no blocking stones for this chamber and the chamber was completely silted, apparently as a result of a single episode of silt flow into the shaft (Figure 8.13). This contributed to the poor preservation of bones (Figure 8.14). The MNI in this chamber is eight, consisting of two adults and six subadults, including a perinatal infant, another infant, and four children.

Burial 1 (including Burial 62 and dental remains marked #23) consists of cranial, dental, and postcranial bones of a child between three and four years of age. There is no evidence of skeletal or dental disease. Burial 1a contains both cranial and postcranial fragments of a five-year-old child. All the elements available for analysis are normal. Burial 2 (including Burial 60) is represented by cranial dental and postcranial material of an adult female 35 or more years of age. The interior surface of the sphenoid sinus has extensive reactive bone formation probably indicative of a chronic sinus infection. Antemortem tooth loss with alveolar remodeling is present. Burial 2a (including dental elements marked as #22) consists of normal cranial, dental, and postcranial bone fragments of a child between one and three years of age. Only cranial fragments of a one- to three-year-old child are included in Burial 2b. The fragments duplicate some of the bones associated with Burial 2a and, for that reason, could not be part of the same burial. All the cranial fragments are normal.

Burial 3 contains cranial, dental, and postcranial elements of an adult male about 25 years of age. Other

Figure 8.13. Tomb chamber A105SE. There was no blocking stone for this chamber, and it was completely silted.

than evidence of dental abscess in the mandibular fragment associated with the left lower third molar, the bones are normal. Burial 20 is represented by dental remains and possibly some cranial fragments. The endocranial surface of the occipital fragments has porous reactive bone deposition indicating an active disorder at the time of death. Infection and metabolic disease are the most likely possibilities. Burial 63 consists of postcranial elements of a perinatal infant. All the elements are normal.

Tomb A105, Chamber Southwest
No blocking stone was identified for this chamber at the time of excavation. Both ceiling fall and silting were substantial, greatly limiting recovery of human remains which were in poor condition (Figures 8.15a, 8.15b). The MNI in this chamber is three, consisting of two adults and a child.

Burial 1 (including postcranial bones B-62) consists of very fragmentary cranial, dental, and postcranial materials of a very gracile female about 40 years of age. Dental attrition is severe. There is also evidence of dental hypoplasia in the left canine and premolar. The dental attrition combined with abnormally thin cortical bone of the long bone fragments suggests that the age estimate may be too young. The thin cortical bone is most likely the result of osteoporosis, a fairly common condition in the shaft tomb remains.

Burial 2 (including postcranial fragments B-60) contains cranial, dental, and postcranial elements of an adult male between 25 and 30 years of age. The left mandibular canine and premolar contain evidence of multiple hypoplastic lines in the enamel, indicative of multiple stress events in childhood. There are two circular depressions in the outer table of the right parietal. There is some porosity within the lesion, suggesting remodeling. The probable cause is blunt force trauma, but sebaceous cysts are fairly common and produce a similar lesion. Cribra orbitalia is present bilaterally and may reflect infection or metabolic disease.

Burial 63 contains fragmentary mandibular, dental, and postcranial elements of a child three to four years of age. There is no evidence of skeletal or dental disease.

TOMB A106

Tomb A106 consists of a single chamber with an unexcavated entryway. Presumably, there was a central

EB I Tombs Excavated in 1979 161

Figure 8.14. Central bone pile of A105SE embedded in silt.

tomb shaft connected to the entryway, but the very poor condition of the tomb chamber made recovery of significant tomb contents unlikely, and further excavation was not attempted. Because of this, there remains uncertainty about the total number of chambers associated with this tomb.

Tomb A106, Chamber South
Preservation of the chamber architecture was so poor that the chamber was excavated from the surface rather than through an excavated tomb shaft. The chamber was completely filled in with a combination of ceiling fall, partial collapse of the chamber wall, and silting. There may have been more than one episode of silting, suggested by the different levels within the chamber at which pottery was found. The bones were shifted with the postcranial bones located near the eastern edge of the chamber (Figure 8.16). The preservation of the human remains was very poor (Figure 8.17). Because of the disturbance and poor preservation of the bones, associations between skeletal elements were less common than in most of the EB IA tombs. The MNI interred in this chamber is 12, consisting of five adults and seven subadults.

Burial 1 consists of cranial fragments and possibly some fragmentary postcranial elements. The individual represented by the bones was an adult female, but a more specific age cannot be assigned because of poor preservation. However, an anterior, right maxillary fragment contains the premolars, which exhibit calculus deposits and severe attrition. The latter, in particular, suggests an age estimate at least later than early adulthood. There is also alveolar resorption that reflects the presence of periodontal disease. There is no other evidence of significant skeletal or dental disease.

Burial 2 contains fragmentary cranial bones and some postcranial elements of an adult female. A more specific age estimate is not possible. The right patella of this burial has slight evidence of osteoarthritis, with a small, well-remodeled lytic focus in the subchondral bone.

Burial 60 includes cranial, dental, and postcranial fragments of a child about eight years of age. The two femora associated with this burial have significant anterio-posterior bowing most likely caused by rickets. Burial 61 is the normal right humeral fragment of a child about 12 years of age. Burial 62 is a normal right humerus of a child about 8.5 years of age that might be associated with Burial 61 (this is uncertain). Burials 63, 64, and 65 are normal femoral fragments of infants between three and six months of age.

Burial 66 consists of robust postcranial elements of an adult male. There may be cranial elements associated with this burial, but this is not certain. In the subchondral bone of both tibiae there is a well-remodeled lytic focus that may be osteochondritis dissecans, a defect probably caused by trauma to the joint.

Figure 8.15a. Tomb chamber A105SW, with pottery between the entryway and the central bone pile.

Figure 8.15b. Detail of the central postcranial bone pile in A105SW demonstrating poor preservation.

Figure 8.16. The south chamber of Tomb A106.

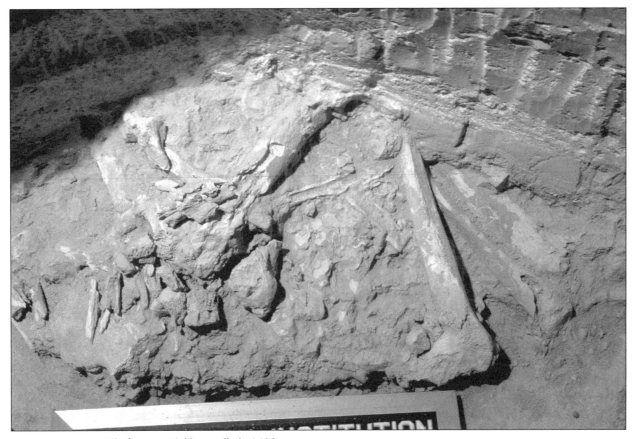

Figure 8.17. Detail of postcranial bone pile in A106.

Burial 67 contains postcranial elements of a male in his mid-20s at the time of death. Several subchondral bone surfaces have an unusual fine porosity. The significance of this abnormality is not known, but the porosity does suggest a mild inflammatory or angiitic (vascular) problem. Burial 68 includes only postcranial elements of an adult male, probably at least 25 years of age at the time of death. There is no evidence of significant disease in the elements available for analysis. Burial 69 is represented by a single right fragment of the femur that is normal.

TOMB A107

This shaft tomb consists of the central shaft and four chambers radiating from the base of the shaft (Figures 8.18a, 8.18b). Blocking stones were either inadequate or had been displaced by silt flow into the shaft (Figure 8.19). Because of this, major silting of all the chambers occurred in antiquity, creating problems in excavation and recovery of tomb contents, particularly the human remains.

Tomb A107, Chamber North

The ceiling was intact in this chamber, but there had been extensive silting in antiquity that had filled some of the pottery and completely encased most of the skeletal remains (Figures 8.20a, 8.20b). The silt was deposited in at least four distinct layers, which suggests four silting events. The arrangement of tomb contents was similar to that of other chambers of the

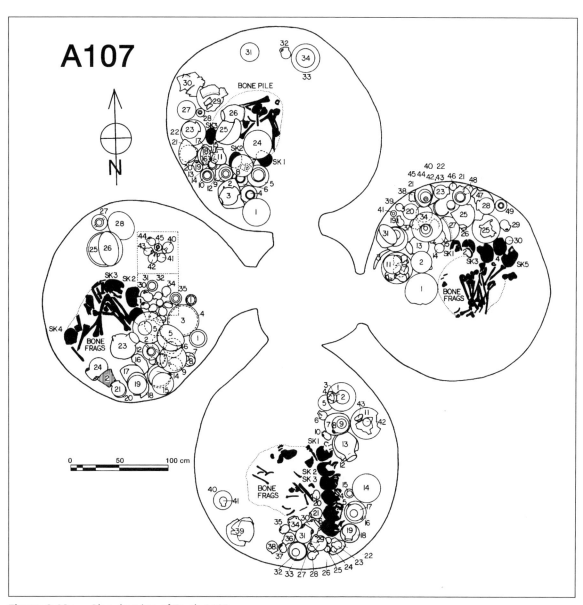

Figure 8.18a. Plan drawing of Tomb A107.

EB I Tombs Excavated in 1979   165

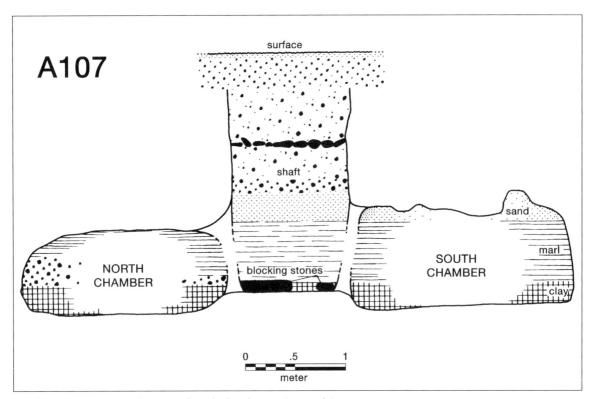

Figure 8.18b. Section drawing of tomb chambers A107N and S.

Figure 8.19. Shaft of Tomb A107 showing silted entryways to the chambers with blocking stones out of place.

Figure 8.20a. Silting in A107N that filled several of the pots and completely encased the burials.

Figure 8.20b. Detail of central postcranial bone pile in A107N embedded in silt.

EB IA period, with a central postcranial bone pile and pottery placed on the periphery of the chamber. However, the skull elements seem to have been placed on or near the central bone pile rather than arranged to the left of the postcranial remains. Despite the silting, the preservation of skeletal elements was relatively good. The MNI interred in this chamber is five, all of which were adults.

Burial 1 is represented mostly by cranial elements. There are a few postcranial elements that possibly may be attributed to this skull (B-60) but this is not certain. The estimated age of this individual is 40 or more years and the sex is female. In the fragments of the maxilla, there is antemortem tooth loss with remodeling of the alveolar bone, which also supports an older age estimate. The cortical bone of the postcranial fragments that likely are associated with this burial are abnormally thin, and this is indicative of osteoporosis, a condition associated particularly with postmenopausal women. This supports an age estimate closer to 50+, but this cannot be confirmed from the other anatomical evidence. The anterior portion of the right tibia has at least two foci of periosteal reactive bone that was remodeling into compact bone at the time of death. Several conditions can cause this type of lesion, but trauma is most likely.

Burial 2a includes both cranial and postcranial bones, and the association between the two seems fairly certain. The estimated age is about 25 years and the sex is male. There is no evidence of significant skeletal disease. Burial 2b consists of cranial elements and postcranial bones that are probably associated with the skull. Although the cranial vault is well preserved, the postcranial bones consist primarily of diaphyseal fragments. The estimated age is 35+ and the sex is female. There is considerable antemortem tooth loss that particularly affects the posterior dentition. Bilateral cribra orbitalia is present in the frontal bone. The lesions are remodeling, so the condition producing the lesions probably was no longer present. The possible causes include infection and scurvy. Similar to Burial 1 from this chamber, the cortex of the long bone diaphyses are abnormally thin, indicative of osteoporosis. The young estimated age of this burial argues against a diagnosis of postmenopausal osteoporosis. Severe malnutrition is likely, although chronic debilitating illness is also possible.

Burial 3 consists of a partially intact cranium, a mandible, and a few postcranial skeletal elements labeled B-62 but probably associated with the cranium. The individual was 35+ years of age at the time of death and the bones are gracile, indicative of female sex. The mandibular alveolar process shows evidence of antemortem tooth loss with some remodeling. There is also evidence of bilateral abscesses associated with the mandibular canines, both of which were lost antemortem. Alveolar resorption surrounding remaining tooth roots suggests periodontal disease that was active at the time of death. Bilateral cribra orbitalia is present, some of which was actively forming when the person died. There is also abnormal porosity around the supraorbital foramen suggestive of an infectious focus in this portion of the frontal bone. The tibiae and femora are both abnormally bowed, and this probably reflects childhood rickets, although osteomalacia is possible. As apparent in other female burials from this chamber, the cortex of the long bones is abnormally thin and porous and indicative of osteoporosis probably associated with severe malnutrition. The subchondral bone of the postcranial elements has evidence of osteoarthritis.

Burial 50 includes skeletal fragments of the upper and lower extremity that can be attributed to an adult male, but a more precise age estimate is not possible. The cortices of the upper extremities are abnormally thin, raising the possibility of severe malnutrition. The diaphyseal fragment of the left ulna contains callus, indicative of healed fracture. This is a common site for parry fractures associated with defending against a blow. However, accidental fracture is possible as well. Fragments of the right femur have abnormal exostoses projecting from the cortical surface. These exostoses contain a trabecular core and are suggestive of a dysplasia such as multiple cartilaginous exostoses. This is a benign congenital condition, although malignant transformation of these lesions does occur in rare cases.

There are a substantial number of bone fragments that cannot be associated with a specific burial, some of which show evidence of significant disorder. At least seven rib fragments have callus, indicative of healed fracture. Subchondral bone associated with several joints has evidence of osteoarthritis. Some of the diaphyseal bone fragments have reactive porous bone formation that most likely is the result of chronic infection or trauma. All these skeletal disorders support the adult age of all five individuals and are compatible with age estimates well into the adult category.

Tomb A107, Chamber East
Similar to the north chamber of this tomb, the east chamber had significant silting but no roof fall. Unfortunately, photographic documentation is not available

for this chamber. The bones available for analysis are in relatively good condition, with many long bones intact. The association between the cranial bones and the postcranial elements is more problematic in this chamber than in others excavated in this time period. Only one adult cranium (Burial 5) is present. However, there are several sets of adult postcranial long bones, particularly femora. The remaining crania are from subadults, and many of these could be associated with postcranial elements. A curious distribution of postcranial bones is present, with many more femora present than other postcranial bones. This implies a specific selection at the primary burial site of bones to be interred. Femora occur more frequently in both adult and subadult remains. The other general point to be made about the subadult burials is that the estimated age of the skulls (largely based on dental eruption) was older than the age estimates based on the postcranial bones. The implication of this observation is that growth of long bones was somewhat delayed relative to the development of the dentition. Dental size and the eruption sequence is less affected by disease and malnutrition than long bone growth. This implies that the children interred in this chamber suffered from poor health and/or malnutrition. The MNI interred in this chamber is 21. This includes nine adults, one adolescent, and eleven subadults.

Burial 1 includes postcranial bones initially identified as Burial 75 and consists of cranial dental and postcranial bones of a child between 18 and 24 months of age. There is porous reactive bone on the endocranial surface of the skull. Cribra orbitalia is present in both orbital roofs. Bilateral fragments of the femora have reactive bone formation associated with the diaphysis. The overall pattern of bone disorders could be caused by either infection or scurvy.

Burial 2 is now associated with postcranial bones formerly labeled as Burial 62. The burial includes most of the major long bones as well as cranial and dental elements. The estimated age based on dental development is about ten years. However, the long bone lengths provide an estimated age of about seven years. The permanent teeth exhibit two hypoplastic lines indicative of two serious episodes of illness or malnutrition during the development of these teeth. Bilateral cribra orbitalia is present. The subchondral bone of the metaphyseal ends of both femora is abnormal, indicating a growth disturbance at the time of death. The most likely cause of the bone lesions that occur in this child is scurvy although systemic infection is also possible.

Burial 3 (ca. nine years) is now linked to postcranial bones formerly identified as Burial 63. The burial comprises cranial, dental, and postcranial bones. Moderate and active cribra orbitalia is present bilaterally. There is a well-healed depression on the right frontal boss that could be from healed trauma to the head, but benign cysts also occur between the skull and scalp, and these also can result in depressed lesions. The alveolar bone of the maxilla is abnormally porous and has reactive bone formation. The permanent dentition exhibits dental hypoplasia indicative of two episodes of severe illness. The right ascending ramus has porous reactive cortical bone. The left side is fragmentary, and the presence of this abnormality cannot be determined. The overall pattern of bone lesions may be the result of scurvy or chronic infection.

Burial 4 includes the remains of a child about nine years of age and consists of cranial and postcranial bones as well as teeth. The only skeletal disorder observable in the bone available for analysis is moderate cribra orbitalia. Infection and scurvy are the most likely causes for this abnormality in this child.

Burial 5 is the skeleton of a robust adult male about 30 years of age. The burial includes cranial, dental, and incomplete as well as fragmentary postcranial bones. The teeth exhibit enamel hypoplasia indicative of two disturbances on the development of the dental enamel. Alveolar bone of the maxilla has evidence of remodeling caused by periodontal disease.

Burial 60 is probably associated with Burial 71 and includes femora, humeri, and ulnae of a child about six years of age. There is no evidence of disorder in any of the bones available for analysis. Burial 64 is represented by the normal right femur of a newborn. Burial 65 consists of the left femur and tibia of an infant about two months of age. The tibia is bowed in the anterio-posterior axis and the cortex is abnormally porous, suggestive of a defect in mineralization of bone protein (osteoid). Both of these abnormalities suggest the presence of rickets in this infant, although this disorder is unusual before the age of six months. It can occur at an early age if the mother was deficient in vitamin D during pregnancy.

Burial 66 is now associated with Burial 74 and consists of a right femur and humerus from an infant about six months of age. There is no evidence of skeletal disease in the bones available for analysis. Burial 68 includes the right femur and both humeri of a newborn. Reactive bone is present on both humeri. Scurvy is unlikely in a newborn, and if the lesions are the result of infection the condition must have been congenital. Burial 76 consists of a normal right humerus from an infant about nine months. Burial 78 is the right humerus of a late-term fetus about 36 weeks (normal gestation is 40 weeks). There is no evidence of skeletal disorder in this bone.

Burial 80 is represented by a right femur of an adult gracile female. The diaphysis is thinner than normal and postmortem damage in the metaphyseal area shows abnormal thickening of a few trabeculae, associated with a general loss of trabecular bone. This abnormality is not specific, but is commonly seen in osteoporosis. Because an accurate age estimate is not possible in this burial, the cause of osteoporosis remains unknown, but postmenopausal changes or malnutrition are the most likely diagnostic options.

Burials 81, 82, and 84 consist of paired femora from robust adult male skeletons. There is no evidence of skeletal disorder in any of the femora. Burials 83, 86, and 87 are also composed of pairs of normal femora from gracile adult female skeletons. Burial 85 is represented by a pair of femora and the pelvic bones of a female between 30 and 34 years of age. There is no evidence of significant skeletal disorder. Burial 89 includes the distal half of a left femur and a pelvic fragment from a normal male between 16 and 20 years of age.

As in other tomb chambers there are a few skeletal fragments or elements that cannot be attributed to a specific burial. In some cases, these fragments show evidence of some skeletal disorder. Most of the evidence is of arthritis, but there is a right rib fragment that has reactive bone formation on the visceral surface. This is likely caused by a chronic infection of the lung or pleura, the most common cause of which is tuberculosis.

Tomb A107, Chamber South
Silting was extensive in this chamber, and there is evidence of at least slight ceiling fall. The postcranial bone pile was completely encased in silt, and there was considerable silt infiltration into the skulls. Bone preservation was unusually good, with association between cranial and postcranial bones fairly certain in most burials. We have seen better than normal bone preservation in several other tomb chambers that were heavily silted. This suggests that the silting is not necessarily harmful to the bones and, in some cases, was beneficial relative to long-term preservation. The MNI associated with this chamber is seven, including two adult females, two adult males, one late adolescent female, one child about 12 years of age, and the bones of a fetus about at the 30-week stage of development. It may be that one of the women interred in this chamber had died during pregnancy and the fetal bones are from this woman.

Burial 1 includes both cranial and postcranial bones. The estimated age is about 12 years and the sex cannot be established with any certainty. The child had cribra orbitalia that appears to be remodeling at the time of death. This implies that the cause of the disorder, infection and scurvy being most likely, was no longer a factor when the child died. The left parietal had an irregular depression that had remodeled. This is most likely the result of trauma that the child survived.

Burial 2 consists of many of the cranial, dental, and postcranial bones of a female between 30 and 35 years of age. There is no evidence of significant skeletal disorder. Burial 3 is the skeleton of a robust adult male between 25 and 30 years of age. The dentition has evidence of at least one episode of stress resulting in enamel hypoplasia. There is a relatively large area of irregular cortical bone in the outer table of the frontal bone extending into both parietals. On the left side of this lesion there is a very sharp margin between normal and abnormal bone. This is suggestive of a scalp disorder, perhaps resulting from injury to the scalp, followed by more general infection. This lesion was remodeling at the time of death, indicating probable healing. The orbits of the frontal bone have bilateral cribra orbitalia that is in the remodeling stage. Vertebrae that are probably associated with this burial have evidence of at least slight scoliosis. There are several causes for this disorder, and the condition is often congenital.

Burial 4 consists of both cranial and postcranial bones of a very gracile female about 17 years of age. There is no significant evidence of skeletal disorder in this burial. Burial 5 includes the cranial, dental, and postcranial bones of a gracile female between 25 and 29 years of age. The teeth have evidence of two stress events resulting in enamel hypoplasia, reflecting childhood disorders at the time the dental enamel was forming on the permanent teeth.

Burial 6 comprises the cranial, dental, and postcranial elements of a robust male with an estimated age between 30 and 39 years of age. The right temporomandibular joint has evidence of marked osteoarthritis. The condyle of the right mandible is deformed, and the dental wear pattern supports an abnormal relationship between the upper and lower teeth. The teeth themselves have two hypoplastic lines, indicative of two serious stress events during childhood. Abscesses occur in the right and left upper molar areas. The vertebrae have subchondral bone changes associated with rather severe osteoarthritis, especially in the mid and lower thoracic vertebrae. The humeri are abnormally bowed in the A-P axis. There is no evidence of rickets or osteomalacia in the other bones, so the most likely explanation for this abnormality is abnormal biomechanical loading, probably beginning in early childhood. There is a suggestion of diminished use of the lower extremities that may be indicative of partial paralysis requiring the use of crutches, which could have caused the abnormalities apparent in the humeri. Both the vertebrae and bones of the lower extremity have evidence of osteoarthritis.

Burial 50 is represented by the distal half of the right humerus. The bone fragment was from a fetus about 30 weeks in utero. There was no evidence of skeletal disorder in this bone.

As with other EB IA tomb chambers, there is evidence of skeletal disorders in some of the bones and bone fragments that cannot be associated with a specific burial. Osteoarthritis is present in some of the vertebral fragments. Four lumbar vertebral bodies have superior rim destruction that is probably caused by anterior herniation of the disk. There are three rib fragments with healed fractures and a right fibula with a healed fracture of the lateral malleolus.

Figure 8.21. Student field supervisor (now Father John Meoska) applying preservative in A107W.

Tomb A107, Chamber West

This chamber had minimal ceiling fall, but silting filled most of the chamber. Fortunately, the silt was easily removed and, although skeletal recovery was poor, the crania were relatively intact (Figure 8.21). Unlike in most shaft tomb chambers, both skulls and postcranial remains were placed in the center (Figure 8.22), with the pottery surrounding the bone pile. The poor recovery of long bones in contrast with the relatively good recovery of cranial elements suggests that skulls were of greater significance to the people of Bâb edh-Dhrâ' in the transition from the primary burial site to the secondary interment in the chambers. As with most of the shaft tomb chambers, the secondary burial tradition made the association of postcranial skeletal elements with a specific skull challenging and often no more than tentative. The MNI identified in this chamber is ten, including six adults, one adolescent, two children, and one infant.

Burial 1 is represented by cranial, dental, and probable postcranial elements of an adult female at least 30 years of age. Both the cranial and postcranial elements were markedly gracile, in contrast with other burials in this chamber, and the association of the postcranial bones was made on this basis. The criteria for estimating age are either absent or inadequately expressed because of the fragmentary nature of the skeletal elements. The severe dental wear, in some teeth involving

Figure 8.22. Tomb chamber A107W, with postcranial bones placed centrally and skulls arranged to the right of the pile.

the entire crown, is suggestive of someone much older than 30 years. An older estimated age is also supported by the significant reduction in bone mass (osteoporosis) that tends to occur in older women following menopause, but can occur at younger ages because of malnutrition or other serious illness.

Burial 2 consists of cranial and dental elements of an adult male between 30 and 35 years of age. There are long bones that probably are associated with this cranium, but the poor recovery and condition of the long bones makes such an association too uncertain. There is some antemortem tooth loss with subsequent remodeling of the alveolar bone. There is a depressed lesion in the frontal bone that may have been the result of blunt force trauma with survival and recovery.

Burial 3 includes only cranial remains and a left radius that is probably associated with the skull. The estimated age is 35 years and the sex is female. Only the right maxilla of this burial is present, and there is an abnormal depression of the canine fossa that is slightly porous. Within the maxillary sinus, there is a mineralized bone mass that may be a benign tumor or reactive bone formation caused by chronic infection.

Burial 4 consists of incomplete and fragmentary cranial elements and some teeth. The bone fragments are robust, suggesting male sex. There are postcranial remains possibly associated with these cranial bones, but there is too much overlap between this and other skulls from this chamber to permit any certainty about the association. Estimation of age is problematic, but the severe dental wear suggests an age in excess of 45 years. There is a depressed lesion of the left frontal bone that is most likely the result of blunt force trauma.

Burial 5 includes the cranium and mandible, some teeth, and the lumbar vertebrae. They are from a female about 25 years of age. There is evidence of antemortem tooth loss, and the remaining teeth have hypoplastic lines indicative of stress events during dental development. Burial 7a is represented by fragmentary cranial bones of an adult who was robust and probably male. There is no evidence of skeletal disease in the elements available for analysis.

The remaining burials are all of subadults. Burial 7 consists of cranial and dental remains of a child about two years of age. Both orbital plates are abnormally porous, and there is reactive bone formation on the

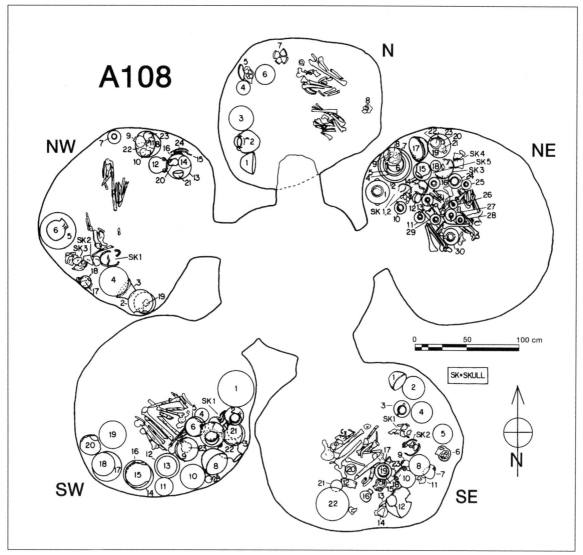

Figure 8.23. Plan drawing of Tomb A108.

endocranial surfaces. These abnormalities could be the result of scurvy, anemia, or infection, but scurvy seems the most likely option. Burial 8 is the cranial and dental remains of a child between three and four years of age. There is extensive porotic hyperostosis with marrow hyperplasia of the parietal and occipital bones. There is also reactive bone formation on the endocranial surfaces of the cranial bones. Anemia is possible in this case, as is scurvy. Burial 60 consists of fragments of a mandible, femora, and a tibia of an infant between six and twelve months of age. There is no evidence of skeletal disorder in the bones available for analysis. Burial 61 includes vertebrae, pelvic bones, and fragments of the tibia of a subadult male about 17 years of age. There is no evidence of significant disease in these skeletal elements. Skeletal disorders that cannot be attributed to a specific burial include a possible compression fracture of the third and fourth lumbar vertebrae, multiple vertebral fragments with syndesmophyte formation suggestive of DISH, and osteoarthritis of several joints.

TOMB A108

Regrettably, the photographic documentation for this tomb is less than ideal, so many of the details about the chamber arrangement and contents will depend on the plan drawing (Figure 8.23) and field notes. Tomb A108 is one of the very few tombs that have five chambers radiating out from the base of the shaft (Figure 8.24). Siltation was extensive in the chambers, but ceiling fall was minimal. The extent of the silting made recovery of the tomb contents challenging, and some damage to the bones occurred during the process (Figure 8.25).

Figure 8.24. Excavated shaft of A108, with some entryways silted but others open.

Figure 8.25. Skull of Burial 3 from A108SW shortly after removal from tomb chamber, showing breakage and silt infiltrate into the endocranium.

Tomb A108, Chamber North

This chamber was entirely silted, with considerable damage to the human remains. There were no intact crania, and the cranial fragments were generally found mixed in the central bone pile. This is in contrast with most of the EB IA tomb chambers, where the crania were usually arranged to the left of the entryway and in a row away from the central postcranial bone assemblage. Burials in this chamber are represented by only a few bones for each burial, and determining association between cranial and postcranial skeletal fragments was not possible. The MNI in this chamber is 14, including two adult females, two adult males, six children from one to seven years of age, and four infants ranging from neonatal to one year of age. Because there were no skulls that could be assigned a specific burial number, the numbering of the burials was done in the laboratory beginning with Burial 50.

Burial 50 consists of a few postcranial bones of a gracile female between 25 and 34 years of age. The lumbar vertebrae associated with this burial have significant osteophyte formation limited to the right lateral bodies. This could be a manifestation of DISH, although it is unusual for osteophyte formation to be limited to the right side in lumbar vertebrae. The sacrum probably associated with this burial contains reactive spicules in the neural canal. This is an abnormality most often associated with chronic infection, but little is known about infectious involvement limited to the sacrum. This abnormality also occurs in Burial 51, and it may be that we have not noticed it before as we have studied previous burials. A more careful search for evidence of this abnormal feature may be a worthwhile exercise.

Burial 51 comprises only a few postcranial bones from a gracile female between 35 and 39 years of age. The left humeral head exhibits subchondral bone destruction and periarticular lipping. The right humerus is missing, but right and left scapulae probably associated with this burial have similar subchondral bone erosion and lipping on the margins of the joint, indicating that this arthropathy was bilateral. The bony surface of the neural canal contains fine spicules, a feature also present in Burial 50. The cause of this abnormality is unknown. Vertebrae probably associated with this burial also have eburnation, marginal lipping, and cyst-like lesions at the margins of the diarthroidal joints. This is a manifestation of osteoarthritis.

Burial 52 is represented by a few postcranial bones and fragments. The individual was a somewhat gracile male between 35 and 44 years of age. Several of the subchondral surfaces have evidence of bone abnormalities associated with osteoarthritis. The bone between the condyles on both distal femora is markedly eroded, which may reflect remodeling associated with torn cruciate ligaments. The subchondral bone of the proximal tibia shows significant subchondral destruction, cyst-like lesions in the margins, and evidence of repair. This would certainly be expected if the cruciate ligaments were not maintaining the relationship between the two components of the knee.

Burial 53 consists of a few postcranial bones of a somewhat gracile male about 25 years of age. There is almost complete erosion of the subchondral bone of the left acetabulum. The right acetabulum was not recovered. The cause of this abnormality is not known, but some type of erosive arthropathy is possible. The burial number 54 was not used. Burial 55 includes a few postcranial elements of an infant about six months of age. The diaphyses of the tibiae are abnormally porous, but the cause is unknown.

The following burials of subadults showed no evidence of skeletal disease in the bones available for analysis: Burial 56, a child of two to three years of age; Burial 57, a child between 1 and 1.5 years of age; Burial 58, a child also between 1 and 1.5 years of age; Burial 59, a child between six and seven years of age; Burial 61, an infant about three months of age; Burial 62, an infant about three months of age; Burial 63, a child between 1 and 1.5 years of age; and Burial 64, a child between three and four years of age.

Burial 60 is from an infant between six and twelve months of age. There are plaques of abnormal woven bone on the left humerus. The right humerus is not available for analysis. The right proximal femoral diaphysis has abnormal reactive bone formation that may have been the result of traumatic strain on the gluteal muscles.

Skeletal disorders in bones that cannot be associated with a specific burial include a left tibia with callus formation presumably the result of trauma. A similar lesion is present on a left fibula that appears to be associated with the tibia. The tibia is also abnormally bowed in the mediolateral axis, and this may have been caused by rickets.

Tomb A108, Chamber Northeast

The northeast chamber of Tomb A108 was completely silted at the time of excavation. Skeletal element preservation is poor, which made the association between bones and a specific burial difficult and often impossible. An unusual feature of the bone assemblage was the presence of nonhuman bones in the bone pile. Poor preservation did not permit the identification of the animal bones. The MNI interred in this chamber is 14, including five adults, six children, and three perinatal burials. Field notes indicate the presence of four crania, but there was some commingling of cranial elements that was noted during laboratory analysis.

Burial 1 consists of the highly fragmented cranial and postcranial elements of a child between two and five years of age. Poor preservation prevents a more specific age estimate. Fragments of the frontal bone have bilateral reactive porosity of the intracranial supraorbital plate. No bone formation is associated with this abnormality. There is reactive bone formation on the orbital roofs that is particularly noticeable in the right orbit.

Burial 2 is very incomplete and fragmentary with less than 10% of the skeletal elements represented. There are no cranial bones that could be associated with this burial. The estimated age of this burial is about 35 years. The gracility and morphology of the pelvic bone fragments indicate that the individual was female. There is no significant evidence of skeletal disorder in this burial.

Burial 3 is also very incomplete, with less than 10% of the postcranial elements. Cranial fragments labeled Burial 4 at the time of excavation are most likely associated with this burial based on a similar estimated age, about 35 years, and general morphology of the bone fragments. The sex is male on the basis of the robust postcranial long bone fragments and the anatomy of the pelvic fragments. The acetabula exhibit subchondral bone degeneration as well as marginal cysts and osteophytes. This indication of significant joint disease, probably osteoarthritis, is supported by enthesopathies and joint margin deterioration in some of the other bones present. Burial 4 does include a few cranial fragments, not linked to Burial 3, and some postcranial elements of an adult male. There is no evidence of skeletal abnormality in this burial.

Burial 5 consists of incomplete postcranial fragments of a very gracile adult female; a more specific age estimate is not possible. The diaphyses associated with this burial have abnormally thin cortical bone suggestive of osteoporosis. Burial 5a includes most of the long bones, pelvic girdle, and maxillary as well as dental elements of the skull. The estimated age is about nine years, and sex cannot be determined. There is no evidence of skeletal disorder in the elements available for study. Burial 6 is the very incomplete postcranial skeleton of a female 35 or more years of age. Joint margins available for study exhibit considerable evidence of osteoarthritis.

Burial 50 includes the incomplete postcranial bones of a child between two and five years of age. Poor preservation prevents a more specific age estimate. There is no evidence of skeletal disease in the elements available for analysis. Burial 51 includes only very incomplete postcranial bones of an infant about one year of age. The tibiae have more than normal anterioposterior bowing, most likely caused by vitamin D deficiency. This long bone abnormality, probably caused by the same vitamin deficiency, is also present in the right tibia of Burial 52, which also has an estimated age of about one year. The right tibia also has abnormal woven bone added to the cortical surface. The left tibia is not available for analysis.

Burials 53, 54, and 55 consist of the very incomplete remains of neonatal infants with no evidence of skeletal disorders. Burial 56 is represented by a single right femoral fragment that is not associated with another infant burial. The fragment is normal. Unattributed vertebral fragments from adult skeletons exhibit a compression fracture, degenerative disk disorder, and marginal osteophyte formation.

Tomb A108, Chamber Southeast
Major silting and ceiling fall covered the burials in this chamber. There is evidence of damage to the bones in antiquity that was exacerbated during excavation. An MNI of six were identified from this chamber, including three adults, two children, and one neonatal skeleton. Poor preservation makes the association between cranial and postcranial bones of a burial problematic.

Burial 1 is represented by a fragmented and incomplete cranium and postcranial bones of an adult male between 25 and 29 years of age. There is no evidence of skeletal disorder in the bones available for analysis. Burial 2 includes less than 10% of the skeletal elements of both the skull and postcranial skeleton. The individual represented by this burial was a female at least 30 years of age at the time of death. The skeletal elements of this burial exhibit enthesopathies and evidence of slight to moderate osteoarthritis.

Burial 50 is poorly preserved and very incomplete, including clavicular, femoral, tibial, and pelvic fragments of a child between six and ten years of age. There were subadult teeth found in the bone pile that may be attributable to this burial. The clavicle has reactive bone formation on the cortex. A right upper canine tooth probably associated with this burial has two hypoplastic lines, indicative of two stress events that occurred during the development of the tooth.

Burial 51 consists of the incomplete and poorly preserved postcranial skeleton of a child about five years of age. There is no evidence of skeletal disease in the few bone fragments available for analysis. Burial 52 includes a right distal fragment and the left zygomatic bone of a perinatal infant. These fragments are normal. Burial 53 includes only the poorly preserved pelvic fragments of an adult between 25 and 29 years of age. The anatomy of the pelvic fragments indicates they are probably from a male, but this cannot be determined with certainty. Both acetabula are more shallow than normal, but there is no evidence of chronic subluxation or dislocation on either side.

Evidence of skeletal abnormalities that cannot be attributed to a specific burial includes tibial and fibular fragments with evidence of chronic infection. Another long bone fragment has remodeled woven bone covering most of the original cortical surface. Fragments of right and left orbits, possibly from the same individual, have slight cribra orbitalia that was in the healing stage. The possible causes of this abnormality include chronic infection, scurvy, and anemia.

Tomb A108, Chamber Southwest
No blocking stone was present for this chamber, so it is not surprising that it was completely silted. The burials in the chamber were poorly preserved. The cortical bone is degraded and cracked. Although only one skull is mentioned in the field notes, laboratory analysis indicates the presence of at least four incomplete and fragmentary skulls. The extensive silting created major problems during excavation, which resulted in some breakage. However, considerable breakage occurred in antiquity, presumably during the movement of the burials from the primary to the secondary burial site. Poor preservation and incomplete burials made the association between cranial and postcranial elements impossible in most cases. The MNI from this chamber is eight, including four adults and four subadults, three of which were infants and one of which was in very early childhood. The poor preservation also made the assignment of estimated ages very difficult, although the unusual absence of evidence of osteoarthritis in the adult remains argues for young adulthood.

Burial 1 consists of an almost complete skull, including the maxilla and mandible. No association can be made with any of the postcranial burials. The estimated age of the skull is about 25 years. The sex criteria are somewhat ambivalent, but more indicative of female than male. In the dentition, there are two hypoplastic lines caused by two major illnesses during the development of the permanent teeth.

Burial 2 is also a cranium with mandible for which there is no likely association with postcranial remains. The moderate dental wear and general appearance of the skull supports an age between 30 and 35. The sagittal suture fused prematurely, causing a condition in which the skull becomes abnormally elongated as abnormal growth occurs in the sutures that are not prematurely fused. As is common in cases of scaphocephaly, the inner table contains multiple depressions from the pressure caused by the normally developing brain and the inadequate intracranial space. Individuals with this abnormality regularly attain a normal life span.

Burial 3 consists of a calvarium lacking the maxilla and mandible. The estimated age based on the open cranial sutures is young adult, and the morphology of the skull supports the likelihood that the individual was female. There is no evidence of significant disease.

Burial 50 is represented by less than 10% of the postcranial bones. There is no evidence of joint disease, and a young adult age estimate is appropriate. The bones available for analysis are normal. Burial 51 is also very incomplete and includes only postcranial skeletal elements. As with Burial 50, the lack of joint abnormalities supports an estimated age in the young adult range; the individual was female. There is no significant evidence of skeletal disorder.

Burial 52 consists of less than 10% of the postcranial bones. The sex criteria, including the robusticity and femoral head diameter, indicate this was a male. There is slight evidence of joint disease bilaterally in the acetabula, but generally the age is most likely in the young adult range, although possibly at the upper end of the range. There is no significant evidence of skeletal disease in the elements available for analysis. Burial 53 includes the right and left proximal femora of what is probably a female adult. A more specific age estimate is not possible. There is no evidence of abnormality in the two bone fragments.

Burial 54 consists of the normal left humerus of a neonatal infant. Burial 55 is an infant between birth and six months of age and represented by both femoral diaphyses and the right tibial diaphysis. There is no significant evidence of bone abnormality. Burial 56 includes less than 10% of the cranial and postcranial bones of a young child between one and one and one-half years of age. Evidence of reactive woven bone is present on the medial diaphysis of the right tibia. The left tibia was not recovered. The endocranial cortex of the right greater wing of the sphenoid is abnormally porous, with woven bone deposits on the cortical surface. Both scurvy and infection can cause this type of abnormality, but a specific diagnosis is not possible. Burial 57 comprises a single right femoral diaphysis of an infant between birth and six months of age. There is no evidence of significant abnormality in this bone.

Tomb A108, Chamber Northwest
This chamber was heavily silted, but there was minimal evidence of ceiling collapse. The bones were in poor condition in antiquity, and this problem was compounded during excavation by the silting. Curiously, the small bones of the hands and feet of at least two burials were remarkably intact, but generally bone preservation is poor. The condition of the bone made association of the cranial bones with postcranial elements impossible in most burials. However, the estimated MNI (11) was the same for cranial as well as postcranial bones. There were five adults buried in the chamber, including three females, one male, and one adult whose sex could not be determined. Six subadults were interred in the chamber ranging in age between midterm fetus and about twelve years.

Burial 1 is represented only by cranial bones, and preservation is poor. The bones are from a gracile adult female; a more specific age is not possible. There is an apical abscess associated with the left mandibular second premolar. The destructive focus is surrounded by unremodeled woven bone, indicative of active inflammation at the time of death.

Burial 2 is one of the few burials in which cranial and postcranial bones could be associated with some confidence. This is because the age of the child represented by the bones is seven to eight years, and there was no other burial of that age in the chamber. The dentition of this child had at least five episodes of stress during development that resulted in multiple enamel hypoplastic grooves. The child also had cribra orbitalia in the active stage at the time of death. The cause of this abnormality cannot be stated with certainty, but the most likely options include scurvy and infection. There is no evidence of marrow hyperplasia in this burial, so anemia is very unlikely.

Burial 3 includes only the poorly preserved cranial bones of a female about 30 years of age at the time of death. The left maxilla has evidence of a dental abscess associated with the first molar, which was lost antemortem. Burial 50 consists of a right temporal and some postcranial fragments of an infant between six and twelve months of age. There is no evidence of skeletal disorder in this burial. Burial 51 includes both cranial and postcranial bones of a child about two to three years of age. There is no significant evidence of skeletal disease. Burial 52 is represented by both cranial and postcranial bones of a normal late-term fetus.

Burial 53 consists of both cranial and postcranial bones of a midterm fetus. There is evidence of possible developmental defects in the cranial bones. The significance of these possible abnormalities is uncertain, but they raise the question about whether fetal burials are the result of abortion by the mother or death of the mother, and whether, if the latter is true, the fetal bones were later gathered when the bones of the mother were collected for reburial in the chamber. Burial 54 includes only the left temporal bone of an adolescent about 12 years of age. The sex is unknown, and there is no evidence of significant bone disorder in this burial.

Burial 55 consists of postcranial bones of a male between 25 and 29 years. Osteoarthritis is apparent in some of the postcranial bones. The left distal tibia and fibula have reactive woven bone formation in both active and remodeling stages. This abnormality is probably the result of trauma, although there is no evidence of fracture. The right acetabulum has two lytic lesions in the acetabular notch at the margins of the subchondral bone. These lesions occur at the attachment of the ligamentum teres femoris. This ligament connects with the fovea capitis on the femoral head, and this fovea is abnormally deep. These right hip abnormalities are suggestive of at least a partial dislocation of the joint, although this likely was reduced long before the death of the individual.

Burial 56 is the incomplete skeletal remains of an adult female. A more specific age estimate is not possible. The only significant evidence of skeletal disorder is some evidence of osteoarthritis. Burial 57 is represented by a right femoral head and neck and probably a pair of tali. Little can be said about this burial other than it was an adult of unknown sex. Burial 58 is represented only by a right temporal bone of a gracile adult female. There is no evidence of disorder in this bone. There is one long bone fragment that is not associated with any burial that has a lesion of woven bone that was being remodeled at the time of death. Trauma and infection are the most likely diagnostic possibilities.

TOMB A109

Tomb A109 includes a central shaft and three tomb chambers radiating from the lower end of the shaft

(Figure 8.26a). A single blocking stone was associated with the south chamber but was not in place when the shaft was excavated. An unusual feature of this tomb is the antechamber between the shaft and the east chamber (Figures 8.26b). An antechamber is associated with one of the tombs excavated in 1977 (Chapter 6, Tomb A86). The photographic documentation of the tomb chambers was not obtained at the time of excavation, so observations of the tomb chamber contents at the time of excavation are based on the field notes.

Tomb A109, Chamber North
At the time of excavation there was extensive evidence of ceiling fall in this chamber but little silting. Much of the damage to the human remains occurred in antiquity, when bones were moved from the primary burial site. Damage was also caused by the ceiling fall and during excavation and subsequent shipping. Preservation of the bone is poor. The burials tend to be represented by cranial fragments and a few of the major long bones, although some of the small bones of the hands and feet were placed in the chamber and were well preserved. The MNI interred in this chamber is six, including two adults, one child, two infants, and one mid- to late-term fetus.

Burial 1 is represented by both cranial and postcranial bones of a child about 11–12 years of age. The sex cannot be determined. Healed porous lesions of the

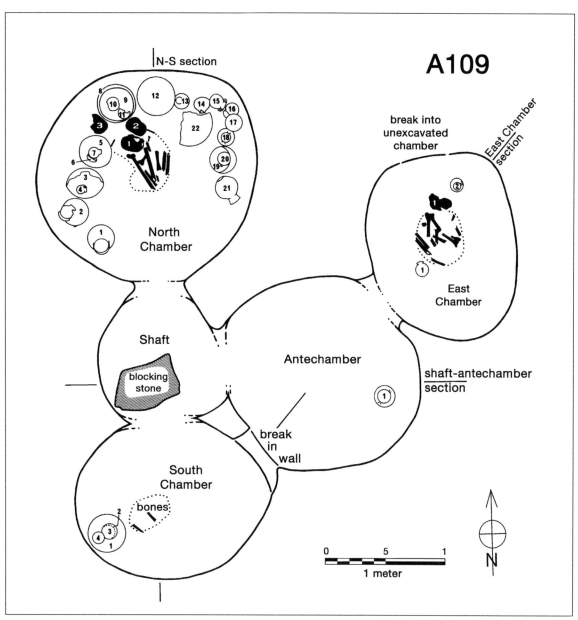

Figure 8.26a. Plan drawing of Tomb A109.

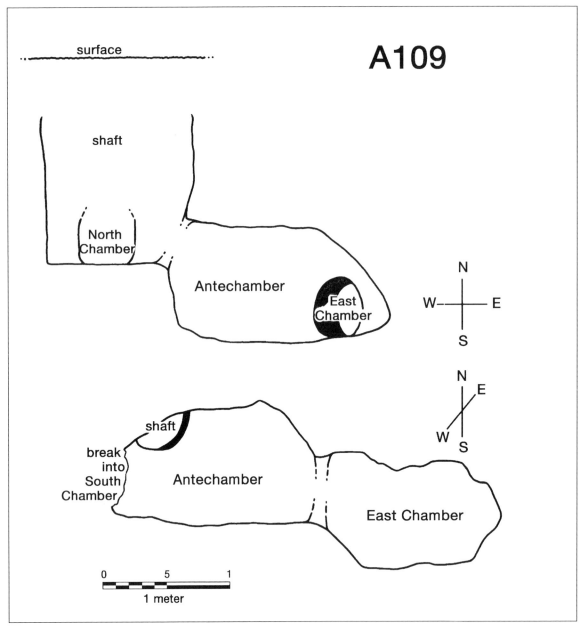

Figure 8.26b. Section drawing of Tomb A109E.

orbital roof occur in both orbits but are most severe in the left side. The remodeling of these lesions is indicative of recovery from the disease that caused the porosity. Scurvy, infection, and anemia are possible causes. There is a depressed lesion with remodeling on the external table of the frontal bone. The presence of active remodeling supports a diagnosis of trauma rather than a cyst.

Burial 2 includes a partially intact skull and postcranial elements that are plausibly associated with the skull. The remains are from an adult female about 30 years of age. There is evidence of temporomandibular degeneration unilaterally on the left mandible, with subchondral bone abnormality and cyst-like excavation of the juxta-articular bone.

Burial 3 consists of a partially intact skull and some postcranial skeletal elements that are probably but not certainly associated with the skull. On the basis of dental wear, the individual associated with these remains was an adult and probably at least 30 years of age. The sex is female based on the gracile morphology of the bones. The long bone cortices are thinner than normal and also abnormally porous, as revealed in the broken cross-sections. The tibiae have lesions of remodeled woven bone on the mid-diaphyses. Infection and trauma are common causes of this type of lesion. The bilateral nature of the lesions favors a diagnosis of a systemic infection, since bilateral and symmetrical trauma, although certainly possible, is uncommon.

Burial 50 consists of a few cranial and postcranial bones of a young infant between birth and three months of age. There is no evidence of skeletal abnormality in the bones available for analysis. Burial 51 includes cranial fragments and postcranial bones from an infant about one year of age. The bones are normal. Burial 52 is represented by both cranial and postcranial bones of a mid- to later-term fetus. The preservation and recovery of the very small and fragile bones of a fetus testify to the care taken in recovery from the primary burial site and interment in the tomb chambers. There is no evidence of abnormality in the bones available for analysis.

Tomb A109, Chamber East
Preservation of chamber contents is very poor, and little can be said about the human remains. The chamber was completely silted at the time of excavation, and this resulted in damage to the fragmentary remains, but the burials interred in the chamber were poorly represented in antiquity as well. The MNI associated with this chamber is seven, including three adults and four subadults. There is no evidence of significant abnormality in any of the bone fragments available for study, but the burials are so incomplete that the lack of skeletal abnormality contributes little to our understanding of the health of the people interred in this chamber. Following excavation, Burial 1 was assigned to some of the fragments. Unfortunately, this assignment included two adult burials, and because of this we have renumbered all the burials from this chamber during laboratory analysis.

Burial 50 consists of both cranial and postcranial fragments of an adult whose age cannot be estimated, although the good condition of the subchondral bone suggests a young adult age. The bone fragments are relatively robust, indicating the likelihood that the burial is a male. There is no significant evidence of skeletal disorder in the fragments available for analysis.

Burials 51 and 56 include bone fragments of two adult females of unspecified age. The acetabulae of Burial 51 are somewhat shallower than normal, which could have been associated with hip dysplasia and chronic subluxation of one or both hips. Otherwise, there is no evidence of skeletal abnormality in the fragmentary, very incomplete, and poorly preserved remains of either burial.

The remaining burials from this chamber are from infants or young children. None of the very incomplete and fragmented bones exhibit any evidence of skeletal abnormality. Burial 52 consists of the petrous portion of both temporal bones and corresponding femoral fragments of a child between one and three years of age at the time of death. Burial 53 includes both petrous portions of the temporal and a fragment of the left femoral diaphysis from a child between one and two years of age. Burial 54 consists of another pair of the petrous portion of the temporal bones and some fragmentary postcranial elements that may but cannot with certainty be linked with the temporal bones. Burial 55 is represented by another pair of petrous fragments of the temporal bone of a mid- to late-term fetus.

There are a few unassociated bone fragments that exhibit slight abnormality, including a fragment of a right adult frontal bone with healed cribra orbitalia. Chronic eye infection such as trachoma is one diagnostic option, but this infection is unlikely to have healed. Scurvy can cause these lesions, and if recovery occurs the lesions will remodel and be similar to what occurs in this fragment. Anemia followed by recovery could have resulted in remodeling of the lesion. A fragment of the left greater wing of an adult sphenoid has a destructive lesion with marginal sclerosis that is indicative of a chronic disorder. A proximal phalanx of the hand has a destructive lesion of the subchondral bone with marginal sclerosis that may have been caused by trauma to the joint.

Tomb A109, Chamber South
This chamber was completely silted in, with no evidence of ceiling collapse contributing to the fill. The burials were very incomplete, fragmentary, and poorly preserved. Part of this condition is attributable to degradation before excavation, but silting made excavation very difficult and additional damage occurred during excavation of the chamber and removal of the remains. Both cranial and postcranial elements are present, but the poor preservation makes an association between these elements problematic. An MNI of five were interred in this chamber, including two adults of unknown age who probably were females. There were three subadults, including a child about nine years of age, an infant about one year of age, and a perinatal infant.

Burial 50 consists of a right humeral fragment and both patellae of a gracile adult of unknown age who was probably female. The right patella has a lytic focus in the subchondral bone that has incomplete marginal sclerosis. The cause of these lesions remains uncertain, but trauma to the joint is one possibility. Burial 51 is the other adult female and consists of a pair of humeri, a right femoral diaphysis fragment, and a left patella. Because of the poor preservation, the association of these skeletal elements is uncertain. The age of this person at death is unknown. There was no evidence of skeletal disorder in the bones present.

Burial 52 includes a left humeral fragment, a left ulnar fragment, the left ischium, the left femoral head epiphysis, and a fragment of the left distal femur. This is the only burial of a nine-year-old child in this chamber, so the association of these bones is certainly more certain than the association of bone fragments in the adult burials. Burial 53 is represented by a femoral diaphysis. The size of this fragment suggests an age of about one year. Burial 54 consists of a fragment of the proximal right femur and proximal tibia. The size of these elements indicates a neonatal age estimate. None of the subadult skeletal elements available for study show any evidence of abnormality.

CHAPTER 9

The Tombs and Burials of the EB I People: Tombs Excavated in 1981

BRUNO FROHLICH, EVAN M. GAROFALO, AND DONALD J. ORTNER

EXCAVATIONS DURING THE LAST of the three field seasons at Bâb edh-Dhrâʿ cemetery took place during the summer of 1981. The field crew completed three tasks: (1) prepared a topographical map of the burial ground, (2) tested geophysical equipment to the evaluate potential for identifying the location of below-the-ground man-made structures (Frohlich and Ortner 1982; Frohlich and Lancaster 1986), and (3) excavated and retrieved human remains and associated burial objects from tombs identified by the geophysical equipment.

Although the 1981 excavations were largely in the A area of the cemetery, the area surveyed and excavated was considerably different from those excavated in the 1977 and 1979 seasons. In 1977 and 1979 the tombs excavated were located on the banks and adjacent areas of the wadis and depressions traversing the burial ground area. These areas were used in earlier excavations by Paul Lapp (1965 to 1967) and provided a substantial number of tombs, including those excavated in 1977 and 1979 (Ortner 1981; Schaub 1981b).

During the 1981 season we elected to focus on the flat areas between the wadis to explore the usage made of areas some distance from the wadis. There were several reasons for this approach. First, our early testing of the geophysical electromagnetic equipment (Geonics EM-31 conductivity meter) suggested that we would be less successful in identifying tombs in areas with substantial slope distances. Second, we wanted to see if the distribution and density of tombs was similar to the tomb concentrations closer to the wadis. During excavations of burial grounds in Bahrain, it became evident that during the transition from a nonsedentary to a more sedentary society, the structural and architectural aspects of mortuary traditions tended to stay unchanged (Frohlich 1986). However, the location of burial structures would be significantly different. Thus, the location of tombs related to nonsedentary populations tended to be associated with geological and geographical features such as wadis and hills, while more sedentary populations selected locations close to the sedentary housing structures that may have included a collection of semimovable structures such as tents or more permanent mudbrick and stone buildings (Frohlich 1986; Frohlich and Ortner 2000). This change in tomb location might also be reflected in the distribution of pottery types and especially in the positioning of the human remains and the degree of skeletal disarticulation. This tendency has also been observed by Schaub and Rast (1989:27–29), who suggested that one subdivision of the EB IA period is based on the presence or absence of articulated skeletons within the burial chambers.

A total of six tombs were identified in 1981 containing 15 chambers (Frohlich and Ortner 1982). Except for one tomb (C11), all were identified by the geophysical equipment (EM-31), and excavations were initiated knowing the exact location of the shafts leading to the chambers, as well as the degree of silting that might have taken place within the chambers over the last five millennia. This was complicated by disturbance of the shaft tombs that occurred both in antiquity and within the past two hundred years. The architectural features and biological contents will be described for all chambers in Tombs A110, A111, A114, and C11, and architectural features only will be described for Shaft Tombs A112 and A113, which were robbed and most of the human remains destroyed.

Five of the tombs excavated in 1981 (A110, A111, A112, A113, and A114) were located on a flat and undisturbed surface east of the previously excavated A65 to A68 group, south of the A1, A3, A4, A11, and A12 group, and southwest of the A42, A43, A45, and A47 group excavated by Paul Lapp from 1965 to 1967 (Schaub and Rast 1989; Figure 1.1). The area surveyed by the EM-31 and including the A110 to A114 shaft tombs measured 60 meters by 30 meters. The precise location of each of the five tombs was identified using the EM-31 equipment (Figure 3.6). One tomb located in the C Cemetery (C11), about 550 meters west-northwest of the excavated A tombs and adjacent to the east-west road leading from the site to the town of Kerak, had earlier been observed by Sami Rabadi of the Jordanian Department of Antiquities. Erosion had exposed the upper end of the shaft, making it noticeable from the road (Frohlich and Ortner 1982).

Of the six excavated tombs, four were undisturbed (A110, A111, A114, and C11), one was disturbed by thieves (A112), and one included a collapsed chamber (A113). Human remains were found in all of the tombs except for the one with a collapsed chamber (A113). Of the remaining five tombs, two included four chambers (A111, and A112), one included three chambers (A110), and two contained one chamber (A114 and C11). The excavated chambers (N = 13) included four that were severely disturbed and nine that were undisturbed. Of the nine undisturbed tombs, seven were unsilted and two were partly silted (20% and 95%). The 13 chambers were all associated with five well-defined shafts (Frohlich and Ortner 1982).

One additional tomb, belonging to an unexcavated shaft tomb system, was identified from within one of the four tombs belonging to the disturbed shaft tomb (the north chamber of A112). Apparently, thieves searched for adjacent burial chambers by forcing sharp spikes into the chamber walls in the hope that they would penetrate the thin walls separating chambers belonging to different tombs, and in this case they succeeded (Frohlich and Ortner 1982).

Each excavated shaft in the A Cemetery was identified by distinguishing isolated and circular high readings using the EM-31 conductivity meter (Frohlich and Ortner 1982; Frohlich and Lancaster 1986). Based on our survey of the 60 by 30 meter square, it is likely that an additional eight to ten tombs are present within this square. Thus, a potential 13 to 15 shafts exist within the 1,800 square meter area.

TOMB A110

This tomb consisted of a vertical shaft leading to three burial chambers that extend outward from the lower end of the shaft. The chambers were connected to the shaft by squared entrances, all covered by one or more blocking stones. The seal between shaft and tombs had further been enhanced by the application of mortar completely sealing the entryway.

The center of the shaft was identified by a high conductivity reading, and identification of the upper levels of the shaft fill was carried out by excavating a two meter by one meter square with the two meter side passing through the center of the high conductivity reading. In this way we could identify the border between fill material and undisturbed adjacent soils. This technique worked well for all shafts identified by the EM-31, and in each case the shaft center as determined by the EM-31 readings was less than 50 centimeters from the center of the shaft determined by excavation.

The upper levels of the shaft fill included about two centimeters of small rocks and stones giving way to softer but still very coarse soils. The upper level of the shaft occurred 33 centimeters below the present ground surface. The upper shaft fill consisted of a soft brownish layer mixed with small stones and a few pottery fragments. At approximately 58 centimeters below the present surface, the northeastern half of the shaft yielded a two centimeter layer of darker brownish soils followed by a two centimeter layer of gravel. The remaining shaft appeared to be relatively homogeneous in color and density.

The maximum depth of the shaft as measured from the current and 5,000-year-old ground surface to the lowest point of the shaft was 148 centimeters; maximum and minimum upper diameters were, respectively, 125 centimeters and 118 centimeters. At the shaft's lowest level, the maximum and minimum diameters were, respectively, 114 centimeters and 91 centimeters. The upper surface of the entrances leading into the chambers was found between 88 centimeters and 94 centimeters below the original surface, thus between 121 centimeters and 127 centimeters below the present surface. The shaft extended between 6 centimeters and 10 centimeters below the lowest points of the three entrances.

The shaft fill was, in general, a homogeneous, intentional fill with similar consistency and density throughout all levels. We conclude that the fill was a result of a single use pattern; thus, the three chambers most likely were constructed and used at the same time. The shaft and all three chambers were cut into a soil matrix consisting of various layers and isolated concentrations of hard-packed clay, sand, and gravel. No marl layers were present.

Each chamber is described separately. The text is divided into two areas: (1) information pertinent to field observation that would include the architecture, measurements, description of soil types, location of objects, and human remains in each tomb and (2) description and analysis of the human remains based on the analysis of the remains in our laboratory facilities at the Smithsonian Institution.

Tomb A110, Chamber Northeast

The chamber had an almost circular, flat floor, and a dome-like wall and ceiling structure. The direction of the chamber axis from the center of the shaft through the center of the chamber is 26° from magnetic north. The entrance was covered by one large blocking stone measuring 50 by 54 centimeters and five to six centimeters deep. A few additional but smaller stones were used in combination with a mud mortar mix to seal off the chamber from the shaft. The mud mortar mix covered the entire external surface of the blocking stone. The entrance measured 41 centimeters by 39 centimeters and is 19 centimeters deep. The lower part of the entrance is located about six centimeters above the lowest part of the shaft. The floor of the chamber was located six centimeters below the lower margin of the entrance and 178 centimeters below the present surface (Figure 9.1).

The northeast chamber measured 175 centimeters from the entrance to the back of the chamber and is 184 centimeters wide. The maximum height was 88 centimeters. No ceiling fall or silt was present in the chamber. The walls consisted of light brown hard-packed clay mixed with gravel. Tool marks produced during the construction of the tomb were clearly visible.

Human remains were found at the center of the chamber. The adult remains had been placed on reed matting. The remains included a total of three individuals: one almost completely articulated adult fe-

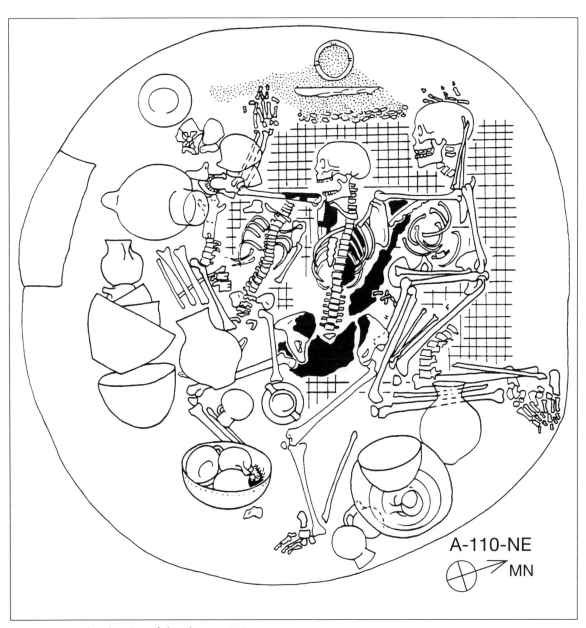

Figure 9.1. Plan drawing of chamber A110NE.

male aged 30 years (Burial 3), one partly articulated male adult aged 25 years (Burial 4), and one subadult aged six to nine years (Burial 1). Some of the skeletal elements of the male skeleton (Burial 4), including ribs, vertebrae, and long bones, appeared to have been placed in locations not corresponding with the expected anatomically correct positions; thus, those elements were, in all likelihood, completely disarticulated before placement in the chamber (Figure 9.2). All three individuals were placed in semiflexed positions, with the skulls located to the left (north) of the entrance and the lower extremities to the right (northeast) of the entrance. The subadult was only about 50% articulated (Figure 9.3). The three burials have been assigned numbers 1, 3, and 4. Burial 2 was originally assigned to a small cluster of bones that later proved to be part of Burial 4.

All three burials were secondary, although the time between primary and secondary interment must have been brief. We argue that the nonsedentary population inhabited a large geographical area and only visited the burial ground infrequently during the construction of tombs and the final interment of the deceased members of the group. The interval between visits was variable, probably ranging between a few months and several years. The degree of skeletal articulation found in the chambers is a measure of the time between the occurrence of death and primary burial and the time of the secondary burial in the shaft tomb. The presence of almost fully articulated skeletons is a function of death occurring just before a planned visit to the burial ground, or perhaps death occurring during a visit to the burial ground (Frohlich and Ortner 1982, 2000).

The remains from this chamber are very well preserved, although with slight damage. The remains are fragile and brittle, and this has resulted in some damage before excavation and removal of the remains. There is also some notable damage that occurred in early antiquity that is observable on the Burial 1 cranium. The nature of some of the breaks as well as the coloration provide the evidence of damage in antiquity. For all three of the burials, the majority of the skeletal elements are present, exceeding 90% for each individual. None of the soil from the primary burial site was found on or within the burials. The total number of individuals in this chamber is three, identified as Burial 1, Burial 3, and Burial 4.

Figure 9.2. View of chamber A110NE. Three burials include one male, one female and one child.

Figure 9.3. View of chamber A110NE. Partly articulated child in the foreground. Note clay figurine between child's skeletal elements.

Burial 1 is a subadult skeleton that was placed between the entrance and the female skeleton (Burial 3) with the skull resting on the female's right arm (Figures 9.1, 9.3). The skeleton is 90–95+% complete, with most bones present. The cranial vault is somewhat fragmented. The facial elements are intact. Because of the disarticulated nature of the child's cranium and the stacked placement of the individual bones it is likely that the child's cranial bones were disarticulated when placed on and next to the woman's (Burial 3) right lower arm. The trunk was partly articulated, with both upper and lower extremities located adjacent to the body, suggesting that the bones had been disarticulated either by natural means or by intentional removal of the extremities from the trunk (Figure 9.1). The latter option is supported by the presence of a transverse cut in the distal and posterior end of the right femur (Figure 9.4).

The age at death was determined by the dental and vertebral development as well as the development and mineralization of secondary centers of ossification (epiphyseal development). The osteometrics of the long bone diaphyses provided an age-at-death estimate of between five and seven years.

Observation of the permanent dentition indicated evidence of multiple stress events expressed as enamel hypoplasia. The increased porosity of the maxillary alveolar bone is likely associated with the active dental development and eruption. Bilaterally, but not symmetrically, subchondral bone lesions are present on the posterior aspect of the lateral femoral epiphyseal condyles (possibly osteochondritis). The lesion on the right distal epiphysis is well remodeled, indicating that this process was in the remodeling stage well before the time of death. Within the lesion on the left, however, the central nodule of sequestered bone is still present, which suggests that this abnormality occurred closer to the time of death.

Burial 2 consists of two groups of articulated vertebrae located between Burial 1 and Burial 3. After the removal of the burials, we determined that all elements originally designated as Burial 2 belonged to Burial 4.

Burial 3 consists of a female skeleton placed supine and parallel to the male skeleton (Burial 4). The lower extremities are slightly flexed, with the knee regions closer toward the entrance. The left upper extremity

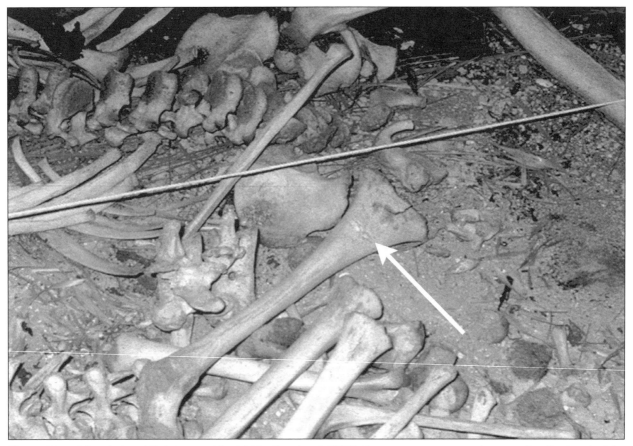

Figure 9.4. Disarticulated limb bones of Burial 1 in Chamber A110NE. Note transverse cut on the posterior-distal end of the right femur (arrow).

was placed partly on top of the male skeleton, and the right upper extremity was flexed about 90 degrees and toward the entrance. The placement of the female's right arm below the subadult's (Burial 1) cranial fragments and the female's left arm on top of the male skeleton (Burial 4) suggests that the order of placement in the tomb was according to the following: first the male (Burial 4), followed by the female (Burial 3), and completed with the subadult (Burial 1). This is further supported by the observation that the male is placed closest to the wall farthest from the entrance and the subadult is closest to the entrance.

The female skeleton is 95+% complete, with elements of the upper and lower extremities missing, including the distal right tibia, midshaft of the left tibia, midshaft and proximal end of the left fibula, and approximately 10% of the hand and foot bones. Most of the female body was covered by shrouds, most likely composed of animal hide and textiles (Figure 9.2).

The age-at-death determination is based on the auricular surfaces of the ilium and the pubic symphysis morphology. These two methods yielded an age range of 25 to 35 years. Dental attrition is advanced. The sex characteristics all indicate a determination of female. However, the sacral morphology is atypical for a female; the alae are not as broad as would be expected.

Multiple stress events are linked to the presence of at least three enamel hypoplasias in some teeth. There is a likelihood of antemortem tooth loss of the left M_3 with complete alveolar remodeling. The left M^3 displays extremely slight polishing on the occlusal surface that would indicate that these two teeth did occlude for a period of time before the lower molar was lost. All of the teeth show secondary eruption resulting from attrition. The right M^1 and M^2 were also lost antemortem, and there is evidence of apical abscess drainage buccally between the two. Antemortem loss of the right M^2 likely occurred close to the time of death, as the socket is in early stages of remodeling. A large caries is present on the lingual cusps of the right M_1, which has destroyed the lingual aspect of the crown. Additionally, there is evidence on the buccal side of complete crown attrition. There is also a buccal apical abscess associated with this tooth. The margin of this defect is rounded and slightly remodeled, with mild porosity. The tooth has shifted buccally out of alignment with the rest of the dental arcade. The alveolar bone of the molars on the right side of the mandible is receding from the roots. This is likely a secondary reaction to the pus drainage. The crown of the left M_1 is completely destroyed by attrition, exposing the pulp chamber. There is also evidence of apical abscess drainage. The margins of this defect are well rounded and remodeled. This occurred some time before death.

The posterior aspect of the left glenoid fossa exhibits minor ossification of the labrum. This is unilateral, and there is no evidence of degeneration or lipping at the left humeral head. Bilaterally, the medial margins of the radial heads show subchondral destruction and repair. There is no evidence of similar lesions on the ulnae or the humeri. Bilaterally, the lateral humeral epicondyles are strongly excavated and exhibit some lipping at the attachment for the common extensors. This is more marked on the right. The anterior body margins of C5 to C7 show evidence of intervertebral disk degeneration (Figure 9.5). This is also visible on T11 to L5. Additionally, the lumbar vertebrae display some lateral body lipping. This is more marked on the left side.

The enthesopathic activity at the pelvic girdle is above normal for an individual this age. The enthesopathies are most visible at the ischial tuberosity (the attachment site for semitendinosus, semimenbranosus, and biceps femoris), the pectineal line (pectineus), the anterior interior iliac spine (reflected head of rectus femoris), the transverse ligament attachment, the iliosacral ligament attachment area, and the margin of the acetabulum. This is bilateral. In addition to these enthesopathies, the proximal femora also display an unusual amount of enthesophytes. Both the greater and lesser trochanters (gluteus medius, gluteus minimus, iliacus, and psoas major) and the attachment for gluteus maximus have enthesopathies. Atypically, the linea aspera is smooth. This severe enthesopathy is likely due to rigorous physical muscle activity. The cortex is thinner than normal and the trabeculae of the metaphyses are reduced in number. Those that are present are thicker than normal. This may be evidence of the early stages of osteoporosis.

Burial 4, including Burial 2 elements, is a male skeleton that was located closest to the rear wall of the chamber (Figure 9.1). The body was placed on its right side with the right lower extremity extended toward the eastern wall and the left lower extremity bent in approximately 80°, resulting in the left foot being placed adjacent to the north-northeastern wall. The left arm was bent 90°, so that the hand was positioned in front of the innominate bones. The right arm was

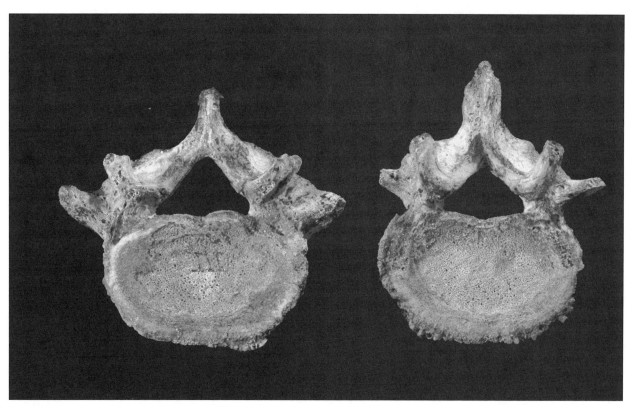

Figure 9.5. Intervertebral disk degeneration. Burial 3, chamber A110NE.

placed forward in the direction of the entrance (southwest). The body is 95+% complete, with only two left ribs and approximately 10% of the hand and foot bones missing.

There had been some disturbance resulting in misplacement of the bones. For example, bones of the upper extremities were found between the lower extremities (Figure 9.6) and T6 to T12, and L1 and L2 were found between the female skeleton and the disarticulated subadult skeleton. Considering the fact that most of the remaining thoracic vertebrae and the innominate bones were found in the normal anatomical location, we argue that the eight vertebrae were moved from their original position after the body became disarticulated during the process of normal decay. At this time, we have no explanation for this misplacement.

Age at death of Burial 4 is estimated at 25 years. This estimate is based on dental attrition and the morphology of the auricular surfaces and pubic symphysis. The medial clavicle is completely fused, but there is a remnant of an epiphyseal line on the posterior proximal medial tibiae. The manubrium and sternum have not yet completely fused. All sex characteristics support assignment to the male sex.

There is a large but well-healed depressed fracture present on the left frontal boss. The depression measures approximately 5 millimeters in depth, with a maximum length of 32 millimeters and width of 34 millimeters. The maximum diameter is 39 millimeters. The depressed area affected the inner table. This lesion is the result of trauma.

The cranial vault is asymmetrical, with a deviation to the right. The right parietal is more bulbous. The nasal bones are deviated to the left, possibly indicating a deviated septum. It is likely that the abnormal cranial shape is antemortem, but postmortem deformation is possible. The deviation of the nasal bones, however, is certainly antemortem.

There are four enamel hypoplasias indicative of significant growth cessation. T7 displays a minor Schmorl's node on the caudal end plate. L4 is present and has well-remodeled, complete laminar spondylolysis; the lamina is not present for observation. The ligamentum flavum attachment sites in the thoracic vertebrae show mild enthesophytes. Unilaterally, the right distal medial femoral condyle contains a destructive lesion caused by osteochondritis dissecans. The abnormality also affects the medial tibial plateau.

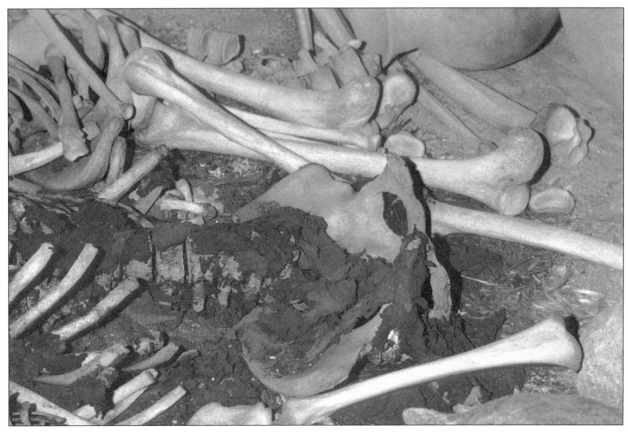

Figure 9.6. Burials 3 and 4, chamber A110NE. Note the displaced humerus between the lower leg bones.

The subchondral surface of the medial tibial plateau displays some hyperostosis. Bilaterally, the tibial tuberosities display mild enthesophytic activity. Cyst formation and repair is visible bilaterally on the medial articular surface of the patellae. This may be evidence of early degenerative joint disease.

Tomb A110, Chamber Northwest
The construction features and general architecture of the A110NW chamber (Figure 9.7) was similar to what was observed in the A110NE chamber. The direction of the axis from the center of the shaft and through the midpoint of the tomb was 320°. The opening was covered by a single blocking stone measuring 50 centimeters by 48 centimeters, and about 30 centimeters thick. Mud mortar was found in all spaces between blocking stone and chamber walls surrounding the entrance. The sealing prevented major incursions of silt and/or water into the chamber from the shaft. The entrance was square, measuring 44 centimeters by 44 centimeters and with a wall thickness of 18 centimeters. The lowest part of the entrance is about 7 centimeters above the floor of the shaft, and the height of the step leading from the entrance to the floor of the chamber is 29 centimeters. Thus, the floor of the northwest tomb was located approximately 22 centimeters below the lowest part of the shaft, and 203 centimeters below the present surface.

The chamber had a roughly circular floor that measured 156 centimeters from the doorway to the back and had a maximum width of 175 centimeters. The maximum height was 84 centimeters. A 2 centimeter layer of silt, covering a small area just below the en-

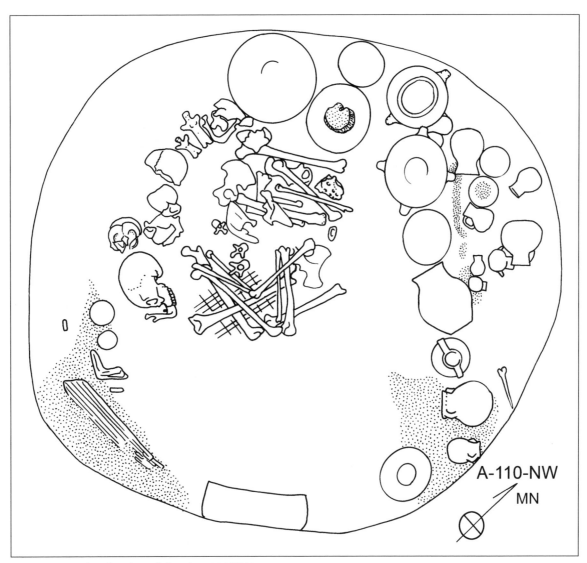

Figure 9.7. Plan drawing of chamber A110NW.

trance, was most likely a result of silt flowing through the entrance. No other silt and/or roof fall occurred.

The MNI is six. This includes two adult male skeletons, two subadults around 1.5 years, one newborn, and one late-term fetus. Preservation is very good, and the adults are nearly complete. Most of the subadult postcranial remains are very well preserved and complete. Only one subadult burial is not well represented (Burial 3B).

Placement of the mostly disarticulated remains followed the pattern found in almost every other EB IA chamber, including the arrangement of the disarticulated remains. The cranial elements were located to the left of the entrance in a row parallel with the left chamber wall. The postcranial elements were located in an organized bone pile at the center of the chamber (Figures 9.7, 9.8).

The skulls of two adult males were positioned at each end of the row. The male cranium and mandible (Burial 1) closest to the entrance was found with all the individual bones articulated and placed on its right side with the face toward the entrance. The adult male cranium (Burial 5) closest to the back wall consists mostly of disarticulated bones that were stacked on top of each other with the mandible on the top. Three clusters of partly articulated subadult cranial bones were located between the two male crania. The three bone clusters included four subadults ranging from fetus/perinatal to 1.5 years at death.

The bone pile located at the center of the chamber includes the almost complete postcranial skeletons of the two adult males and between 10% and 50% of the postcranial skeletons of the subadults. All of the subadult skeletal elements identified within the bone pile matched up with the subadult crania found along the western arc of the chamber.

Most of the postcranial skeletons were disarticulated. However, the lower extremities of one adult were found in a position that strongly suggested that the tibia, fibulae, and patella were articulated by soft tissue at the time of burial. There were no signs of articulation between the proximal ends of the femora and the innominate bones (Figure 9.8).

There were no signs of repeated usage of this chamber. This observation is supported by the fact that all the articulated and disarticulated bone elements within the

Figure 9.8. View of chamber A110NW. Note partly articulated lower long bones in front and two clay figurines between the fragmentary skulls in the back.

bone pile matched up with the cranial elements along the western wall. All adult articulated and disarticulated elements were sorted into two individuals based on size and shape of each bone. Ribs and hand and foot bones could not be linked to a specific burial and were processed as a single entity representing two individuals. Except for some ribs, hand bones, foot bones, and both coccyxes, all bones are present. Of the ribs, 43 out of a total of 48 were identified. Hand and foot bones are remarkably well preserved, with 180 bones present out of a possible total of 212 representing two adults. The missing 32 bones are represented by phalanges from both hands and feet. The completeness of the subadults is not as good as in the adults; only 10% to 50% of the subadult bones are present.

Burial 1 is the skeleton of an adult male about 30 years of age. The age at death for this individual was based on the morphology of the auricular surfaces and pubic symphysis. Furthermore, the jugular plate is unfused, and the sacral bodies are incompletely fused. Dental attrition is mild to moderate and there is little enthesophytic development at the muscle attachment sites.

The sex characteristics for this individual are, in general, rather ambiguous. The pelvic morphology is more indicative of a male; however, the remains are more gracile than is typical for the males in this population. Of the numerous characteristics that were examined, about half are more similar to male and half to female. The pelvic morphology was the primary criterion for the determination of male sex.

Enamel hypoplasias indicate at least four stress events during the growth period. There is evidence of healed probable cranial trauma. The lesion is oriented vertically at the posterior portion of the right parietal. The margins of this lesion are smooth and well remodeled. A small circumscribed depression is present on the left frontal. The cortical bone within the depression is slightly rugose.

There is evidence of subchondral destruction and repair on the left humeral trochlear ridge. This lesion is unilateral, and there is no evidence of corresponding joint disease to the ulna. Evidence of unilateral activity related enthesopathy is present on the left lesser tubercle of the proximal humerus. The anterior bodies of T8 and T9 show lipping and slight deviation to the right side. The anterior margins of the superior end plates of L4 and L5 bear evidence of mild disk degeneration. There is postmortem deformation of the mandible. Bilaterally, the distal linea aspera show deeply excavated muscle attachment sites. This abnormality occurs bilaterally at the origins of the soleus muscle on the tibia.

Burial 2 consists of a 1.5-year-old child and is represented by 25–50% of the skeletal elements. The majority of the long bones and cranial bones are present. Costal and vertebral fragments are present in the chamber but cannot positively be attributed to this individual. A mandible and maxilla may be attributed to this individual, but the presence of another child of similar age and size (Burial 3A) precludes certainty.

Age at death was determined by the diaphyseal lengths of the long bones as well as dental development. The diaphyseal length measurements yielded an age-at-death estimate of 6 months. Given the similarity in size with Burial 3B and the presence of dental remains of 18 months, the age-at-death estimation was refined. There is very slight evidence of porosity bilaterally on the orbital roofs.

Burial 3A is the skeleton of a 1.5-year-old individual and is represented by both cranial and postcranial remains; 25–50% of the skeletal elements are present. A mandible, maxilla, and several other elements (vertebrae, ribs, and pelvic elements) could not differentially be attributed to either Burial 2 or Burial 3A because of similar size and age at death of each child.

The diaphyseal lengths of the long bones yielded an age at death of approximately 6 months. Dental remains, which could not positively be attributed to either Burial 2 or Burial 3A because of similar size and age, refined the age estimate to 1.5 years.

Bilaterally, the tibiae show slight anterio-posterior bowing, and the posterior cortical surface is pumice-like. Pores provide evidence of unmineralized osteoid. Postmortem damage has eroded most of the porous plaques of woven bone. This abnormality may be an indication of a metabolic disease such as rickets. Porous plaques of woven bone are present on the internal surface of the parietals and the occipital.

Burial 3B consists of a six-month-old infant and is represented by approximately 10% of the skeletal elements. Both cranial and postcranial elements are represented. The age-at-death determination was made using the diaphyseal length measurement of the left femur. The cranial development is consistent with this age estimate. No dental remains are present.

Burial 4 includes the bones of a late-term fetus with more than 25% of the skeletal elements, including cranial squama, present. Age at death was determined by using diaphyseal length measurements of the long bones as well as the development of the petrous temporal.

Burial 5 is a 25- to 30-year-old male with nearly complete presence of all skeletal elements. The cranium is almost complete. The basal cranium and midface are broken but present.

Age-at-death determination was made using the auricular surfaces and the pubic symphysis, providing an estimate of 25 to 35 years at the time of death. However, given the incomplete fusion of the annular rings, the mild dental attrition, the unfused jugular plate, and the slight evidence of enthesophytes, the age-at-death estimation was placed in the lower part of the age range. The skeletal elements do not exhibit the robustness that is more typical for the males in this population. However, the innominate morphology is indicative of a male.

Enamel hypoplasias indicate that there were at least three stress events during childhood development of the permanent dentition. The dentition displays crowding in the maxilla. This is most evident on the right side and is most obvious with the canine. Dental crowding is unusual in this population. The right M^3 was lost antemortem; the socket displays evidence of the early stage of remodeling. There is bilateral evidence of subchondral erosion and repair at the superior surfaces of the acetabula. Similar lesions are not observable on the femoral heads. Very mild scoliosis is discernable in the upper and mid thoracic vertebrae. The spinous processes deviate to the left and the anterior bodies to the right. The first sacral element is incompletely fused with the rest of the sacrum at the body and the neural canal. This may be evidence of incomplete lumbarization. Unilaterally, the left distal tibia displays a lesion on the medial aspect of the articular surface. This lesion appears to be the result of osteochondritis dissecans, a condition associated with joint trauma. The articulating talus could not be identified. There is a fracture callus on the right distal ulna that has the location typical of a parry fracture. The repaired cortical bone is encroaching on the medullary cavity. The diaphysis is damaged distal to the callus and therefore not present for evaluation. This abnormality appears to be unilateral.

Additionally, a fragment of a right rib shows a remodeled and irregular hyperostotic visceral surface. Some evidence of remodeling is also present on the external surface. This is likely the result of a healed fracture rather than a pleural infection. It was not possible to allocate this anomaly to either of the two adult males (Burial 1 and Burial 5).

Tomb A110, Chamber Southeast
The general layout and architecture is similar to the arrangement found in tomb chambers A110NE and A110NW. The axis extending from the center of the shaft and through the center of the chamber is 118° from north. The entrance was covered by three blocking stones sealed with mud mortar. The entrance measures 54 centimeters by 46 centimeters and has a wall thickness of 19 centimeters. The lower part of the entrance is approximately 9 centimeters higher than the lowest part of the shaft, and the chamber floor is between 30 and 40 centimeters below the entrance, thus 202 centimeters below the present surface. The chamber measures 172 centimeters from the entrance to the back wall and is 172 centimeters wide. The maximum height is 72 centimeters from the center of the chamber floor to the ceiling.

The floor of the chamber was covered by thin layers of silt, each a few millimeters thick and totaling 40 to 50 centimeters. The silt is formed by substantial amounts of water moving through the chamber over long periods of time, with the silt being added in layers from the chamber floor to the lowest part of the entrance. The presence of this type of silting is supported by two observations. All pots except for three smaller juglets were embedded in the silt. These three juglets had floated in the water and later settled in the top layers of the soft and liquefied silt. The juglets were found about 40 centimeters above the floor level, partly embedded in the silt (Figure 9.9). The second observation is that the pots that were embedded in the silt displayed multiple horizontal layers of silt that corresponded with the position of the pot within the horizontally layered clay matrix, thus reflecting multiple events of silt water and flow (Figure 9.10). Similar horizontal lines of silt deposits were found within the crania, which at the time of excavation were all located in the position typical of EB IA burials on the chamber floor to the left of the entrance.

The origin of the silt is problematic. The events creating the horizontal varve-like sedimentation could have taken hundreds of years and been the result of multiple events of water intrusion into the chamber. The time estimate by one sedimentologist (D. Stanley, pers. comm. 2006) is between 800 and 1,000 years. The water could have entered from the shaft through very small spaces between the three blocking stones. It could have penetrated through the soils between the surface and chamber ceiling. Or it could have penetrated through natural holes, or "heat cracks," found in some chamber walls. Similar conditions were observed in chambers excavated during the 1977 season (Ortner 1981). The hardness and high density of the silt made the excavation and retrieval phases very difficult. This

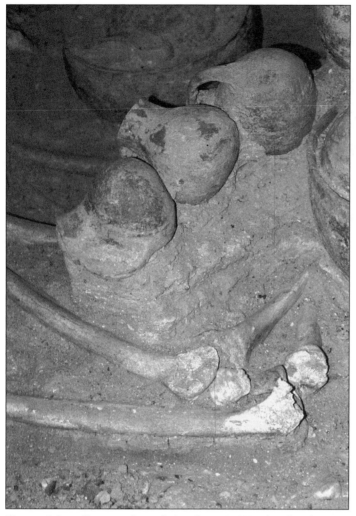

Figure 9.9. Close-up view of chamber A110SE. The chamber was silted about 50%. Note the articulated lower long bones and three pots "floating" in the silt.

was especially true for the smaller bones, such as those of the hands and feet.

This chamber contained ten individuals, including two articulated skeletons and two adults and six subadults represented in a bone pile. The crania were located to the left (southeast) of the entrance, and the majority of the cultural finds were found to the right (south and southwest) of the entrance and toward the back.

Two articulated skeletons, one male (Burial 1) and one female (Burial 2), were placed on their right sides and in a semiflexed position at the center of the chamber, with the crania to the left (east) of the entrance and lower extremities to the right (west). The lower extremities are bent approximately 80 degrees, with the knees toward the entrance. The upper extremities are bent, and the distal elements are positioned in the front of the faces (Figure 9.11).

Three crania (Burial 3, Burial 4, and Burial 5) are positioned to the left of the articulated skeletons (Burial 1 and Burial 2). A bone pile including disarticulated skeletal material is located south of Burial 2 and west of Burial 5. The three clusters of articulated and disarticulated cranial bones include the remains of two adults (one male and one female) and six subadults ranging from prenatal to approximately four years old.

The preservation is very good, even with the chamber badly silted prior to excavation.

Figure 9.10. Pots from chamber A110SE filled with silt (left) and with horizontal silt lines from floating (right).

Figure 9.11. View of chamber A110SE after the majority of silt had been removed. Note articulated bodies in front and "floating" pots at center/right.

Three of the four crania are largely intact, and the majority of the adult and disarticulated postcranial remains found in the central bone pile can be associated with the two adult crania found along the eastern wall. The subadult burials are also fairly well preserved. All remains were cleaned after removal; however, some silt remains in the marrow space and sinuses, particularly the maxillary sinuses.

Burial 1 is an articulated 30- to 35-year-old male skeleton. The body was placed in a flexed position on its right side, with the head toward the east and the face toward the entrance. The skeleton is almost complete, about 95+%, with only four cervical vertebrae, one thoracic vertebra, the sternum, the coccyx, and about 15 hand and foot bones missing. It was impossible to attribute hand and foot bones to each adult individual, thus such bones have been evaluated based on the total number of hand and foot bones represented by four adults. Thus a total of 60 hand and foot bones are missing, which would be an average of 15 for each of the four adult individuals. This approximates the number of missing hand and foot bones in both the northeast and northwest chambers. The recovery process during excavation for the silted chamber A110SE was as successful as in the two unsilted tombs.

The age at death was established by using the auricular surfaces and comparing the extent of dental attrition to other individuals within our sample. The majority of the classic sex criteria are indicative of a male, however, the greater sciatic notch is wider than typical for a male.

Caries and apical abscess drainage are evident. For example, the left M^1 lingual cusps are entirely destroyed by caries. Apical abscess drainage from chronic infection is associated with this carious lesion. The drainage pathway perforates the maxillary sinus wall. The maxilla itself is intact, thus the interior surface of the sinus cannot be evaluated for possible evidence of chronic infection. The left M_1 was lost antemortem, and remodeling was progressing at the time of death. Interproximal caries is present on the left M_3. It does not appear to have affected the left M_2. The right M^3 was lost antemortem. Apical abscess drainage is associated with this tooth.

Evidence of enamel hypoplasia indicates at least five events of stress affecting the teeth during development. There is evidence of unilateral chronic subluxation of the right temporomandibular joint (TMJ) with destruction of the mandibular eminence on the temporal bone (TMJ dysplasia). Cyst formation is evident, but lipping is not. The mandibular condyle is not present for observation. Two very slight circumscribed depressions are observable on the right posterior parietal. One depression is more circular and has very slight porosity at the center. The other is more oblong and shallow in the cortex and also exhibits slight porosity at the center. Neither communicates with the inner table of the cranium. The circular lesion may be the result of benign fatty/sebaceous, fibrous cysts, or old and remodeled depressions resulting from blunt force trauma. The oblong lesion is most likely the result of trauma. Enthesophytes and macropores are present bilaterally on the humeral tubercles. These are asymmetrical.

Lesions are present on the lesser tubercle (associated with subscapularis muscle) on the left humerus and on the greater tubercle (associated with supraspinatus, infraspinatus, and teres minor) on the right humerus. These enthesopathies indicate severe strain of the rotator cuff for the right and left arm. Enthesophytes are also evident bilaterally on the humeral epicondyles at the attachment for the common flexors. This may indicate habitual overuse of the common flexors. These enthesophytes are symmetrical.

The right distal radius exhibits a healed fracture, probably the result of accident rather than interpersonal violence. The distal end of the radius is damaged, which obscures much of the fracture. The fracture ends were not perfectly aligned. A minor spur probably was caused by mineralization of adjacent connective tissue—a condition known as myositis ossificans. The ulnar head exhibits some lipping that is likely associated with the trauma and repair. Bilaterally, the curvature of the ulna is more pronounced lateroposteriorly than is typical. This abnormality is more marked on the left ulna, which is also slightly shorter in the diaphysis than the right (about 5 millimeters). However, the radial curvature appears to be within the normal range.

A callus is present on the angle of a right rib that cannot be associated with a specific burial. The fracture appears well remodeled. Mild scoliosis is evident in the bodies of T1–T5 vertebrae that cannot be associated with a specific burial. The apophyseal joints on the right side of T3–T5 exhibit lipping, although the left joints appear normal. Postmortem damage to the spinous process of these vertebrae precludes the observation of deviation. The neural arch of the first sacral element shows evidence of trauma, possibly a herniation of the neural sheath. This kind of trauma would not cause tremendous dysfunction this low in the spinal column. Bilaterally, the surfaces and margins of the superior aspects of the acetabula are interrupted at the junction of the anterior inferior iliac spines. A small segment of the acetabular rims is partially separated from the subchondral surfaces. This may be the result of some kind of trauma around the time of pelvic fusion (12–16 years old). This expression is more marked on the left.

Burial 2 includes the articulated skeleton of a 25-year-old female and was located just south of and parallel to the articulated male skeleton (Burial 1). The body has been placed in a similar position to Burial 1. The individual is represented by both cranial and postcranial remains and is virtually complete in element presence, with better than 95% of the bones present. The cranium is intact and the postcranial remains are in a very good state of preservation. One cervical vertebra, the coccyx, and about 15 hand and foot bones are missing.

The age-at-death estimate was based on the auricular surfaces, the pubic symphysis, and the presence of some residual lines of epiphyseal fusion. Additionally, dental attrition is relatively moderate. The sex estimate is based on characteristic female attributes in the cranium, innominate bones, and the femora.

Dental anomalies and caries are present. The presence of enamel hypoplasia indicates that this individual survived at least three major stress events during development. Supernumerary teeth are present in the maxillary arcade. One supernumerary tooth originally located between the right lateral incisor and the canine was lost postmortem. As both the lateral incisor and the canine are accounted for, this socket indicates that either a second lateral incisor or canine was present or a deciduous tooth was retained. A tooth fitting the socket was not recovered among the loose teeth. The right M_3 exhibits intercuspal caries. This caries does not appear to involve the pulp chamber. The left M_3 is not present in the dental arcade. There does not appear to be evidence of remodeling, so it is possible that this tooth either did not erupt or is agenetic. The occluding M_3 shows no wear or polishing.

The right distal radius exhibits a very well-remodeled fracture. The location and alignment of the fractured segments is typical of a Colles' fracture usually associated with a fall. However, the realignment of the fracture segments is near perfect and the callus is very well remodeled. The presence of the healed fracture is not immediately observable, and there is minimal displacement of the two segments. This may indicate that this fracture was set and immobilized. The distal ulna is not present for observation. A small pit occurs unilaterally on the superior margin of the right acetabulum. The pit measures no larger than five millimeters in diameter and approximately a similar depth. The margins of the defect are smooth and rather rounded. The surface bone of the defect appears to be similar to subchondral bone. The etiology of this pit defect is unknown.

Burial 3 is the remains of a 35-year-old female. The individual is represented by both cranial and postcranial bones. The cranium was placed to the left of the entrance and the postcranial skeleton was placed in an orderly fashion in the bone pile mixed with postcranial elements from another adult (Burial 5) and various skeletal elements representing six subadults. Approximately 75% of the skeletal elements are represented. Seven cervical vertebrae, five thoracic vertebrae, the coccyx, and about 15 hand and foot bones are missing.

The age-at-death estimate was made using the auricular surfaces and pubic symphysis. The moderate dental attrition is compatible with this age estimate. The sex characteristics indicate a female. The bones of this individual are gracile, but with marked muscle attachment sites.

The expected weight of the skeletal elements is less than normal relative to other burials from this tomb. The completeness of the long bones precludes cross-sectional observation. Bilaterally, the surfaces of the acetabular notches are slightly thinner than is expected, which may indicate a decrease in bone density. Also, thin and translucent bone is present on the orbital roofs. The vertebrae were not observed for micro fractures associated with osteoporosis in the bodies as they are also not fragmented. The reduction in bone density is probably osteoporosis resulting from malnutrition.

Dental pathologies/anomalies are present. The presence of enamel hypoplasia indicates that this individual survived at least three events of stress that led to temporary growth cessation of the dental enamel. The buccal side of the left M^1 is completely obscured by a plate of calculus. The third molars have not erupted and may not have developed. The left M_2 was lost antemortem shortly after eruption and occlusion. The occluding left M^2 shows very mild polishing on the occlusal surface. A large caries on the mesial cusps of the left M_1 has destroyed much of the crown. Apical abscess drainage on the buccal side of the mandibular ramus is associated with the cavities. There is excess space in the mandibular arcade that has allowed for the rotation of the PM_4s bilaterally. There is very little evidence of enthesopathies with the exception of the left intratrochanteric notch. The pelvic inlet appears to be quite narrow for a female, and the sacral curvature is more marked than is expected. This morphology may have caused difficulty during labor.

Burial 4 is represented by the cranial fragments of a two-year-old child. It was located between Burial 3 and Burial 5. About 50% of the postcranial elements were found in the bone pile. The age-at-death estimate was based on a combination of the dental development stage and the diaphyseal lengths of the measurable long bones. The length of the left femur is within the 1.5–2.5 year range. The iliac breadth, however, measured in the range for 0.5–1.5 years.

The endocranial frontal and parietals exhibit abnormal woven bone plaques. On the frontal, the woven bone is largely focused within the convolutions stemming from pressure erosion. This expression is indicative of increased internal pressure but also possible evidence of localized nodules of infection that are bundled within the connective tissue (granuloma lesions). Tuberculosis is one disease that may result in these lesions. An infection affecting the dura could also explain these lesions.

A woven bone lesion is present bilaterally on the distal half of the medial tibiae. The woven bone is almost completely remodeled into normal cortical bone, which may indicate a periosteal insult from which the child recovered during infancy. This lesion is repeated bilaterally on the distal posterior humeri. Observation of the maxillary alveolus revealed increased porosity as well as some plaques of woven bone that are partially obscured by soil matrix. Some of this reactive bone may be related to the process of dental development and eruption or to a pathological process such as scurvy. Increased incidence of porosity is also present on the mental eminence and the lingual surface of the ascending ramus. The squamous portions of the sphenoids are not present for observation, and there are no lesions evident on the orbital roofs.

Burial 5 is from a 21- to 25-year-old male. The skeleton is completely disarticulated, with the cranial elements placed east of the bone pile and south of the crania of Burial 2 and Burial 4. All postcranial elements are found in the bone pile and represent approximately 90% of the complete number of elements. One scapula, one thoracic vertebra, the coccyx, and about 15 hand and foot bones are missing. All postcranial elements are placed in the bone pile in an orderly manner, but mixed with the adult postcranial elements of Burial 3 and the majority of the subadult skeletal elements. The estimate of age at death for this burial was determined using the pubic symphysis, fusion of the epiphyseal plate of the medial clavicle, and dental development and degree of attrition. This individual is markedly robust, and all of the sex characteristics are indicative of a male.

There is probable postmortem damage to the posterior left parietal. The damage is circular, with spalling on the endocranial table. However, because the cranium was fragmented and repaired, it is at least possible that this damage was perimortem and due to a blow to the head. There is some evidence of enamel hypoplasia, and the dental attrition is very slight. Unilaterally, the left lateral clavicle shows an exostosis projecting medially on the caudal surface from the margin of the acromioclavicular joint. Postmortem damage precludes observation of the acromial component of the joint. This abnormality may have been the result of trauma. The right humeral medial epicondyle exhibits an unusual groove between the trochlea and the epicondyle. This may be due to overuse of or trauma to the common flexors. This condition is unilateral. The coronoid fossa of the left distal humerus shows pressure erosion. The trabecular structures are visible, and the margin of the defect is smooth. This condition is unilateral and is likely activity related, due to hyperflexion of the elbow. The sacrum shows abnormal mediolateral curvature, deviating to the right. This probably is a developmental anomaly. There is no evidence of scoliosis in the other vertebrae. Schmorl's nodes are present in association with T6–T12. Additionally, the anterior margin of the superior end plate of T10 exhibits a lesion that may indicate trauma to the annulus fibrosis, perhaps a localized avulsion of the connective tissue. The inferior end plate of T9 is not affected. This may have occurred near the time of or during the same event that resulted in the Schmorl's nodes, as the amount of remodeling is similar.

Burial 6 depicts an approximately 50% complete skeleton of a four-year-old child. All bones were disarticulated. The cranial elements were found within the cluster of cranial fragments, including Burial 5 and other subadult individuals. The postcranial elements were predominantly found within the bone pile. Age estimate is based on dental development (3–5 years old) and the diaphyseal length of the long bones (2.5–3.5 years old). The breadth of the ilium measured within the range of 1.5–2.5 year olds. Dental development was weighted more heavily in the estimation of age at death.

Porous plaques of woven bone as well as increased porosity are observable in the maxillary and mandibular alveoli. Both locations of the condition may be associated with the active dental development and eruption. Increased porosity is also observable at the mandibular ascending ramus. Similarly, the etiology of this is unclear because of the active dental development.

The remaining four burials all represent subadults in the range from late prenatal to newborn. One fragmentary newborn cranium was found within two of the clusters along the eastern wall. This cranium cannot be positively attributed to any single burial. However, based on estimated age at death, it may be attributed to the postcranial remains of Burial 50, Burial 51, or Burial 52.

Cranial fragments exhibit the following pathologies. Layers of porous lamellar bone are present on fragments of the endocranial surface. These do not appear to be related to the normal developmental process and likely are the result of a pathological process. Possibilities include infection and scurvy; however, scurvy is unlikely because of the young age.

Burial 50 is the postcranial remains of a newborn with about 10% element presence. Age at death is estimated to between 40 intrauterine weeks and newborn, based on diaphyseal lengths of intact long bones. However, subadult age estimates based on long bone lengths tend to underestimate age in the EB IA skeletal sample.

Burial 51 is represented by normal postcranial remains, with approximately 10% element presence. Age at death is estimated to 38 intrauterine weeks and is based on the diaphyseal lengths of the intact long bones.

Burial 52 is a late-term fetus. Age at death is based on the diaphyseal lengths of intact long bones. Ten percent of the postcranial elements are present. There is no evidence of skeletal disorder.

Burial 53 is represented by a single left tibia. The diaphyseal length suggests an age at death of between two and three years. Remodeling woven bone is present on the posterior-medial diaphysis. This may indicate recovery from infection or trauma.

TOMB A111

Tomb A111 was located 8 meters northeast of Tomb A114 and 16 meters southwest of Tomb A113. The tomb was identified by a centralized, circular, high conductivity reading surrounded by significantly lower conductivity, suggesting unsilted chambers surrounding a shaft (Frohlich and Ortner 1982; Frohlich and Lancaster 1986). This condition proved to be partly true. Four chambers were associated with A111, of which three had been used for burials and one chamber had not been used. The three chambers including burials were all sealed off from the shaft by blocking stones and mud mortar and did not include any silt or deposit. The fourth and unused chamber did not include any blocking stones and was 95% silted (Figure 9.12).

The shaft refill was slightly lighter in color and less dense than the surrounding soils. No sign in the fill stratigraphy suggests that the tomb had been used more than once. The upper end of the shaft was located between 49 and 52 centimeters below the present surface, and the shaft had a depth of 258 centimeters from the present surface and 202 centimeters below the original surface. The upper shaft diameter is 129 centimeters, and the lowest shaft diameter is between 76 centimeters and 84 centimeters. The shaft fill is homogeneous and includes a few pottery and bone fragments. One large pottery fragment was found 137 centimeters below the present surface in a small alcove or step measuring 28 centimeters wide and 15 centimeters tall (Figure 9.13). The shaft extended between zero centimeters and 22 centimeters below the lowest parts of the chamber entryways.

Tomb A111, Chamber North

The chamber is circular with a round, dome-like ceiling. There is a minute amount of roof fall deposited on the chamber floor and within some of the pots. The entryway is rectangular, with well-defined round corners measuring 49 centimeters wide, 55 centimeters high, and 26 centimeters deep. The lowest part of the entryway is 10 centimeters above the base of the shaft. One large blocking stone measuring 53 centimeters wide, 67 centimeters high, and 7 centimeters thick covered the entryway. Sealing of the entryway was enhanced by added mortar, which succeeded in protecting the chamber from intrusive material from the shaft. The chamber measured 149 centimeters from the doorway to the back and 164 centimeters side to side, and the maximum height of the ceiling was 86 centimeters. The walls of the

Figure 9.12. View of the shaft of Tomb A111. Note the entrances to four chambers, with blocking stones covering three chambers only.

Figure 9.13. Large pot fragment found embedded in the shaft wall of Tomb A111, 137 cm below the present surface.

chamber are made up of a light brown matrix of dense clay mixed with pockets of gravel. No signs of tool marks were observed.

A total of 12 individuals have been identified. This includes four adults (two males and two females) and eight subadults, including four children and four infants. All remains are disarticulated, and contrary to all other chambers with undisturbed human remains, none of the burials in this chamber adds up to more than approximately 25% of the total number of elements in a complete burial (Figure 9.14).

Four crania are placed to the left of the entryway and parallel to the wall. This includes two adults (one male and one female) and two children (Figure 9.15). One of the crania is disarticulated, and some of the elements are stacked. Postcranial elements and disarticulated cranial elements of some of the subadults are placed in an unorganized bone pile at the center of the chamber (Figure 9.16).

The remains in this chamber were completely disarticulated. The elements that are present are relatively well preserved. However, element presence is poor, with less than 25% representation for most of the individuals. The majority of the crania are largely intact, with teeth present. One cranium (Burial 5) was badly damaged in antiquity and is represented by little more than fragmentary elements. This specific cranium was disarticulated and located within the central bone pile. The subadult remains are fairly well preserved. One additional adult was identified. The adult MNI is confirmed by the presence of four adult manubria. Age-at-death estimates for the subadults with dentition present show a discrepancy between the dental age and the postcranial osteometric age in which the dental age is older (Burials 1 and 2). Additionally, the Burial 1 cranium may be attributed to the postcranial material of either Burial 51 or Burial 52.

Burial 1 is a child with nearly complete cranial and dental material. The attribution of postcranial material is quite tentative and could include elements from either Burial 51 or Burial 52. The age at death is estimated to be three years, based on dental development. The osteometric long bone age determinations of both Burial 51 (ca. 2.5 years) and Burial 52 (1.5–2.5 years) are younger than Burial 1. Sex has not been established because of the young age at death. Bilateral cribra orbitalia is present. There is some postdepositional

Figure 9.14. View of chamber A111N. Note the unusual disorganized placement of disarticulated skeletal elements at the center of the chamber.

damage to both orbital roofs, so the severity of the lesions is obscured (Figure 9.17). We cannot determine if the lesions were active or remodeling at the time of death. The mental eminence and the maxillae show increased cortical porosity but no plaques of woven bone (Figure 9.18).

Burial 2 is an individual represented by dental, cranial, and postcranial remains. Approximately 15% of the skeletal elements are present. The attribution of the postcranial remains is based on similarities between age-at-death determination of cranial and postcranial elements. The age-at-death estimate is six to seven years and is based on dental development. The long bone osteometric age determination underestimates this. The lengths of long bones indicate an individual between 4.5 and 5.5 years old. No sex estimate is possible because of the young age. Evidence of mild bilateral cribra orbitalia is present.

Burial 3 is an adult female including mostly postcranial remains. The cranium is not fully intact, but the postcranial elements are well preserved. Approximately 25% of the elements are present, including a few vertebrae. The age-at-death estimate is between 30 and 35 years and is based on the auricular surfaces; it is consistent with a moderate dental attrition. This individual is female based on the standard anatomical characteristics. There is no evidence of skeletal disease in the bones available for analysis.

In Burial 3 there is bilateral evidence of maxillary sinus infection. The individual survived this condition, and remodeling is evident within the sinus, but it appears as though the infection perforated the posterior sinus. Porous reactive bone is present, with evidence of remodeling on the external surface of the maxillary sinus. The condition of the right side is more marked and less remodeled, suggesting an active disease process at the time of death. Lobular, porous bone formation is present within the sphenoid sinus. This also appears well remodeled. This may be related to the maxillary sinus infection. The right M_2 and possibly the M_3 were lost antemortem, and the alveoli are nearly completely remodeled. The left M_3 either was lost antemortem or was agenetic. The alveolus is damaged on the right so bilaterality cannot be determined. There are no data available in regard to enamel hypoplasia. The left parietal and left frontal exhibit lesions that may be the result of healed cranial trauma or subcutaneous cysts.

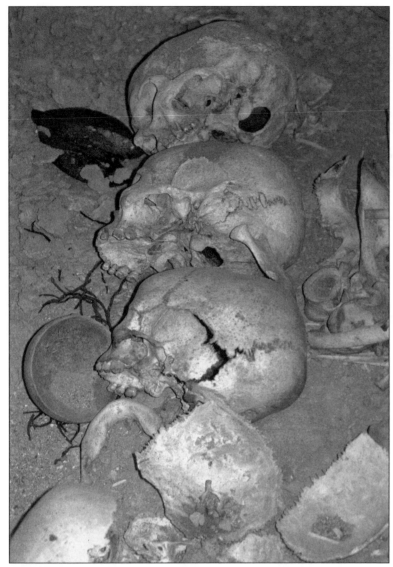

Figure 9.15. Close-up view of four skulls and crania located to the left of entrance in chamber A111N.

The costal facet for the left first rib on T1 shows deformation and lipping with cyst formation. The hyperplasia extends to the lateral inferior end plate of C7 and is associated with focal destruction and cyst formation. This is likely the result of trauma. Bilateral, small focal destructive lesions are also present on the inferior lateral end plate of T1. Additionally, the spinous process shows slight deviation to the left, which might be linked with slight scoliosis. There is evidence of degenerative joint disease or overuse trauma on the left glenohumeral joint. The glenoid shows some periarticular reactive bone formation and subchondral damage. The humeral head also has periarticular bone formation and enthesopathy of the lesser tubercle. The contralateral elements are not present for observation. The acromioclavicular joint appears to be unaffected.

Bilaterally, the anterio-medial margins of the radial heads exhibit cyst formation and repair. The radial tubercles are mildly enthesopathic.

Evidence of osteoarthritis is present on the left patella, with subchondral cyst formation and nodular bone formation as well as periarticular lipping. The distal femur is not present for observation. The right tibia shows moderate mediolateral bowing. Because the left is not present for observation, we cannot determine whether this is the result of a healed, poorly aligned fracture, rickets, osteomalacia, or another process.

Burial 4 is an adult male including cranial, postcranial, and dental remains. Approximately 20% of the skeletal elements are present and in good condition. The cranium is largely intact. The age-at-death estimate is between 35 and 44 years, based on the morphology of the auricular surfaces. The left portion of the pubic symphysis is available for observation, but was not used for aging, as it was affected by severe trauma. The dental attrition is consistent with an individual of this age. The standard sex characteristics are indicative of a male.

Cribra orbitalia is present unilaterally on the right orbit. The majority of the teeth, especially the molars, show significant attrition and destruction of much of the crown. Shoulder arthropathy is observable in the right shoulder complex. The left is not present. The right humeral head shows nodular subchondral depositions. This is especially marked on the anterio-medial aspect of the head. The lesser tubercle is enthesopathic. Unilaterally, the left lateral clavicle exhibits exostoses. The glenoid fossa shows cyst formation outlining the posterior margin and exhibits subchondral bony projections. The acromion process is not present for observation. This arthropathy is likely related to a traumatic event. A remodeled fracture is present on the left distal radius. Some anterio-posterior deviation from the normal axis is present. The distal ulna is not present for observation. L5 exhibits complete spondylolysis of the lamina, which is not present. The anterior margin of

Figure 9.16. Center bone pile in chamber A111N. Disarticulated and incomplete skeletons of 12 individuals were identified.

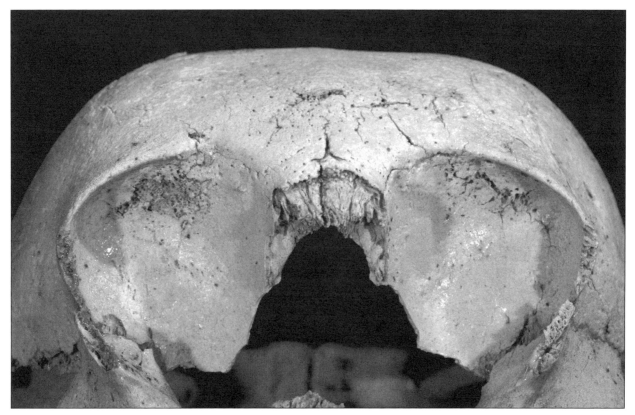

Figure 9.17. Case of bilateral cribra orbitalia. Chamber A111N, Burial 1, representing a 2.5-year-old child.

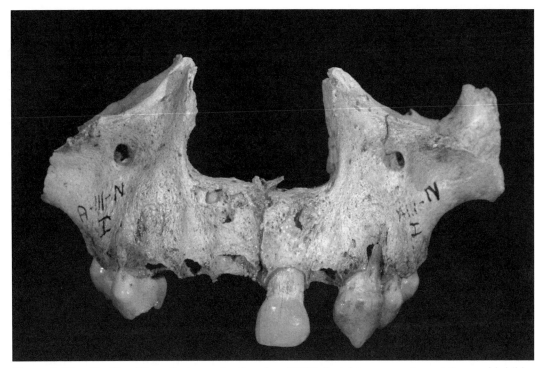

Figure 9.18. Maxilla with cortical porosity. Chamber A111N, Burial 1, representing a 2.5-year-old child.

the inferior end plate shows evidence of syndesmophyte formation and possible rupture of the annulus fibrosis. A perforating lesion located at the center of the interior end plate of L3 may indicate the presence of a compression fracture with callus formation. This may also be indicative of herniation of the intervertebral disk. A cluster of three lytic lesions of unknown etiology are present on the superior end plate of this vertebra. The left os coxae exhibits a well-healed fracture with morphological deformity and enthesopathies. The fracture compromised both the ilio-pubic and ischio-pubic rami. The anterior portion of the bone is less angled than is normal, and the rami are shortened. The pubic symphysis is deviated from the normal orientation and is at a 90° angle on the dorsal surface of the pubis.

The right os coxae is not present. The force causing this degree of deformation would have been quite significant, and it is likely that the right os coxae would have shown involvement. Bilaterally, the patellae show cyst formation and subchondral nodular bone formation extending beyond the surface. Periarticular lipping is also present. The right distal femur displays periarticular lipping, but no subchondral damage is evident. The left distal femur is not present.

Burial 5 is an adult female represented by cranial and mandibular fragments and possibly a manubrium. Some postcranial material attributed to Burial 3 may be attributable to this burial. The age estimate is adult, but not enough information is available to allow a more specific estimate. This is based on the absence of most characteristics that are indicative of age at death. Also, the dental matrix is inconclusive. The sex estimate is female, based on most elements showing a gracile individual; the mandible is morphologically feminine.

A remodeled lesion is present on the endocranial surface of a possible frontal fragment. Reactive bone is present on the ectocranial surface directly overlaying this lesion. It is possible that this was induced by a blow to the head resulting in an endocranial hematoma. The dental attrition is slight. The M_3 is not observable on the fragmented segment of the right mandible. This may be indicative of agenesis, antemortem tooth loss, or immature dental development.

Burial 51 is represented by a right femur that may be attributable to the Burial 1 cranium. The age-at-death estimate is 2.5 years and is based on long bone lengths. Sex is unknown. No pathologies are present.

Burial 52 is a young child and includes the majority of the major long bones. It is possible that some of the elements are attributable to Burial 1. The age at death is between 1.5 years and 2.5 years and is based on the length of long bones. Sex is unknown. Bilaterally, the tibiae display anterio-posterior bowing and the left femur exhibits slightly increased curvature. This may be an indication of rickets.

Burial 53 is a young child represented by approximately 10% of the postcranial remains. The age at death is 2.5 to 4.5 years, based on long bone lengths. The sex is unknown. No pathologies were observed.

Burial 54 is an infant and includes a pair of femora and a pair of humeri. Additional cranial and postcranial remains are present but cannot be positively attributed. The age-at-death estimate is between perinatal and six months and was made using long bone lengths. The sex cannot be determined.

Bilaterally, the tibiae show plaques of woven bone on the anterio-medial side of the diaphyses. Unilaterally, the left femur shows a localized lesion of woven bone on the anterior diaphysis. Scapulae are present, with lesions of periosteal woven bone on the infraspinous and supraspinous fossae. These periosteal lesions may be indicative of scurvy or another systemic disorder such as infection.

Burial 55 is an infant and includes humeri, femora, and tibiae. Additional cranial and postcranial remains also may be present but cannot be positively attributed. The age-at-death estimate is perinatal to six months old and is based on long bone lengths. No sex determination is possible because of the young age. Circumferential woven bone plaques are present bilaterally on the femoral and humeral diaphyses. The anterio-medial surface of the tibial diaphyses also displays woven bone plaques. Scapulae are present with lesions of periosteal woven bone on the infraspinous and supraspinous fossae. These lesions are indicative of a systemic disorder. Given the distribution of the lesions, scurvy is the most likely diagnosis, but infection is also possible.

Burial 56 is an infant represented by some postcranial elements. Additional cranial and postcranial remains were recovered but cannot be positively attributed. The age-at-death estimate is perinatal to six months and is based on long bone lengths. Sex is unknown. The left tibia exhibits a plaque of woven bone on the anterio-medial diaphysis. The right tibia is not present for observation. Bilaterally, the humeri show periosteal woven bone on the posterior distal region of the diaphysis. Scapulae are present with lesions of periosteal woven bone on the infraspinous and supraspinous fossae. These lesions are the result of a systemic disorder such as scurvy or infection.

Burial 57 is an infant and includes humeral, tibial, and femoral elements. Additional cranial and postcranial remains were recovered but cannot be positively attributed. The age-at-death estimate is between perinatal and six months and is based on long bone lengths. Sex cannot be determined for this individual, and no pathologies were observed.

Burial 58 is an adult male identified during the laboratory analysis of the burials excavated in 1981. It is represented by less than 10% of the postcranial elements. The age at death is between 30 and 39 years based on morphology of the auricular surfaces and pubic symphysis. In addition, the epiphysis of the medial clavicle is still distinguishable. The skeleton is male, based on pelvic morphology and the robusticity of the elements. Bilaterally, the acetabula are rather shallow. The superior margin of the right acetabulum displays a pressure erosion notch. This suggests that the joint was loose, perhaps with slight subluxation but not dislocation.

Unattributed pathological skeletal elements include the left greater wing of a subadult sphenoid that displays pathologically increased porosity. This is especially marked in and surrounding the foramen rotundum. The right side could not be identified. The abnormality was probably associated with scurvy. Two adult male left ribs are present with remodeled fractures. The spinous processes of unattributed C6 and C7 deviate slightly to the left. This may indicate mild scoliosis.

Tomb A111, Chamber East

The chamber floor is almost circular, and the chamber has a dome-like ceiling. The square entryway measures 39 centimeters wide, 61 centimeters high, and 29 centimeters deep. The lowest part of the entryway is 22 centimeters above the base of the shaft. The entryway is sealed from the shaft by about 11 smaller blocking stones held together by extensive use of mud mortar and smaller rocks. The blocking stones kept silt and fill from the shaft out of the chamber. The walls included a light brown and very dense clay matrix with a few pockets of gravel and marl. A few tool marks were visible in the isolated marl pockets.

The chamber was not silted but had some pieces of ceiling fall especially along the back wall and partly covering one of the articulated skeleton's foot bones. Also, some roof fall was found within the standing pots. The chamber measures 163 centimeters from the entryway to the back wall, 190 centimeters from side to side, and 89 centimeters at its maximum ceiling height.

Five burials were recovered. This includes two adults (one male and one female) and three subadults. The two adult skeleton are both articulated and placed on their right sides with the heads to the left of the entryway (northwest of the entrance; Figure 9.19). The male skeleton (Burial 3) was closest to the back

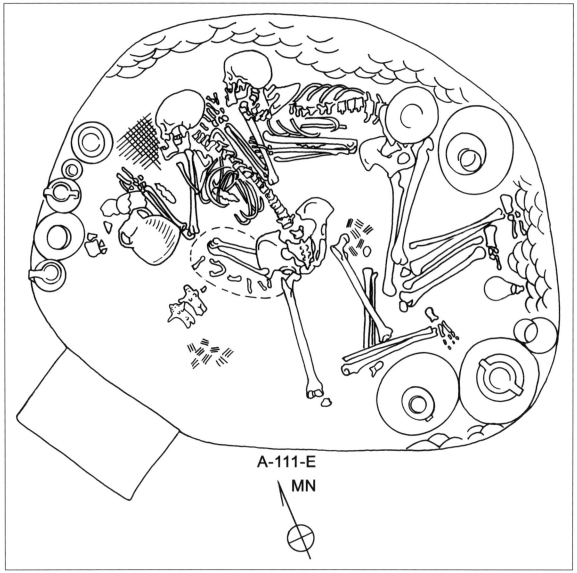

Figure 9.19. Plan drawing of Chamber A111E. The chamber includes one adult male, one adult female (both articulated), two disarticulated infants, and one disarticulated fetus.

wall, and the female skeleton (Burial 2) was placed at the center of the chamber parallel to the male skeleton. The lower extremities of both articulated skeletons were in a semiflexed position at the center and to the right of the entryway and with the knee regions toward the entryway. The upper extremities of the male were flexed, with the hands placed in front of the face. The left upper extremity of the female was bent with the hand in front of the face. The female's right upper extremity was partly disarticulated from the body and placed next to the female's right innominate bone, with the ulna's proximal end articulated with the humerus' distal end, and with the radius disarticulated from the ulna and placed partly below the right innominate bone. The distal end of the ulna was resting on top of the innominate bone (Figure 9.20). The location of individual elements of Burial 2, especially ribs, vertebrae, and some of the bones of the extremities, suggest that this burial was less articulated compared to Burial 3 at the time they were placed in the chamber. Apparently, disarticulated elements were positioned in what was assumed to be anatomically correct position. With some misplacement, it is clear that the cause was an error during interment and not a function of a natural displacement (Figure 9.21).

The cranial elements of two subadults were located between the left side of the entryway and the female's cranium. Both were disarticulated and partly stacked. The disarticulated bones of three subadults were placed in a small bone pile between the female

Figure 9.20. View of chamber A111E. Two clay figurines are located between articulated skeletons and the entrance way.

Figure 9.21. Close up view of A111E showing the disarticulated subadult bone pile. Note the unusual location of lower arm bones from an adult female.

skeleton and the entryway. Part of the bone pile was covered by skeletal elements from the female. Based on the location of skeletal elements, the male skeleton (Burial 3) was the first burial to be placed into the chamber, followed by the disarticulated skeletons of three subadults (Burials 1, 4, and 5), and last by the female (Burial 2).

Bone preservation in this chamber is very good. The adults, in particular, have nearly complete skeletal element presence, although with some damage. Most of the damage is due to the fragile, chalky quality of the bone. The crania are largely intact; repair of those that were broken was easily done in the laboratory. The juvenile skeletal elements are also well preserved, but this varies between the burials. One subadult has an element presence in excess of 75%, including many fragile cranial elements. Another burial has slightly less extensive preservation quality, but is still well preserved and also provides reasonably good evidence in terms of skeletal pathology. Surprisingly, a few small and very fragile remains of a midterm fetus are also present, including parts of the cranial squama.

Burial 1 is an infant and includes cranial, dental, and postcranial remains with 75+% element presence. Vertebral, costal, manual, and pedal elements are also present in the chamber but could not be positively attributed to a specific burial. The age-at-death estimate for this individual is from newborn to six months. This is based on length measurements of long bones, dental development, and skeletal development. The metrics obtained from the ilium underestimate the age at death as compared with the other methods. The sex cannot be determined.

Extensive abnormal porosity in the skeletal elements suggests that this infant suffered from scurvy, although this diagnosis is unusual in an infant younger than six months of age. This may be indicative of maternal prenatal metabolic deficiency. In an infant of this age, it is sometimes difficult to isolate pathological processes from normal growth and developmental changes in the bone tissue. Bilaterally, the tibiae exhibit porous lesions along the diaphyses. Some lesions are plaque-like and appear more active, and others appear more as abnormal porosity in the cortical complex. Also, bilaterally, the humeri show the development of porous plaques of new bone formation at the posterior distal half of the diaphysis. All of the long bones have metaphyseal porosity that extends beyond the porosity associated with normal growth. Abnormal porosity is also present on the scapulae at the fossae and, most diagnostically, on the greater wing of the sphenoids, which is probably a reaction to bleeding from abrasion of the underlying weakened blood vessels by the temporalis muscles. The gluteal muscle attachments on the posterior femora are quite marked. This could be the result of the strain of the tightly flexed position often seen in scorbutic subadults. Periosteal lesions are also present at the ascending rami of the mandible. This porosity is more extensive than normal and is probably additional evidence of scurvy.

Endocranial lesions are extensive in this individual and vary in terms of stages of activity and remodeling. The lesions include fine porosity found on the endocranial frontal, thickened and striated woven bone deposits in the sigmoid sinus, and elevated and sinuous lamellar deposits on the endocranial parietal and occipital, which show some remodeling but also appear to be active. At least one lesion of lamellated porous periosteal reactive bone was observed externally on a parietal fragment. These were likely active lesions at the time of death. These lesions are most likely attributable to metabolic or infectious disorder.

Burial 2 is an adult female and very nearly complete. The skeletal elements are very well preserved, with the exception of a small amount of postmortem damage. The cranium is intact. The age-at-death estimate is between 30 and 40 years and is based on auricular surface morphology (phase 3) and the pubic symphysis (Suchey-Brooks phase 3–4). There is very little enthesophyte development. Dental attrition is slight to moderate, cranial sutures are open, and the subchondral surfaces appear normal. The sex is female, based on very gracile skeletal elements and typical female morphological characteristics, especially in the pelvic bones.

Bilateral enthesopathies of the greater tubercle of the humeri indicate unusual strain on the muscles of the rotator cuff complex. A small septal aperture is present unilaterally on the left humerus.

Dental caries, antemortem tooth loss, abscess drainage, and possible cysts are apparent in the dentition. Bilaterally, the M_2s were lost antemortem as well as the left M_3. Remodeling is nearly complete for the right M_2 and left M_3 sockets. A large nearly spherical cavity is present in the place of the left M_2 that may be evidence of a cyst. The margins are sharp, and remodeling occurred. The crowns of the right third molars, both mandibular and maxillary, have been destroyed by caries. The root of the right M^3 perforates the buccal alveolus. The appearance of the margin indicates that this occurred near the time of death. The right M_3 shows evidence of apical abscess drainage. The left M^3 shows little attrition. It is likely that the occluding molar was lost soon after occlusion occurred. The ligamentum flavum is enthesophytic in the lower thoracic vertebrae. The rest of the vertebrae do not exhibit this. This may indicate that this region of the vertebral column had abnormal mechanical stress. Enamel hypoplasias indicate at least three stress events during dental development.

Burial 3 is a nearly complete adult male. Most of the fragmented cranium has been repaired in the laboratory, and a few of the postcranial remains are damaged. The age-at-death estimate is between 30 and 39 years and is based on the morphology of the auricular surfaces. The dental attrition is moderate. Some enthesopathies are present. This individual is rather slight for a male of this population, but anatomical characteristics indicate a male burial.

Dental caries, antemortem tooth loss and abscess drainage, and attrition anomalies are all found in this burial. The left M_2 exhibits a buccal apical abscess that destroyed the alveolus between M_2 and M_3. The roots were in contact at the time of death. The segment of alveolus that is associated with the right M_1 is recessed and exposing the mesial root of that tooth. The crowns of the maxillary M_2s show significant wear posteriorally, which is likely due to the early antemortem tooth loss of the mandibular M_3s. Remodeling of the sockets is complete. Interproximal caries is present on the posterior of the right M^2. The posterior buccal root of the left M^2 is exposed due to abscess drainage and recession of the alveolus. The right M^1 was lost antemortem, with apical abscess drainage buccally and lingually. Apical abscesses are associated with all three roots of the left M^1. The lingual drainage site is the most marked. Attrition exposing the pulp chamber may have caused this. Woven bone plaques are present at the drainage site buccally, indicating that infection was active at the time of death. An apical abscess is associated with the left I^2. The pattern of attrition on the maxillary incisors is unusual for this population. The anterior surfaces of the crowns are worn.

Enamel hypoplasias are present and give evidence of at least five episodes of stress during dental development. The earliest event appears to have been the most severe. A supratrochlear exostosis is present unilaterally on the right humerus. A unilateral subchondral lesion with remodeling on the anterio-medial margin of the left radial head may indicate osteoarthritis. The left humerus is not involved.

Two right and three left ribs exhibit remodeled shaft fractures. One right rib exhibits a periosteal lesion in the visceral surface. This is usually associated with inflammation of the pleura. No other ribs had similar lesions. Visceral surface lesions on the ribs are most often associated with tuberculosis, but there is no other evidence of this disorder in this burial. Evidence of disk degeneration is present between L5 and S1. The right diarthroidal joint of L4/L5 is destroyed, with some reactive bone growth. As this is isolated in one diarthroidal joint, this may be evidence of traumatic osteoarthritis. L2 and L3 are ankylosed circumferentially at the body. The segments of the body within the neural space are not affected, nor are the diarthroidal joints. The disk space was beginning to be compromised at the time of death. The anterior superior body margin of T1 exhibits some macroporosity. This is indicative of disk disruption.

Burial 4 is an infant represented by cranial, dental, and postcranial remains. Approximately 70% of the skeletal elements are present and in very good condition. The age-at-death estimate is perinatal and was determined using the osteometrics of the measurable long bones. The metrics of the ilium underestimate the age at death compared with the long bones and dental developmental. Sex has not been established because of the young age of the individual.

There is abnormal bone development on the squamous portion of the right temporal. Thickened spicules of woven bone replace normal bone posterior to the zygomatic root. This occurs unilaterally. The etiology is unknown. A lesion consisting of a plaque of woven bone is present bilaterally on the superior surface of the clavicles. This lesion is more marked on the left. Possible causes include trauma, as the clavicular shafts are vulnerable.

Burial 5 is a fetus that includes the cranial squama, a left humerus, and left ilium. It is quite surprising that these were preserved, given the age of the individual. This age-at-death estimate is between 22 and 24 gestational weeks and is based on the length of the humerus and the very well-preserved left ilium. A determination of sex could not be completed, and no pathology was noted.

Tomb A111, Chamber West
The western chamber in Tomb A111 included secondary burials with disarticulated and partly articulated burials. The chamber was unsilted and protected from silt and shaft fill by seven blocking stones held together with extensive use of mud mortar. The blocking stone measured 44 centimeters wide, 85 centimeters high, and 17 centimeters thick. The roughly square entry-

way measured 45 centimeters wide, 55 centimeters high, and 16 centimeters deep. The lowest part of the entryway was located at the same level as the base of the shaft and was lower than the entryways of any of the other three chambers.

The unsilted and almost circular chamber measured 153 centimeters from entryway to back wall, 152 centimeters from side to side, and 83 centimeters at its maximum height. Some roof fall was present along the walls, and some was found within the standing pots, but very little was found among the skeletal elements. The walls consisted of a light brown, high-density clay matrix mixed with isolated pockets of gravel and marl (Figure 9.22).

Seven burials were found in this chamber. This includes five adults (three females and two males) and two subadults. One adult male (Burial 1) was located closest to the entryway, set in a partly articulated position, with the head to the left and lower extremities to the right. Most of the elements were articulated, however, the positions of the extremities, vertebrae, and ribs strongly suggest that the body was placed in its present position by arranging articulated and unarticulated elements in anatomical position (Figure 9.23). Four skulls (three adults and one child) were placed to the left of the entrance in a line parallel to the wall. A well-organized bone pile was located between the articulated burial (Burial 1) and the back wall. The lower legs of Burial 1 were bent approximately 90°, which placed them between the bone pile and the wall to the right of the entrance. The bone pile was well organized, with most long bones in a left to right position (or parallel with the body of Burial 1).

Nearly all of the elements from this chamber are intact. Almost complete sets of hand, foot, costal, and vertebral elements are present for four of the five adults. One additional adult was discovered and is represented by an isolated left radius. The subadult remains are also well preserved. There is no silt infiltration from the primary burial site in any of the remains. An unknown crisp, black, layered paper-like substance is adherent to the inferior ventral aspect of the sacrum attributed to Burial 4. The postcranial elements for the adult burials have been attributed to a specific burial with relatively high confidence.

Burial 1 is an adult female represented by a nearly complete skeleton. The cranium is intact and the majority of the postcranial elements are present. The age-at-death estimate is between 25 and 30 years and is based on the morphology of the auricular surfaces and the pubic symphysis. Dental attrition is consistent

Figure 9.22. View of chamber A111W. Chamber includes a total of five adults and two subadults.

Figure 9.23. Close-up view of skulls, crania, and bone pile in chamber A111N. Apparently, an attempt was made to place some bones in a correct anatomical position.

with this estimate; it is slight, and the polishing on the third molars is minimal. The remains are gracile, and the majority of the sex characteristics, including pelvic and cranial morphology, are indicative of a female.

Caries occurs on the posterior cusps of the left M_3. There is evidence of at least four events of stress during dental development as indicated by enamel hypoplasia. Bilateral cribra orbitalia is present. The porosity is more marked on the left orbit. Enthesophytic lesions found bilaterally but asymmetrically on the humeri are indicative of stressful use of muscles. The lesion on the left is on the lesser tubercle, and the lesion on the right is on the greater tubercle.

T6–T8 vertebrae exhibit hyperostosis of the anterior longitudinal ligament on the right side of the body. T7 has a Schmorl's node on the inferior plate. The uncinate process at C3 is degenerated, with a corresponding lesion for the C2 articulation.

The transverse ligament on the right acetabulum is enthesophytic; this is unilateral. Subchondral destruction is present on the inferior and superior horns of the acetabular surface bilaterally. Active plaques of woven bone are present bilaterally and symmetrically on the distal tibiae and fibulae. Only the anterior portions are unaffected. There is evidence of some remodeling. The cause of this new bone formation is unknown but indicative of a systemic disorder, and infection is possible. Partially remodeled bone is present bilaterally on the fibular midshafts. The healed lesions are largely situated posteriorly. These do not appear to be healed fractures.

Burial 2 is a child represented by most of the postcranial elements, as well as cranial and dental remains. The age-at-death estimate is three years and is based on dental development and long bone measurements. The age estimate based upon femoral length is between 1.5 years and 3.5 years. Sex cannot be established.

Bilateral deposition of woven bone on the endocranial surface of the greater wing of the sphenoids is probably indicative of a chronic low-grade infection; there is very little evidence of remodeling. Woven bone is also present on the internal cranial vault with an occipital focus. These lesions were remodeling at the time of death. Cribra orbitalia is present bilaterally and the lesions are mixed, that is, active and remodeling. Abnormal porosity is also present on the orbital floor and continues to the external maxilla in proximity to the infraorbital foramina.

The mental eminence also exhibits abnormal porosity. Woven bone is present bilaterally on the internal surface of the ascending ramus of the mandible. The postcranial remains exhibit no active or observable pathological process except well-remodeled woven bone on the lateral left tibia. The right is not present for observation.

Burial 2A is an infant represented by most of the major postcranial and cranial remains. The age at death is newborn, based on long bone measurements. The dental remains are not present for observation. Sex is unknown. Plaques of woven bone are present bilaterally on the scapula. The lesions are circumspinal and on the right subscapular fossa. No lesions were noted on the left subscapular fossa. Postmortem damage obscures the area where abnormal woven bone might have been present on the lateral shaft of the left femur.

Burial 3 is an adult female represented by cranial, dental, and the majority of the major postcranial elements. Although the cranium largely is complete, it was broken. The age-at-death is between 30 and 40 years. The estimate is based on morphology of the auricular surfaces and pubic symphysis. The dental attrition is slight. The elements attributed to this burial are gracile. The pelvic and cranial morphology are indicative of a female.

The left M^1 shows antemortem tooth loss with complete remodeling. Cervical vertebrae indicate healed trauma primarily affecting C5 and C6. The inferior bodies are fractured and displaced superiorly on the right anterio-lateral body with focal compression. There is some deformation and reactive hyperostosis. The right uncinate process of C3 and the articulating C2 are also deformed. Remodeling woven bone is present bilaterally on the medio-posterior midshafts of the fibulae. The tibiae are unaffected. The long bones are light in weight relative to those from similar burials.

Burial 3A is an adult represented by an isolated left radius. The single radius does not match up with any of the other burials in this chamber and has been recorded as belonging to one additional individual. The age-at-death estimate is nonspecific adult, since there are no specific aging characteristics that are observable. However, it is likely that this individual is a young to middle-aged adult, as the articular surfaces are normal and the cross-sectional cortical thickness is very good and dense. The radius is quite gracile and most likely belongs to a female. No pathology was observed.

Burial 4 is a young adult male including the majority of the skeletal elements. The cranium is intact and the elements are well preserved. The age-at-death estimate is between 19 and 21 years and is based on the degree of epiphyseal fusion. The epiphyseal lines are visible on the femoral and humeral heads indicating incomplete fusion. The medial clavicle epiphysis is beginning to fuse, and the iliac crests and ischial rami are also incompletely fused. Elements are robust and the pelvic morphology is indicative of a male.

This individual exhibits possible agenesis of all third molars. The canines, mandibular and maxillary, exhibit enamel defects. These appear to be coalesced enamel hypoplasia. Malocclusion of the left central incisors is indicated by a wear facet on the anterior surface of the left I^1 cusp. Bilateral caries are present on the posterior interproximal surface of the M_1s. Cribra orbitalia is present bilaterally. Bilaterally, the retrocondylar foramina are obliterated by bone formation. Woven bone is present bilaterally on the posterior-medial midshaft of the fibulae. This reactive bone appears to be remodeling. The left fibula appears to exhibit anterio-posterior bowing, but this may be taphonomic rather than pathological.

Burial 5 is an adult male including the majority of the cranial and postcranial elements. The elements are well preserved and the cranium is intact. The age-at-death estimate is between 30 and 40 years and is based upon the morphology of the auricular surfaces and the pubic symphysis. Dental attrition is moderate to severe, and the left first molars show a significant destruction of the crown. The postcranial elements are robust and have a classic male pelvic morphology.

This individual exhibits possible agenesis of the third molars. The right M^1 was lost antemortem. The alveolus at the sockets and buccal and labial abscesses were actively remodeling at the time of death, indicating recovery from the abscess following the loss of the tooth. Active woven bone is also present on the palate. The presence of at least 5 lines of enamel hypoplasia indicates at least five events of stress during dental development.

Bilateral cribra orbitalia is present. A unilateral remodeled depression on the right parietal may have been caused by cranial trauma. Bilaterally, the radial heads exhibit lesions caused by subchondral destruction and repair. L4 exhibits complete spondylolysis of the lamina; the lamina is present for observation. L2 and L3 have enthesophytes on the spinous processes and disk degeneration accompanied by syndesmophytes at the anterio-superior margins of the bodies.

Unattributed pathologies and anomalies include a total of five left ribs with fractures that were in the early stages of repair at the time of death. The process of remodeling is at a similar stage for all of the ribs, and it is likely that they can be attributed to the same individual and represent a single traumatic event. Five additional ribs exhibit well-remodeled shaft fractures. These may be from the same individual.

Tomb A111, Chamber South

No blocking stone or stones were found associated with this chamber's entryway (Figure 9.12). A careful search of the excavated shaft silt, especially from the lower part of the shaft, did not yield any signs of possible fragmentary or smaller sections of blocking stones or mortar. This suggests that the tomb was constructed ahead of time and only three chambers were needed at the time of interment.

The chamber was extensively silted. About one-third of the silt was derived from the shaft fill. Observation of silt intrusion into chambers through entryways with no blocking stones shows that only 15% to 30% of the chamber may have been filled with sediments from the shaft. The intrusion may have come to a halt when the shaft refill reached the upper part of the entryway. This is supported by the presence of layered varve-like sediments constituting the upper parts of the chamber fill.

The entryway measured 47 centimeters wide, 63 centimeters high, and 26 centimeters deep. The chamber is circular and with a domed ceiling corresponding in size and shape to all other chambers recorded during the 1981 season. No metric data were obtained from this chamber. No human remains or burial objects were found in the chamber.

TOMB A112

The A112 Tomb was identified by high conductivity readings, suggesting the presence of a shaft. The difference between the high conductivity of the shaft and the reading in the surrounding area was similar to those found in Tombs A110 and A114. Tomb A112 provided us with the best geophysical data, including conductivity, interpretation, and results (Figure 9.24). The conductivity patterns indicated that the shaft was surrounded by four chambers, of which two, those located to the east and west, would be silted and two chambers, to the north and south, would be less silted or not silted. This proved to be correct. The shaft was surrounded by four chambers, of which A112N and A112S were about 50% silted and A112E and A112W were about 90% silted. Unlike all other tombs excavated in the A Cemetery during the 1981 season, the A112 shaft and chambers were cut into a matrix of clay and calcium carbonate or marl.

The tomb had been robbed. There were no signs of robbery at the surface. However, it soon became clear that the shaft we had identified was a thieves' shaft dug within the diameter of the original shaft. Accordingly, we identified two fills: the smaller centrally located thieves' fill, mixed with pottery fragments and bone fragments, and the original fill outside the robbers' access shaft, which showed characteristics similar to those found in the undisturbed fills of Tombs A110 and A114.

The metrics of the thieves' access shaft would include somewhat lower values, especially for the horizontal distances. We are presenting only the original shaft tomb metrics. The upper end of the shaft was found 50 centimeters below the present surface. The maximum and minimum upper diameters were, respectively, 172 centimeters and 152 centimeters. The maximum depth of the shaft was 151 centimeters from the original surface and 201 centimeters from the present surface. The maximum and minimum diameters at the base of the shaft were, respectively, 166 centimeters and 142 centimeters. The upper portions of the entryways were about 60 centimeters below the original surface and 110 centimeters from the present surface. The shaft's base extended downward 90 centimeters below the lowest margin of the four entryways.

The marl layers begin 90 centimeters below the present surface and about 40 centimeters below the original surface. The soil matrix between the present surface and the beginning of the marl includes a mixture of hard-packed clay and gravel. Metric dimensions of the four chambers are similar to those found in A110, A111, and A114. The maximum width (side to side) ranges from 152 centimeters to 183 centimeters, the maximum length (entryway to back of chamber) from 151 centimeters to 180 centimeters, and maximum height from 80 centimeters to 114 centimeters. No blocking stones were found in situ.

Large numbers of fragmentary human bones were found mixed with the silt from the shaft emptying into the chambers when the shaft was filled in. The thieves successfully accessed all four chambers. We argue that none of the chambers were silted at the time of the robbery. All of the silt found in the chambers was similar to the silt found in the shaft, thus the chamber silt had all originated from the shaft and the result of

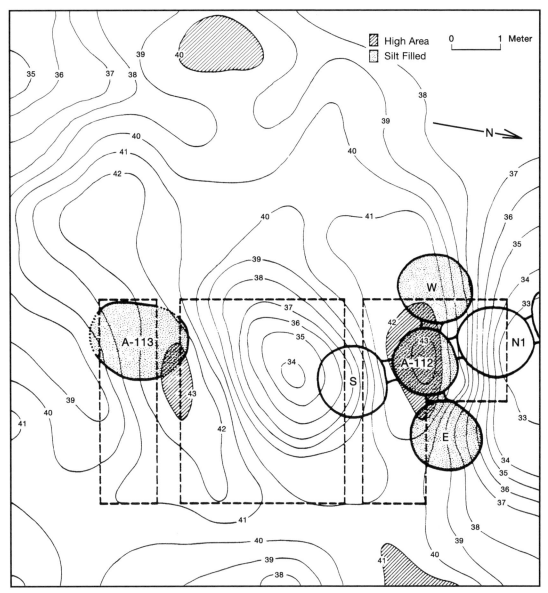

Figure 9.24. EM contour map showing ground conductivity variation pertinent to Tomb A112 and Tomb A113 (see Chapter 3).

intrusion of shaft fill material after the blocking stones had been removed.

The thieves successfully emptied all of the chambers of burial artifacts and objects. Human remains were left behind, although some unintentionally were brought up to the surface. A significant number of bone fragments was found in the shaft's fill.

Earlier experimentation with the EM-31 had shown that shaft refills from recently excavated tombs displayed higher conductivity readings than shafts that had been filled several thousand years ago. The conductivity values found in A112 were very similar to those found in the undisturbed tombs of A110, A111, and A114. Because of this, we argue that Tomb A112 was robbed sometime in antiquity. The robbery of tombs in early antiquity is common. Study of Bronze Age burial mounds in Bahrain has shown that most have been robbed, and most robberies took place during the past 100 to 150 years. However, some can be proven to have been robbed within a few years after the original interment (Frohlich 1982).

During the thieves' exploration and looting of the northern chamber, they found a chamber associated with another shaft tomb south of A112. Subsequently, they broke down part of the thin wall separating the chambers and could have succeeded in robbing an adjacent shaft tomb system. However, and most likely to their great frustration, they found that the chamber had already been robbed and emptied (Figure 9.25). Apparently, earlier thieves had broken into the adjacent tomb and removed the blocking stones before emptying the burial artifacts out of the chamber. The subsequent refilling of the shaft resulted in the intrusion of material from the shaft through the open entryway. Unfortunately, the damage to the human remains was severe, and we were not able to extract useful data.

TOMB A113

Tomb A113 was found through high conductivity readings using the EM-31 conductivity meter. The 250-centimeter shaft had an upper diameter of 136 centimeters and a lower diameter of 110 centimeters. About 35 centimeters of mixed sand and gravel covered the upper part of the shaft. The top 30 centimeters of the shaft fill included several pottery fragments. Three complete medium-size pots and fragmentary blocking stones were found at the base of the shaft. No chambers or entryways leading to chambers could be identified. We argue that Tomb A113 exemplifies a tomb construction running into major problems. Most likely, the excavators of the shaft dug into an old chamber that in the process collapsed, destroying its contents. Apparently, the construction was cancelled and moved to another site. No human remains are associated with this tomb.

TOMB A114

Tomb A114 is located approximately five meters east-southeast from Tomb A110. This tomb consisted of a vertical shaft leading to a single and unsilted burial chamber that extended outward from the lower end of the shaft (Figure 9.26).

The upper shaft was identified by the EM-31 equipment. A significant and circular area of high conductivity indicated the presence of a shaft. This anomaly extended slightly toward the north, suggesting the presence of a silted chamber and the possibility of unsilted chambers to the south, east, and west. However, the excavations yielded only one unsilted chamber to the north (Figure 3.6).

The upper layer of the shaft was found 40 centimeters below the present surface and had maximum and minimum diameters of 139 centimeters and 126 centimeters, respectively. The fill was softer in density and lighter in color when compared to the surrounding undisturbed clay and gravel matrix. The maximum depth recorded from the original surface was 133 cen-

Figure 9.25. Partly silted chamber located adjacent to Tomb A112 as viewed from a break in the wall from chamber A112N. Note the silting process through the undisturbed entrance way.

timeters. Consequently, the lower part of the shaft was 173 centimeters below the present surface. The maximum and minimum diameters at the lowest part of the shaft were, respectively, 96 centimeters and 92 centimeters.

Tomb A114, Chamber North

The entryway leading from the shaft to the chamber was square, measuring 44 centimeters wide by 43 centimeters high. The wall thickness was 19 centimeters. The upper end of the entryway was located 88 centimeters below the original surface and 128 centimeters below the present surface. Unlike in most of the other EB IA burial chambers, the lowest part of the entryway was level with the lowest part of the shaft. The chamber was sealed from the shaft by a single blocking stone measuring 40 centimeters wide, 66 centimeters high, and between 7 and 9 centimeters thick. Small stones were placed in spaces where the single blocking stone was not wide enough to cover the entrance. The sealing was enhanced by the use of mud mortar, which prevented any intrusion of materials from the shaft.

The unsilted chamber measured 192 centimeters (entryway to back) by 208 centimeters (side to side), and the maximum height was 74 centimeters. Chamber stratigraphy showed a compact matrix of clay mixed with isolated layers and pockets of marl and gravel. This mixed layer of marl and gravel represented the lowest 15 to 20 centimeters of the chamber wall. Dense horizontal layers of undisturbed clay constituted the upper part of the walls and roof. Tool marks were visible, especially in the upper part of the wall and roof. Some roof fall was present adjacent to the entryway and along the margins between wall and floor.

The well-preserved human remains of seven individuals have been identified. These includes two adult females, one adult male, one adolescent, one child, one infant, and one prenatal fetus. The burials are secondary, and the majority of the skeletal elements are disarticulated. Five skulls and one cranium were located parallel to the wall to the left of the entryway

Figure 9.26. Shaft leading to chamber A114N. Only one chamber was associated with this tomb.

(Figures 9.27, 9.28). The five articulated skulls include mandibles, and the disarticulated cranium of an infant had been placed in a small pile between two of the articulated skulls. Postcranial elements and the skeletal elements of the prenatal burial were located in a well-organized bone pile at the center of the chamber (Figure 9.29). One complete and articulated vertebral column was located directly on the floor. The direction of the column was approximately east to west, with the cervical vertebrae pointing toward the skulls located to the left of the entryway. A second and partly articulated vertebral column was found parallel to the first but closer to the entryway. The majority of the long bones were placed parallel to the back wall and parallel with the two articulated vertebral columns. All other skeletal elements had been placed randomly between and below the long bones and between the majority of the long bones and the entryway. Sections of textiles were found on the upper surfaces of the bones in the top layers of the bone pile. This suggests that the bone pile, but not the skulls to the left, were partly covered by a shroud of textile. The human remains had been placed on two rectangular reed mats, with the skulls placed on one mat to the left of the entryway and the bone pile placed on a second mat at the center of the chamber (Figure 9.30).

The A114N chamber yielded an unusual amount of associated burial artifacts and objects, including mace heads, unfired clay figurines, and well-preserved wood objects (Figure 9.31; Frohlich and Ortner 1982).

Burial 1 is nearly complete, and the cranium is intact, including the small and delicate bones, such as the lacrimals. The vertebral elements are all present, and the majority of the hand and foot elements are present but cannot be attributed to a specific burial. The costal elements are poorly represented.

The age at death is estimated at 25 to 30 years, based upon the morphology of the articular surfaces and the pubic symphysis. Additionally, the dental attrition is slight and there is little evidence of age-related degeneration and enthesophyte formation. The remains attributed to this individual are very gracile. The classic female sex characteristics in the pelvic and cranial morphology, in addition to the very gracile skeletal elements, such as the femoral and humeral head diameters, are all indicative of a female.

The cranium of this individual shows metopism and multiple ossicles. The dentition has enamel hypoplasia

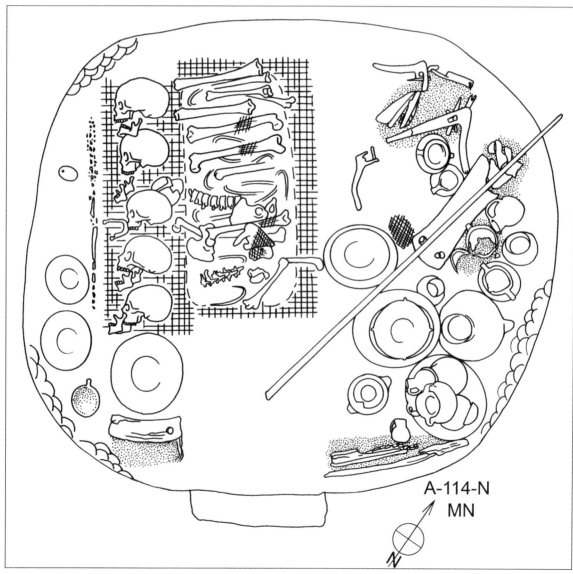

Figure 9.27. Plan view of chamber A114N. Note two reed mats, the textile cover on the bone pile, and the wood tools and stick to the left and right of the entry way.

Figure 9.28. View of chamber A114N. Note the wood stick to the right and mace head to the left.

Figure 9.29. Skulls, crania, and postcranial elements in the center bone pile of chamber A114N. Note the pieces of textile fragments covering the bone pile and articulated vertebrae.

Figure 9.30. Reed matting covering the floor of chamber A114N. Two separate matting pieces were used for skulls and for postcranial elements.

Figure 9.31. View of chamber A114N to the right side of the entrance. Note the presence of well-preserved wood objects.

indicative of three stress events during dental development (Figure 9.32). There is very slight unilateral evidence of cribra orbitalia on the left orbit. The left posterior parietal shows a remodeled lesion, likely due to trauma, approximately one centimeter in diameter. In addition to this, a linear, horizontal depression is present near the midline of the right side of the frontal. The margin of this defect is smooth. If this depression was caused by trauma, it is likely that the trauma occurred long before death. A dark, tar-like, granular matter is present in the neural cavity and is adherent to the internal superior surface. Early periodontal disease is evident and is especially associated with the anterior dentition.

There is a slight deviation to the right of the spinous processes of the midthoracic vertebrae, which likely reflects mild scoliosis. The right humerus shows a unilateral supratrochlear exostosis. Unilaterally, the muscle attachments of the left proximal humerus are quite marked, as is the right radial tuberosity. This may indicate habitual biomechanical stress. There is possible preservation of cartilage on the articular surfaces of the femoral head, the distal femora, and the tibial plateaus. Bilaterally, cystic destructive subchondral lesions are present on the superior horns of the acetabular articular surface.

Burial 2 is nearly complete. The cranium is intact and all of the vertebrae are present. The majority of the hand and foot remains are present but not necessarily positively attributed, due to mixture with other burials in the central bone pile. Costal remains are not well represented.

The age-at-death estimate is 40 years and is based on morphological changes in the auricular surfaces and the pubic symphysis. The dental attrition is very substantial and is consistent with an individual over 40 years old. Skeletal characteristics, including pelvic and cranial morphology and overall gracility of the remains, are all indicative of a female.

The dental attrition is severe and regular. This is likely a contribution of age and coarse diet. There is bilateral degenerative joint disease at the temporomandibular joints. The subchondral surfaces show slight deformation and cyst formation. The gonial angle of the mandible is unusual and very close to 90°.

There are multiple dental pathologies present. Antemortem tooth loss is evident for both maxillary second molars. The process of remodeling of the alveoli shows that tooth loss occurred first on the left and just prior to death on the right. Apical abscesses are present bilaterally in association with these teeth. The left

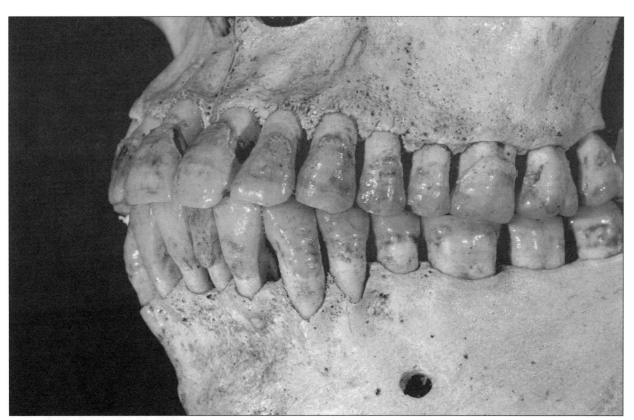

Figure 9.32. Enamel hypoplasias in Burial 1, chamber A114N. At least three events of stress-related growth cessation caused these anomalies.

M³ was lost antemortem, and the alveolus is resorbed. The right M³ shows interproximal caries. A buccally located apical abscess is associated with the right M_1. This is likely due to the severe attrition of this crown and the exposure of the pulp chamber. Enamel hypoplasia provides evidence of at least three stress events during the dental development.

There is evidence of trauma on the left frontal boss, but this is remodeled. The skeletal elements are lighter than normal and may indicate osteoporosis. Given the age at death of this individual, the cause of reduced bone mass could be hormonal changes associated with menopause. Bilaterally, the humeral tubercles are enthesophytic, as are the radial tuberosities. This is likely an indication of stressful muscle activity. Finely woven textile is adhered circumferentially to the right femoral midshaft and a number of articular surfaces, including the right proximal ulna and right auricular surface. The ischial tuberosities are enthesophytic, bilaterally. This is generally considered age related. Osteoarthritis is present bilaterally on the acetabular surface (Figure 9.33). Cyst formation is present on the superior and inferior horns as well as the superior aspect of the acetabular rim. Additionally, the transverse ligament attachment is enthesophytic. This is bilateral, but more marked on the right. Osteoarthritis is also present bilaterally on the femoral/patellar articulation. There is evidence of cyst formation and repair, with marginal lipping, on the subchondral surfaces of both elements. Cyst formation is also present unilaterally on the right lateral tibial plateau. Bilaterally, the cruciate ligament attachments, primarily on the tibiae, are enthesophytic.

Vertebral pathology is extensive. There is evidence of extensive osteoarthritis of the cervical and thoracic vertebrae. On the left, the articulating apophyseal joints and uncinate processes are destroyed and show cyst formation from C3 to C5. The anterior base of C6 is compressed. Syndesmophytes are present. C7–T1 and T4–T5 show cyst formation and joint destruction on the right articulating apophyseal joint. T1–T6 bear similar evidence of osteoarthritis on the left apophyseal joints. T10–T11 display osteoarthritis on the articulating apophyseal joints bilaterally. Syndesmophyte formation is present on T6–T12. L2–L5 display degenerative disk disease at the superior anterior margins of the bodies, with syndesmophytes developing from the superior rim. L5 also displays degenerative disk disease on the inferior anterior plate. Bilaterally, L4 shows abnormally large nutrient foramina on the

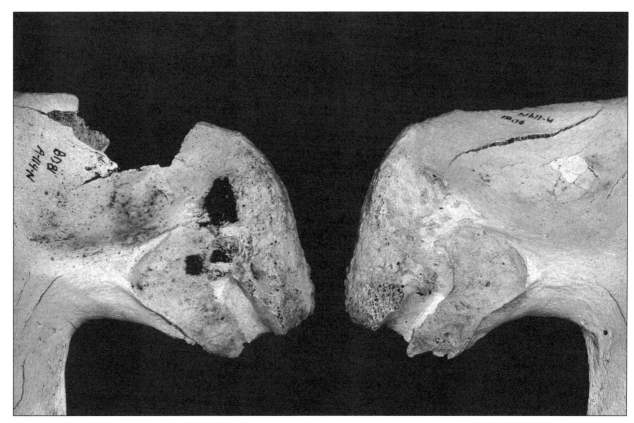

Figure 9.33. Bilateral osteoarthritic developments on the left and right acetabular surfaces, Burial 2, chamber A114N.

anterior body. Unilaterally, the left iliac blade exhibits an abnormal morphology of unknown etiology. The curvature of the blade and rim, especially at the posterior portion, is exaggerated. A longitudinal depression is present perpendicular to the iliac crest, on the center of the external surface of the blade. Additionally, there is evidence of enthesophyte formation of the sacroiliac ligaments. This may be the result of a very well-remodeled fracture with muscle displacement or gait problems.

Burial 3 is a child represented by dental, cranial, and postcranial remains. Most of the major long bones and cranial elements are present, as are the majority of the elements of the pelvic and shoulder girdles. The costal, vertebral, and hand and foot remains, however, are not as well represented for this individual.

The estimated age at death is three years and is based on dental development. The length measurements of the long bones yielded a complementary age of two to three years old. However, as is typical for this population, the iliac breadth metrics indicate an age of 1.5–2.5 years old. No estimation of sex was possible because of the young age. There is no evidence of skeletal pathology in the bones available for study.

Burial 4 is a fetus including cranial and postcranial elements. The costal, vertebral, and hand and foot remains are not well represented. The age-at-death estimate is late-term fetus/perinatal (ca. 38–40 gestation weeks) and is based on the lengths of the long bones as well as cranial development. No sex estimate is possible because of the young age, and no pathology was noted.

Burial 5 is a child or adolescent with an intact skull and dental remains. There are postcranial remains that also may be attributed to this individual. The first cervical vertebrae articulated relatively well with the occipital condyles, but not well enough to be certain, so the attribution remains tentative. The postcranial remains are well represented and include most of the major epiphyses. Vertebrae and hand and foot remains are also associated with this burial. Costal remains are also present, but not complete.

Age at death is estimated to between 11 and 15 years, based on an evaluation of the dental and postcranial remains. The dental development is indicative of an individual 11 years to 15 years old. The postcranial metrics and epiphyseal development, however, are indicative of an individual ranging from as young as 6 years old (iliac breadth) to 8–12 years old (distal femoral epiphyseal development). The majority of the epiphyseal development and metrics indicate an individual of approximately 8–10 years old. This disparity is not unusual for this population. The remains, including the dentition, appear gracile; however, the skeletal immaturity precludes certainty regarding the sex.

At least four events of growth cessation are marked by moderately deep enamel hypoplasia. The right scapula displays reactive woven bone on the inferior margin of the scapular spine and the medial aspect of the subscapular fossa with abnormal and variably-sized porosity. The left scapula is too damaged to determine bilaterality. No other evidence of pathology was observed.

Burial 6 is nearly complete. The skull is intact, and most of the major postcranial elements are present, with the exception of the left femur. The hand and foot remains are largely present, and the vertebral elements are all accounted for. The costal elements are not as well represented.

The age-at-death estimate is between 35 and 44 years. The auricular surfaces are indicative of an individual 40–44 years old, while the pubic symphysis indicates an individual closer to 35 years old. The dental attrition is similar to that of an individual in the mid- to late 20s, and there is little evidence of age-related enthesopathies. The sex is male, based on the morphology and robusticity of the skeletal elements.

At least four events of childhood stress are evident from enamel hypoplasia. The nasal bones of this individual were fractured but healed. The elements are partially fused. There is enthesophyte formation unilaterally on the coracoid process of the left scapula. This was likely caused by trauma or habitual biomechanical stress. The medial margin of the right radial head shows subchondral destruction and repair. This condition is unilateral. Remodeled bone deposition is present within the olecranon fossa of the right humerus (Figure 9.34). This is unilateral and likely the result of habitual stress.

T3–T4 show hyperostosis of the right side of the anterior longitudinal ligament with near fusion of the two vertebral bodies. The lumbar vertebrae show degeneration of the anterior margin with syndesmophyte formation. This is most marked on L4 and L5 superior margins.

The proximal articular facet of the right fibula shows slight joint degeneration with subchondral deformation. The facet of the tibia is also deformed, with some marginal lipping. This is unilateral, although both tibiae show mediolateral flattening and thickening of the posterior crest. This evidence of joint disease may be related to an abnormal gait.

Burial 7 is an infant with the major cranial and postcranial elements present. The costal, vertebral, and hand and foot remains are not well represented. The age-at-death estimate is three to nine months and is based largely on the dental development. The measurements of the postcranial remains are consistent with this determination. The long bone lengths indicate a newborn to six months old. No sex estimate is possible because of the young age.

This individual shows evidence of scurvy. Reactive woven bone lesions are present bilaterally on all long bones, the suprascapular fossae, mildly on the external wings of the sphenoid, markedly on the temporal squama, the mandibular rami, and gonial angles. Woven bone is also present on the internal occipital. Marked gluteal attachments are present on the posterior femoral shaft, and there is formation of a sharp crest at the attachment sites.

Unattributed elements from this chamber have evidence of pathology/anomalies. At least three left ribs exhibit fractures. The callus is well formed, but not entirely remodeled. Bilateral tali exhibit cyst formation on the medial margin of the superior articular surface. These may be associated with Burial 2, but as the tibiae do not exhibit similar lesions, the attribution is quite tentative.

TOMB C11

This tomb was found on a steep bank just south of the east-west road leading from the Lisan peninsula to the town of Kerak (Figure 1.1). The tomb was discovered earlier in the spring of 1981 by Sami Rabadi of the Jordan Department of Antiquities (Frohlich and Ortner 1982). Apparently, a washout of the bank had exposed part of the entryway, revealing the tomb. Before we began excavation, we completed a series of EM-31 conductivity readings with the purpose of finding associated chambers and exploring whether or not other tombs were in the area. Because of the steep bank and the potential water runoff, we did not succeed in obtaining conductivity readings that could be associated with the known shaft or any other possible tombs.

The tomb consisted of a shaft leading from the bank toward the entryway. The shaft was not vertical but extended obliquely into the bank at about 60°. The shaft was cut into a matrix of hard clay. Eighty-six centimeters below the present surface the clay matrix was replaced by a homogeneous matrix of marl. The vertical depth of the slanted shaft was 129 centimeters. The diameter was 105 centimeters in the upper shaft and decreased to 78 centimeters approximately 90 centi-

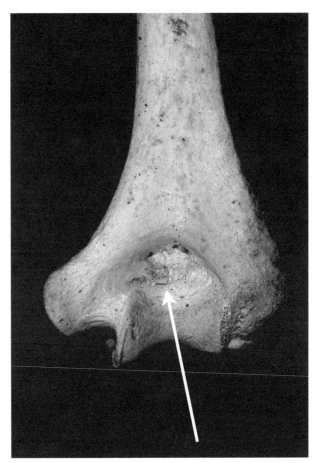

Figure 9.34. Unilateral remodeled bone deposition on the right humerus' olecranon fossa, Burial 6, chamber A114N.

meters below the present surface. The floor diameter at the base of the shaft was 57 centimeters.

The chamber was cut into marl. The entryway's width and height were 119 centimeters and 43 centimeters, respectively. The chamber was sealed from the shaft by a vertical wall constructed from more than ten smaller rocks, all secured by extensive use of mud mortar. A small man-made step was constructed on the border between the shaft and chamber. The floor of the chamber was 26 centimeters below the step. The chamber measured 83 centimeters from the back wall to the entryway and 110 centimeters side to side. The maximum height between floor and ceiling was 67 centimeters. No silt or ceiling fall was observed.

Six burials were identified in the chamber. Burials were secondary, and all skeletal elements were disarticulated (Figure 9.35). The six burials included two adults (one male and one female), one child, one newborn infant, and two perinatal fetuses. The skulls of the two adults and one child were placed in a line to the left of the entryway. All postcranial elements and the cranial elements of the infant and two perinatal burials were found in the bone pile located at the center of the chamber. No reed matting was used in the burial process. Most burial artifacts and objects were located to the right of the entrance. The cultural burial objects, although somewhat visibly different from similar objects found in Tombs A110, A111, and A114, have all been assigned to the EB IA period.

Preservation of the human remains is very good to excellent, with all the individuals disarticulated and the majority of the remains intact and well preserved. Nearly all of the adult elements are present and could be attributed positively, with the exception of the costal elements. The vertebral columns are present in their entirety. Hand and foot remains are also well represented. Both adult crania are intact, with minimal damage. The mandibles are slightly warped and do not articulate well.

The subadult remains are also well represented. The major long bones of the perinatal burials and infant could be attributed to a skull. The major cranial elements are present for all subadults, but positive individual associations are difficult to determine due to the similarity in age for Burials 50 and 51.

Burial 1 is a child with more than 50% of the skeletal elements present, including cranial, dental, and postcranial remains. The elements are largely intact, but the vertebral, costal, and hand and foot remains are incomplete and fragmentary.

The estimated age at death is five to six years, based upon dental development. The postcranial length measurements yielded an age that was significantly younger than the estimate derived from dental eruption. The long bone ages ranged from 2.5 to 4.5 years, and the iliac breadth measurement indicated an age at death of 1.5 to 3.5 years. No sex has been established because of the individual's young age.

Bilaterally, there is evidence of mild and remodeling cribra orbitalia. The medial shaft of the left tibia has a woven bone lesion, possibly caused by infection. The right tibia is damaged, making the determination of bilaterality of the lesions impossible.

Burial 2 is an adult female nearly complete with elements that are, for the most part, intact. The costal elements cannot be positively attributed. The age-at-death estimate is between 30 and 40 years and is based on the auricular surfaces and pubic symphysis. It is likely that the age of this individual was on the lower end of the stated range. The dentition shows moderate attrition, and the majority of the muscle attachments lack enthesophytic activity. Female sex has been determined based on pelvic and cranial morphologies. Additionally, the remains are gracile.

Bilaterally, the ischial tuberosities exhibit marginal enthesopathies. Other muscle attachment sites appear gracile and lack similar osseous indications of muscle over-use. The anterior bodies of T11–L5 display minor marginal syndesmophyte formation. They are most marked on the lower thoracic vertebrae.

The alveolar bone of the right maxilla displays a near-spherical cavitation, with sharp margins and sclerotic walls in proximity to the apex of C^1. This may be due to an apical cyst rather than an abscess. The tooth

Figure 9.35. View of chamber C11. Three skulls are placed to the left, and the bone pile includes skeletal elements from six individuals.

was lost postmortem. The buccal alveolar bone of the mandible is receded at the roots of both first molars.

Burial 3 is a 40-year-old male and nearly complete. The elements are well preserved and largely intact, with the exception of the fragmented tibae. Costal elements cannot be positively attributed to this individual.

The age-at-death estimate is 40+ years and is based on observation of the auricular surfaces and the pubic symphysis. Dental attrition is quite marked. Dental crowns are worn to the root, and the pulp chambers of some teeth are exposed. The extent of joint degeneration is compatible with this age estimate. The sex is male based on pelvic and cranial morphologies. Other skeletal elements are also quite robust.

This individual likely survived a blow to the head. The evidence of skull trauma is large (ca. 35 millimeters long and 20 millimeters wide) and displays significant remodeling. The blow may have perforated the inner table and dislodged a portion of the right parietal vault. The medial margin of the defect is straight and the lateral is slightly curved, as would be the case from trauma caused by an axe cutting through the parietal at an angle.

Bilateral M_1s display apical abscess drainage with evidence of remodeling. This is likely due to the exposure of the pulp chambers because of severe attrition of the crowns. The left M_3 was lost antemortem. The socket is remodeled. The buccal roots, and the buccal crowns, of bilateral M^1s were lost antemortem. The alveolus, associated with the M_2, exhibits evidence of remodeling. Apical abscess drainage is evident in association with the right M^2 and M^3 unilaterally.

Bilaterally, the glenoid fossae exhibit marginal lipping. The humeral heads are unaffected, but the greater and lesser tubercles show cyst formation and enthesophytes bilaterally. This is likely due to biomechanical stress.

There is evidence of arthropathy. Unilaterally, the right posterior trochlea shows subchondral destruction and cyst formation. The articulating ulna displays either the formation of an osseous node or an adhesion of osseous material of the subchondral bone. This may be the result of remodeling or osteochondritis dissecans. The left ulna shows cyst formation and subchondral destruction at the proximal trochlear articular surface. A similar lesion is not present on the humerus. Bilaterally, there is evidence of radio-ulnar osteoarthritis. Both elements display cyst formation, subchondral destruction with some evidence of remodeling, and marginal lipping, all indicative of moderate to severe osteoarthritis. Bilaterally, the patellae are affected with cyst formation and lobular bone deposition or repair. The evidence of this on the femoral surfaces is slight, but they do exhibit marginal lipping. Bilaterally, the acetabula are rather shallow, with marginal lipping along the rim and transverse ligament enthesopathies. Bilaterally, a small lesion of cyst formation and subchondral destruction is present on the superior acetabular surface in proximity to the anterior inferior iliac spines. This is more marked on the right side. The right calcaneus displays a destructive cystic lesion at the medial portion of the talar articulation. The corresponding talus does not display a similar lesion. The left calcaneus is not present for observation.

Bilaterally, the mineralized joint cartilage associated with the first rib is hyperostotic. Enthesophytes are present on the greater trochanters of the bilateral femora, with marked gluteal attachments. Enthesophytes are also present bilaterally at the radial tuberosities and ischial tuberosities.

There is marked hyperostosis of the anterior longitudinal ligament at the lumbar. This hyperostosis is demarcated and present on the left side of the anterior bodies. The ossified ligament is nearly continuously fused from L1 to L3. Diffuse idiopathic skeletal hyperostosis (DISH) is a diagnostic option. The nonvertebral enthesopathies are not as diffuse or as evident at regions such as the olecranon processes, the patellae, and the tibial tuberosities. The L5–S1 intervertebral body articulation displays evidence of disk degeneration that is severe. Proliferative bone formation is encroaching on the intervertebral disk space and macroporosity is present on the end plates. C5–T1 display evidence of disk degeneration as well. The end plates exhibit macroporous cyst formation with marginal syndesmophytes. A centralized lesion is present, perforating the superior body of T1. The margin of the nearly spherical defect within the trabeculae displays sclerosis.

Burial 50 is a perinatal skeleton and includes cranial and postcranial remains. Because Burials 50–52 are all approximately the same age, the specific association of skull elements and postcranial elements cannot be made. Nevertheless, long bones of three neonatal skeletons are present. Woven bone is present along the diaphyses of the long bones. This may be an expression of an infectious process, but all bone formed during the first postnatal year is woven; distinguishing normal growth from bone reaction to pathology is not always possible.

Burial 51 is a perinatal skeleton including cranial and postcranial remains. The age-at-death estimate is perinatal and is based on length measurements of the long bones.

Burial 52 is a newborn with cranial and postcranial remains. Most of the major long bones are present but not as well preserved as those in the other burials. The epiphyses are badly damaged. As the long bones are not measurable for length, the robusticity of the diaphyses was observed for the age-at-death estimation. The robusticity of the diaphyses, in comparison with Burial 50 and Burial 51, indicates that death occurred within the first months of life. No sex has been established. No anomalies and pathologies have been observed.

There are also some unattributed pathological elements included with the collection of remains. Fragments of perinatal/infant cranial squama are present, with variably sized porosity and pumice-like woven bone lesions intracranially. One fragment, likely a parietal or a frontal, exhibits evidence of a progressively remodeling intracranial lesion. These fragments are associated with Burial 50, 51, or 52. Three costal fragments are present and have well remodeled fractures. It is not clear whether they are associated with the same individual.

CHAPTER 10

The Osteology of the EB IA People

BRUNO FROHLICH, DONALD J. ORTNER, AND ALAIN FROMENT

THIS CHAPTER PRESENTS THE OSTEOLOGICAL data, both metric and nonmetric, to provide a general view of the physical aspects of the Bâb edh-Dhrâʿ people and to make some basic comparisons with other contemporary and noncontemporary skeletal populations in the Middle East. We also hope that the data will provide a baseline for research being conducted on human remains from other sites in the Middle East and North Africa that are dated to approximately the same time period. We will not provide raw data in this book. However, the complete data set, high resolution CT (computed tomography), images of crania and mandibles, and field and laboratory photographs are available in electronic format at the Smithsonian DAM (Digital Access Management) server at: anthropology.si.edu/babedhdhra/.

The Bâb edh-Dhrâʿ skeletal collection and the associated archaeological and architectural information are unique. It is one of a very few collections where it has been possible to control information about the collection from the time of excavation to the time of data analysis. Because of this control, we can distinguish between variation in burials due to conditions at the time of excavation versus variation due to curation. Also, the skeletal preservation is relatively good and, although some of the silted tombs were a serious challenge for the excavators, we succeeded in identifying and retrieving the majority of the skeletal elements that had been placed in the tomb chambers during interment. Knowledge of the field conditions enables us to understand that variation in the recovery of chamber contents was due to slightly different conditions at the time of primary interment and not the result of differences in excavating methods. This control and knowledge of the means of collection, in combination with the size, age, and sex distributions of the sample, are extremely important and make our skeletal collection as good a representation of the original population as an archaeological skeletal collection can be.

This chapter is divided into three sections: (1) general description of the burial landscape and burials, including sex and age-at-death estimates; (2) Bâb edh-Dhrâʿ burials, including metric and nonmetric data, stature estimates, age distribution, and sexual dimorphism; and (3) comparative analysis with other contemporary Middle Eastern skeletal collections.

GENERAL DESCRIPTION

The burials are divided into two groups with one dated to the EB IA period and the other to EB IB period (Table 10.1). The EB IA burials are all associated with shaft tomb systems, while the EB IB burials are associated with both shaft tombs and a charnel house. Because of the size of the cemetery, it was divided into identified areas (Schaub and Rast 1989). The archaeological excavations of the Bâb edh-Dhrâʿ cemetery took place during two major periods: three seasons from 1965 to 1967 directed by Paul Lapp and three seasons from 1977 to 1981 conducted by Donald Ortner and Bruno Frohlich (Frohlich and Ortner 1982; Schaub and Rast 1989). The 53 burial structures excavated by Paul Lapp include 38 shaft tombs and 9 charnel houses and have been described by R. Thomas Schaub and Walter Rast (1989); the associated skeletal remains were analyzed by Wilton Krogman (1989). The skeletal burials collected by Paul Lapp were stored in Jerusalem. Unfortunately, they were destroyed during a storm in 1968 (Schaub and Rast 1989) and could not be included in our research.

From the excavations in 1977, 1979, and 1981, a total of 29 burial structures have been dated to EB I (shaft tombs and charnel houses). This includes 28 shaft tombs and one charnel house (Table 10.1). Twenty-

Table 10.1. Distribution of EB I Shaft Tombs and Charnel Houses in Cemeteries A, C, and G

	Shaft Tombs		Charnel Houses		Total
	EB 1A	EB 1B	EB 1A	EB 1B	
Area A	22	1	0	0	23
Area C	3	0	0	0	3
Area G	1	1	0	1	3
Total Tombs	26	2	0	1	29
Total Burials	578	19	0	115	712

three shaft tombs were found in Area A, three shaft tombs were found in Area C, and two shaft tombs and one charnel house were found in Area G (Table 10.1). Twenty-six shaft tombs with 578 burials, including 22 in Area A, three in Area C, and one in Area G, are associated with the EB IA period (Table 10.1). Those associated with the EB IB period include two shaft tombs housing 19 burials in Area A and Area G and one charnel house structure in Area G holding 115 burials. The number of burials identified in the EB IB charnel house is an estimate based on the maximum and minimum number of skeletal elements identified within each age cohort. Thus, the total number of burials assigned to the EB IB period is 134 (Table 10.1). Four burials, all dated to EB IB, have been found in a reused EB IA chamber in the A area. Burials from EB IB period are described and analyzed in Chapter 7.

EB IA BURIALS

The skeletal sample from the EB IA shaft tombs at Bâb edh-Dhrâ' consists of 578 secondary burials. A large portion of these burials include skeletal elements in excellent condition and relatively complete, however some of the burials were not preserved well enough to provide all the data included in our data protocol. There are 239 (41%) adult skeletons and 339 (59%) subadults in the sample (Table 10.2). In the current reanalysis, we have also concentrated on identifying evidence of human skeletal disease, and this has provided more complete data on the prevalence of skeletal diseases than reported in earlier publications. A discussion of this aspect of the research is provided in Chapter 12.

Generally, preservation of the EB IA burials is good to excellent, although the bone is fragile, the protein component of bone degraded by taphonomic processes in both the primary and, particularly, the secondary burial environments. Preservation and recovery was also related to the degree the shaft tomb chambers had remained intact; the best bone preservation is seen in those chambers where there was no ceiling fall or silting. All the EB IA burials were secondary and based on an inventory of all skeletal elements; completeness ranged mostly between 50% and 95%. However, some burials are represented by single skeletal elements. Furthermore, there was some breakage and a likely further loss of bones in antiquity during the transition to the secondary burial. Bone loss particularly affected the recovery of the smaller bones and bone fragments.

A significant factor affecting the completeness of burials is the length of burial at the primary site (Frohlich and Ortner 2000). When only a short time had elapsed between primary and secondary burial, the bones, particularly those of medium or large size, were held in place by soft tissue. However, even in cases where the time in the primary burial site allowed for an almost complete disarticulation of all skeletal elements, we find that the skeletal inventory of each individual in the secondary burial chamber is as much as 95% complete. This high level of care in transferring many of the skeletal elements from the primary burial location to the secondary burial site is remarkable. In some chambers where the burials were embedded in a concrete-like matrix, recovery of all skeletal elements from the original interment may have been incomplete, but we are confident that, in most cases, this has not affected our estimated minimum number of individuals interred in the chamber. In some cases, the variation in the completeness of the burials may be a function of changing burial practices over time and within the EB IA period. For example, burials excavated during the 1981 season and that may belong to a later phase within the EB IA period appear to include more articulated and complete skeletons than burials excavated during the 1977 and 1979 seasons, which may be associated with an earlier subphase within the EB IA period. Nevertheless, the recovery of late fetal bones in many chambers, and more than 90% of hand and foot bones in several chambers, testifies to the care taken by the EB IA people in recovering and interring the remains of individuals of all ages and sexes at the secondary burial sites.

Tomb Statistics (EB IA)

Twenty-six EB IA tombs were excavated during three seasons, 1977, 1979, and 1981 (Table 10.3). This does

Table 10.2. Basic Distribution of Sex and Age

Sex	Number of Burials per Age at Death					
	Prenatals	Infants	Children	Adolescents	Adults	Total
Males	0	0	0	6	120	126
Females	0	0	0	12	112	124
Unknowns	60	98	152	11	7	328
Total	60	98	152	29	239	578

not include two tombs (A112 and A113) excavated in 1981 that were excluded because of previous disturbance, robbery, and destruction (Frohlich and Ortner 1982).

Fifteen tombs containing 275 burials were excavated in 1977, seven tombs containing 247 burials were excavated in 1979, and four tombs (excluding two robbed and destroyed tombs) including 56 burials were excavated in 1981 (Table 10.3). The tombs excavated in 1977 and 1979 were located on relatively flat land but close to the seasonal wadis connecting the drainage from the escarpment east of the cemetery to the Dead Sea. Tombs excavated in 1981 were located midway between wadis, also on relatively flat land (Frohlich et al. 1986; Frohlich and Lancaster 1986; Frohlich and Ortner 1982).

Chamber Statistics (EB IA)

Sixty-three chambers were associated with 26 tombs, with an average of 2.4 chambers per tomb. Tombs included from one to five chambers (Table 10.4). Twelve tombs (46%) had one chamber, six tombs (23%) had three chambers, seven tombs (27%) had four chambers, and one tomb (4%) had five chambers. The majority of tombs included one, three or four chambers; none included two chambers, and only one tomb had five chambers (A108; Table 10.4).

Burial Statistics (EB IA)

A total of 578 burials have been identified in 26 tombs including 63 chambers. The average number of burials in each tomb is 22.2, with a range from four burials in Tomb A103 (one chamber only) to 59 burials in Tomb A105 (including four chambers; Table 10.5). Only three tombs, all excavated in 1979, included more than 50 burials (Table 10.3). Obviously, the total number of burials in each tomb correlates with the number of chambers within each tomb (Tables 10.5 and 10.6). The number of burials averages to 9.2 per chamber and ranges from two burials in chamber A101E to 27 burials in chamber A105NE (Table 10.5).

The orientation of tomb chambers may have cultural significance. For some tombs, the axis was recorded as a function of a line through the center of the chamber entryway and the center of the shaft and classified as north, northeast, east, southeast, south, southwest, west, or northwest (Table 10.5). Analysis of these data did not identify any consistent pattern in the orientation of chambers. Furthermore, the distribution of chambers and burials within each tomb does not indicate any preference for a specific combination of architectural variants, including the number of chambers, the direction of each chamber, or the number of burials within each chamber. One plausible hypothesis is that the number of chambers in each tomb and the number of burials in each chamber are a function of the number of people dying within a specific kinship group during a specific time period and are unrelated to variation in location or the number of chambers in a tomb.

Age Distribution (EB IA)

The accuracy of age-at-death estimates is affected by whether the burial is a subadult or an adult. Subadult age estimates are based on the growth and development of different skeletal elements, while adult age estimates are based on changes in skeletal anatomy that are much slower to develop. These adult changes have greater variability within specific age categories; this variability increases with age. The age-at-death estimates during growth and development are in general very accurate (Bass 1995; Krogman 1973; Saunders 1992). The length of individual long bones in the perinatal skeleton correlates exceptionally well with gestational age, and long

Table 10.3. Tombs and Burials Excavated in 1977, 1979, and 1981

Tomb	Year Excavated			Total
	1977	1979	1981	
A78	28	0	0	28
A79	47	0	0	47
A80	23	0	0	23
A86	12	0	0	12
A87	5	0	0	5
A89	30	0	0	30
A91	13	0	0	13
A92	5	0	0	5
A100	46	0	0	46
A101	12	0	0	12
A102	0	56	0	56
A103	0	4	0	4
A105	0	59	0	59
A106	0	12	0	12
A107	0	43	0	43
A108	0	55	0	55
A109	0	18	0	18
A110	0	0	19	19
A111	0	0	24	24
A114	0	0	7	7
A120	6	0	0	6
A121	9	0	0	9
C10	15	0	0	15
C11	0	0	6	6
C9	11	0	0	11
G4	13	0	0	13
Total Tombs	15	7	4	26
Total Burials	275	247	56	578

Table 10.4. Basic Chamber and Burial Statistics in Tombs with One, Three, Four, and Five chambers

	Number of Chambers in Each Tomb					Total
	1	2	3	4	5	
Tombs in each group	12	0	6	7	1	26
Chambers in each group	12	0	18	28	5	63
Total number of burials in each group	106	0	131	286	55	
Average number of burials in each chamber	8.8	0	7.3	10.2	11	9.2

Note: No tombs included two chambers only.

Table 10.5. Distribution of Burials in Chambers with Different Direction

Tomb	Chamber Direction									Tomb		
	North	Northeast	East	Southeast	South	Southwest	West	Northwest	N/D	Chambers	Burials	Bur/Cha
A78	0	3	0	0	2	11	0	12	0	4	28	7.0
A79	17	0	13	0	8	0	9	0	0	4	47	11.8
A80	3	0	3	0	6	0	11	0	0	4	23	5.8
A86	0	2	0	3	0	7	0	0	0	3	12	4.0
A87	0	0	0	5	0	0	0	0	0	1	5	5.0
A89	0	5	0	13	0	0	0	5	7	4	30	7.5
A91	0	0	0	0	0	0	0	0	13	1	13	13.0
A92	0	0	0	0	0	0	0	0	5	1	5	5.0
A100	0	0	24	0	11	0	11	0	0	3	46	15.3
A101	3	0	2	0	7	0	0	0	0	3	12	4.0
A102	0	10	14	0	17	0	15	0	0	4	56	14.0
A103	0	0	0	0	4	0	0	0	0	1	4	4.0
A105	0	27	0	8	0	3	0	21	0	4	59	14.8
A106	0	0	0	0	12	0	0	0	0	1	12	12.0
A107	5	0	21	0	7	0	10	0	0	4	43	10.8
A108	14	15	0	6	0	8	0	12	0	5	55	11.0
A109	6	0	7	0	5	0	0	0	0	3	18	6.0
A110	0	3	0	10	0	0	0	6	0	3	19	6.3
A111	12	0	5	0	0	0	7	0	0	3	24	8.0
A114	7	0	0	0	0	0	0	0	0	1	7	7.0
A120	0	0	0	0	6	0	0	0	0	1	6	6.0
A121	9	0	0	0	0	0	0	0	0	1	9	9.0
C10	0	0	15	0	0	0	0	0	0	1	15	15.0
C11	0	0	0	0	0	0	0	0	6	1	6	6.0
C9	0	0	0	0	0	0	0	0	11	1	11	11.0
G4	0	0	0	0	0	0	0	0	13	1	13	13.0
Burials	76	65	104	45	85	29	63	56	55		578	
Chambers	9	7	9	6	11	4	6	5	6	63		
Burials/Chamber	8.4	9.3	11.6	7.5	7.7	7.3	10.5	11.2	9.2			9.2

Note: Chamber directions within each tomb can be north, east, south, and west OR northeast, southeast, southwest, and northwest, only.

Table 10.6. Basic Distribution of Age and Sex Divided into Chambers

Chamber	Prenatal Unknown sex	Infants Unknown sex	Children Unknown sex	Adolescents			Adults			All
				Males	Females	Unknown sex	Males	Females	Unknown sex	
A100E	3	5	4	0	0	1	9	2	0	24
A100S	1	2	2	1	1	0	2	2	0	11
A100W	1	6	0	0	0	0	2	2	0	11
A101E	0	0	0	0	0	0	1	1	0	2
A101N	0	0	1	0	0	0	1	1	0	3
A101S	1	2	2	0	0	0	1	1	0	7
A102E	0	1	6	0	1	1	4	1	0	14
A102NE	1	1	5	0	0	0	1	2	0	10
A102S	2	8	4	0	0	0	2	1	0	17
A102W	2	6	2	1	0	1	2	1	0	15
A103S	0	0	1	0	0	0	1	2	0	4
A105NE	5	8	5	0	0	0	4	5	0	27

Table 10.6. continued

Chamber	Prenatal Unknown sex	Infants Unknown sex	Children Unknown sex	Adolescents Males	Females	Unknown sex	Adults Males	Females	Unknown sex	All
A105NW	4	3	9	0	0	0	2	3	0	21
A105SE	0	2	4	0	0	0	1	1	0	8
A105SW	0	0	1	0	0	0	1	1	0	3
A106S	1	3	2	0	0	1	3	2	0	12
A107E	1	5	5	0	0	0	5	5	0	21
A107N	0	0	0	0	0	0	2	3	0	5
A107S	1	0	1	0	1	0	2	2	0	7
A107W	0	1	2	1	0	0	3	3	0	10
A108N	0	4	6	0	0	0	2	2	0	14
A108NE	0	4	6	0	0	0	2	3	0	15
A108NW	2	1	2	0	0	1	1	4	1	12
A108SE	1	0	2	0	0	0	2	1	0	6
A108SW	0	3	1	0	0	0	1	3	0	8
A109E	1	1	2	0	0	0	1	2	0	7
A109N	1	1	2	0	0	0	0	2	0	6
A109S	1	0	2	0	0	0	0	2	0	5
A110NE	0	0	1	0	0	0	1	1	0	3
A110NW	1	1	2	0	0	0	2	0	0	6
A110SE	2	1	3	0	0	0	2	2	0	10
A111E	1	2	0	0	0	0	1	1	0	5
A111N	0	4	4	0	0	0	2	2	0	12
A111W	0	1	1	0	0	0	2	3	0	7
A114N	1	1	1	0	0	1	1	2	0	7
A120S	0	0	1	0	1	0	1	3	0	6
A121N	0	2	1	0	0	0	4	2	0	9
A78NE	0	1	1	0	0	0	0	1	0	3
A78NW	5	2	2	0	0	0	0	2	1	12
A78S	0	0	0	1	0	0	1	0	0	2
A78SW	3	1	2	0	0	0	3	2	0	11
A79E	3	3	5	0	1	0	1	0	0	13
A79N	1	2	9	1	1	0	2	1	0	17
A79S	1	1	1	0	1	0	2	2	0	8
A79W	3	1	1	0	0	1	0	2	1	9
A80E	0	0	0	0	0	0	2	1	0	3
A80N	0	1	0	0	0	0	2	0	0	3
A80S	1	0	2	0	0	0	2	1	0	6
A80W	1	1	4	0	0	1	2	2	0	11
A86NE	0	0	0	0	0	0	1	1	0	2
A86SE	0	0	0	0	0	0	2	1	0	3
A86SW	0	0	2	0	0	0	3	2	0	7
A87SE	0	0	1	1	0	1	2	0	0	5
A89NE	1	0	2	0	0	0	1	1	0	5
A89NW	0	0	2	0	0	0	2	1	0	5
A89SE	2	1	3	0	0	0	5	2	0	13
A89, n/d	0	0	3	0	1	0	1	2	0	7
A91, n/d	1	1	4	0	1	1	1	2	2	13
A92, n/d	1	0	1	0	0	0	2	1	0	5
C10E	2	0	4	0	1	0	5	3	0	15
C11, n/d	0	3	1	0	0	0	1	1	0	6
C9, n/d	0	0	5	0	0	1	2	2	1	11
C9, n/d	1	1	4	0	2	0	1	3	1	13
Total	60	98	152	6	12	11	120	112	7	578

bone length, completion of epiphyseal fusion, and the development of dental elements correlate significantly with age at death in infants, children, and adolescents (Bass 1995; Krogman 1973; Mays 1998; McKern and Stewart 1957; Ortner 2003; Saunders 1992; Stewart 1968; Stewart and Trotter 1954; Tanner 1981; Ubelaker 1989). However, various methods appear to be differentially affected by extrinsic factors such as nutrition, trauma, stress, and pathology. The age-related development of dental elements, especially the formation of teeth, is less affected by extrinsic causes and probably a better focus for estimating age. Nevertheless, the criteria can vary in different human populations, so careful attention is needed in selecting the appropriate standard. Additionally, long-bone growth and epiphyseal fusion can be influenced by extrinsic factors, making these features variably accurate, depending upon the circumstances (Gustafson and Koch 1974; Jantz and Owsley 1994; Mays 1998). Unfortunately, most of our knowledge of age-related changes in both adults and subadults is based on modern Western population studies. Applying such methods to archaeological burials assumes that the correlation between anatomical change relative to age has not changed significantly over time. We argue that such differences are unlikely to have a significant effect in analyzing and interpreting our data.

Following the excavations at Bâb edh-Dhrâ', 578 EB IA burials have been recorded and the following age cohorts have been defined: adults, equal to or older than 18 years, include 239 individuals (41%), and subadults, younger than 18 years, include 339 individuals (59%; Tables 10.2 and 10.7).

The subadult burials have been divided into four age cohorts: prenatal, infant, child, and adolescent.

Prenatal (N = 60, 10.4%) examples include all burials from time of gestation (or fertilization) to birth. It is important to stress that virtually all fetal bones were from fetuses more than 30 weeks of age. Infants (N = 98, 17.0%) are defined as burials from birth to one year of age at the time of death. Children (N = 152, 26.3%) include burials from the age of one year to 12 years. Adolescents (N = 29, 5.0%) are individuals from 12 years to 18 years of age at the time of death. Finally, the adult age cohort (N = 239, 41.3%) includes remains that appear to be older than 18 years of age at the time of death. The latter is an arbitrary age category based on an average onset of sexual maturity at about 16.5 years and the first birth at around 20 years of age (Howell 1976).

The number of subadults is higher than evidenced in most previous reports of skeletal samples (Acsadi and Nemeskeri 1970; Angel 1971; Saunders et al. 1995; Weiss 1973). However, Judith Littleton (1998) found similarly high frequencies of prenatal and infant skeletons in the excavations of Hellenistic burial grounds in Bahrain. She argues that this higher number is a more accurate reflection of prenatal and infant deaths in most archaeological human populations than other published frequencies, which tend to be about 50%

Table 10.7. Basic Distribution of Age and Sex Divided into Tombs

Tomb	Prenatal Total	Infant Total	Child Total	Adolescent			Adult			Male Total	Female Total	Total
				Male	Female	Total	Male	Female	Total			
A78	8	4	5	1	0	1	4	5	10	5	5	28
A79	8	7	16	1	3	5	5	5	11	6	8	47
A80	2	2	6	0	0	1	8	4	12	8	4	23
A86	0	0	2	0	0	0	6	4	10	6	4	12
A87	0	0	1	1	0	2	2	0	2	3	0	5
A89	3	1	10	0	1	1	9	6	15	9	7	30
A91	1	1	4	0	1	2	1	2	5	1	3	13
A92	1	0	1	0	0	0	2	1	3	2	1	5
A100	5	13	6	1	1	3	13	6	19	14	7	46
A101	1	2	3	0	0	0	3	3	6	3	3	12
A102	5	16	17	1	1	4	9	5	14	10	6	56
A103	0	0	1	0	0	0	1	2	3	1	2	4
A105	9	13	19	0	0	0	8	10	18	8	10	59
A106	1	3	2	0	0	1	3	2	5	3	2	12
A107	2	6	8	1	1	2	12	13	25	13	14	43
A108	3	12	17	0	0	1	8	13	22	8	13	55
A109	3	2	6	0	0	0	1	6	7	1	6	18
A110	3	2	6	0	0	0	5	3	8	5	3	19
A111	1	7	5	0	0	0	5	6	11	5	6	24
A114	1	1	1	0	0	1	1	2	3	1	2	7
A120	0	0	1	0	1	1	1	3	4	1	4	6
A121	0	2	1	0	0	0	4	2	6	4	2	9
C10	2	0	4	0	1	1	5	3	8	5	4	15
C11	0	3	1	0	0	0	1	1	2	1	1	6
C9	0	0	5	0	0	1	2	2	5	2	2	11
G4	1	1	4	0	2	2	1	3	5	1	5	13
Total	60	98	152	6	12	29	120	112	239	126	124	578

adults, 35% children and adolescents, and 15% prenatals and infants.

There are several potential reasons for reports of low subadult mortality rates and, especially, low prenatal and infant mortality rates. Prenatal and infant skeletal elements may not be found during excavation due to their more fragile bone architecture, which is more vulnerable to some burial environments (Angel 1969; Gordon and Buikstra 1981; Mays 1998; Saunders 1992; Weiss 1973). Also, the interment of subadults, and especially prenatal and infant individuals, may reflect different mortuary practices than those accorded adults; thus, fewer subadults may be included in the burial distribution investigated by the excavators, even when careful field methods are utilized (Mays 1998; Molleson 1991). Disfiguring diseases such as consumption (tuberculosis) and leprosy, and in some cases mental diseases, could result in exclusion from interment in the general cemetery (Frohlich 1999; Gorecki 1979; Mays 1998; Ucko 1969). Finally, in some excavations, the collection of prenatal and infant burials may not have been part of the original research design, resulting in the excavator focusing on adult and older subadult burials.

Dental estimates of subadult age are less subject to extrinsic factors than are estimates based on the development of long bones. This difference between estimates provides insight regarding factors that adversely affected the growth and health of infants and children (Thomlinson 1965). Saunders and Barrans (1999) argue that the variation in age estimates based on dental development and long bone growth may allow for inferences about related causes of mortality. Saunders et al. (1995) note that variation in long bone growth between individuals with similar age at death as determined by dental development could be the result of acute causes of mortality, including poor nutrition. This observation is based on studies of skeletal collections from a 19th-century Canadian church cemetery, associated parish registers, and similar records (Saunders 1992; Saunders et al. 1995; Saunders and Barrans 1999). In general, mortality data derived from church records and well-controlled burial ground excavations have suggested that neonatal mortality (first four weeks after birth) most often is a function of intrinsic problems, whereas postnatal mortality (from one month to one year) is caused by factors such as poor sanitation, infection, malnutrition, and trauma.

However, the hypothesis that neonatal mortality may exceed postneonatal mortality appears to be contradicted by studies conducted by Saunders (1992) and Forfar and Arneil (1978). These studies support the argument that postnatal mortality far exceeds neonatal mortality and suggest that this assumption can be extrapolated to historic and prehistoric populations. Saunders et al. (1995) report 74 postneonatal skeletons and 26 neonatal skeletons in the 19th-century Canadian church cemetery. Similar trends are reported by John Knodel (1988) in a study of 14 German village populations dating to the 18th and 19th centuries.

The estimated neonatal and postneonatal mortality rates from Bâb edh-Dhrâ' are, respectively, 42% and 58%. These figures are similar to those reported by Saunders et al. (1995) and Knodel (1988). However, the methods used for age-at-death estimates for skeletal remains may not be accurate enough to entirely rely on such comparisons; it is also unknown how well and how accurate the divisions between prenatal, perinatal, and neonatal are in church burials and records. It is noticeable, however, that most mortality rates dating from the early 20th century and earlier suggest a similar trend in which postneonatal mortality is significantly greater than neonatal mortality.

Our research suggests that the Bâb edh-Dhrâ' population, especially the infants, were adversely affected by various pathogens that caused a high infant mortality rate. Finally, the higher percentage of postneonatal mortality when compared to neonatal mortality as observed in the Bâb edh-Dhrâ' material may support the hypothesis that the population was exposed to malnutrition and pathogens. This argument is further discussed in Chapter 12.

Observation of cranial sutures or suture closure is possibly the aging method reported earliest in the literature. For example, Celsus (de Medicina) describes variation in suture closure in the first century AD (Krogman 1973). The method used today is basically based on the research by T. W. Todd and D. W. Lyon (1925) and further developed by T. W. McKern and T. Dale Stewart (1957) and Wilton Krogman (1973). However, the suture closure method more recently has been described as highly unreliable; it has been suggested that, if possible, it should not be used alone as the means to age skeletal remains (Mays 1998; Masset 1989; McKern and Stewart 1957; Ortner 2003).

As an alternative to observation of the cranial sutures, pubic symphysial morphology is probably the most reliable "nondesctructive" method for estimating age at death applied to the adult skeleton. However, from an archaeological point of view, the pubic bone is also the bone in the human body that is most likely to be damaged or destroyed either by taphonomical processes or during the excavation process (Krogman 1973; Mays 1998; McKern and Stewart 1957; Ortner 2003; Ubelaker 1989). Dental wear has been described by Brothwell (1981) and used in the age determination of burials where the diet has been partly known. Other methods, including osteon remodeling and dental microstructure analysis, have shown excellent results; however, these methods appear to be less accurate, due to the inability of researchers to use similar methodologies (Ahlquist and Damsten 1969; Kerley 1965; Kerley and Ubelaker 1978; Ortner 2003). Overall, the age-at-death estimate of any burial greatly improves with the increasing number of methods applied to the sample.

As stated, the majority of age-at-death techniques have been developed by using recent skeletal collections where the sex, race, and age at death are known entities. Such techniques have been applied to archaeological populations assuming that age-related skeletal variation found in modern populations can be applied to human populations deriving from archaeological excavations. Shelley Saunders (1992) has shown that some of the applied techniques developed by using modern skeletal populations with known profiles performed poorly when applied to earlier populations with known birth and death records. However, we hypothesize that variation between different methods of determining age in subadults does provide a relative index of subadult health in archaeological skeletal samples.

Sex Distribution

The difference between male and female skeletons in size, skeletal element proportions, and anatomy varies between human populations. Thus sexual dimorphism is best expressed in the pelvic area, the skull, and the femoral head. Accuracy in determining sex depends on the preservation of the remains, the number and type of skeletal elements that can be used in the determination, and the degree of sexual dimorphism in the skeletal sample. As a single element, the pelvic bone is the most reliable when determining the sex of a skeleton. When both the skull and the pelvic elements are used, the accuracy of the sex estimate may be as high as 95% to 99% (Bass 1995; Brothwell 1981; Cox and Mays 2000; Krogman 1973; Mays 1998; Ortner 2003).

In the Bâb edh-Dhrâʿ collection, 29 burials have been assigned to the adolescent age cohort, and it has been possible to assign a sex to 18 of these. Of the 239 burials assigned to the adult age cohort, 232, or 97%, have been assigned a sex (Tables 10.2 and 10.7). Combining adolescents and adults, it has been possible to assign sex to 250 burials. Of these, 126 (47%) were male and 124 (46%) were female. Eighteen burials (7%) could not be assigned a sex (Tables 10.6 and 10.7).

Each shaft tomb includes adult and subadult burials (Table 10.7). One tomb (A87) has only males and subadults of unknown sex (Table 10.7). All other tombs include males, females, and subadults. Forty-nine chambers include burials, including males, females, and subadults (Table 10.6). All the data about age and sex of burials within the chambers support the conclusion that the EB IA skeletal sample is a reasonable approximation of the living population. We argue that the distribution of males, females, and subadults in each tomb and each chamber may represent members of the same extended family; this hypothesis is further developed in Chapter 13.

METRIC AND NONMETRIC DATA COLLECTION AND ANALYSIS

Our analysis of metric and nonmetric data has several goals in which the fundamental objectives are to reconstruct as much as possible of the biology of the EB I people of Bâb edh-Dhrâʿ and to define the relationship of these people to other Middle Eastern groups dating to approximately the same time period. Our data provide information on stature, sexual dimorphism, robusticity (muscularity), and other variables. These variables can be used in comparison with similar data from other Middle Eastern bioarchaeological human skeletal samples.

Data collection was initiated shortly after the human remains from the 1981 excavation arrived at the Smithsonian Institution. Our first objectives were to reconstruct from the mixed burial assemblages as many individual burials as possible. Once this was done, we recorded, as allowed by the preservation of the remains, both metric and nonmetric data that had been chosen as representative of the osteological biology of the people. More recently, we conducted a detailed analysis of every skeletal element to determine the age distribution of the samples and the minimum number of burials included in each tomb.

In the analysis of these data, we used different statistical programs and procedures to ensure the integrity of the data set. The data protocol and analysis allow us to define with considerable accuracy various subpopulations, such as sex and age cohorts, and to control and verify the basic statistical description of all groups.

Metric variables were selected to ensure as much comparability with data from other Middle Eastern skeletal samples currently available or that will become available as additional excavation is completed. Variables have been assigned an alphabetic designation to facilitate analysis with standard statistical software packages.

When defining metric and nonmetric variables, we have relied on descriptions by Martin (1928), Hrdlička (1920), Howells (1973), and Bass (1995). We used two statistical software packages in the analysis. SAS statistical software version 9.1 for Microsoft Windows, produced by the SAS Institute, North Carolina, USA, has been used in all our descriptive and analytical procedures. The other software package is Systat Version 10, marketed by SPSS, Inc. This software has been used as a control to ensure that descriptive and analytical statistical procedures are congruent.

Systat and SAS have different strategies for data input as well as different options for output following various types of analysis. Because of these differences, the two packages provide a useful method for checking both the accuracy of the data set and the appropriateness of the methods of data input and analysis.

Definitions and Descriptions of Metric Variables

A total of 41 metric variables were recorded. This includes 26 variables related to the cranium and mandible and 15 variables related to the postcrania. Definitions of cranial variables follow Martin (1928), Hrdlička (1920), Howells (1973), and Bass (1995). We selected the following variables:

CRANIAL VARIABLES

- Maximum length of the calvarium (CRMAXL): the maximum length in the median sagittal plane from glabella to the most posterior point on the occipital bone.
- Maximum width of the calvarium (CRMAXW): the maximum distance between the two parietal bones, measured perpendicular to the maximum length of the cranium.
- Biauricular diameter (BIAURIC): the minimum distance across the roots of the left and right zygomatic processes.
- Basion bregma diameter (BASBREG): the distance between basion and bregma.
- Basion nasion diameter (BASNAS): the distance between nasion and basion.
- Basion prosthion diameter (BASPROST): the distance between basion and prosthion.
- Circumference (CIRCUM): the maximum circumference of the calvarium measured through the glabella and the most posterior point of the occipital bone.
- Minimum frontal breadth (FRONTB): the minimum distance between the left and right temporal lines using the left and right fronto-temporale.
- Bizygomatic diameter (BIZYG): the maximum distance between the zygomatic arches between left and right zygion.
- Facial height (FACIALH), or upper facial height: the distance between nasion and alveolare.
- Nasal height (NASALH): the distance between nasion and nasospinale.
- Nasal width (NASALB): the maximum width of the pyriform aperture perpendicular to the midsagittal plane.
- Interorbital width (INTORBIT): the minimum distance between the left and right dacryons.
- Orbit width (ORBITB): the maximum width of the orbit measured from dacryon to the middle of the lateral orbital border (ectoconchion).
- Orbit height (ORBITH): the maximum distance of the orbit measured perpendicular to the orbit width.
- Palate length (PALATEL): the distance between orale and staphylion.
- Palate width (PALATEB): the distance between the left and right endomalare.

MANDIBULAR VARIABLES

- Maximum mandibular length (MMAXL): the maximum horizontal distance in the median plane between a line connecting the most posterior points of the left and right condyloid processes, and the most anterior point of the mandibular body (chin).
- Vertical ramus height (VERTICH): the vertical distance between the most superior part of either condylar process and a horizontal line extending posteriorly from the base of the mandibular body.
- Mandibular body length (BODYL): the maximum horizontal length in the median plane from the most anterior point of the mandibular body (chin) to the midpoint of a line connecting the left and right gonion.
- Oblique ramus height (OBLIQUEH): the distance from the most posterior part of the condyloid process to the gonion.
- Gonial angle (GONIALA): the angle measured between the base of the mandibular body and the ascending ramus.
- Bigonial diameter (BIGONIAL): the maximum distance between the left and right gonion.
- Bicondylar diameter (BICONDYL): the maximum distance between the two most lateral points on the condylar processes.

Symphysial height (SYMPHYS): the distance between infradentale and gnathion.

Minimum ramus width (RAMUSW): the minimum distance between the anterior and posterior borders of the ascending ramus.

POSTCRANIAL VARIABLES

Humerus maximum length (HMAXL): the maximum distance from the proximal end of the head to the medial margin of the trochlea at the distal end of the bone.

Humerus epicondylar width (HEPICOND): the maximum width of the distal end of the bone.

Ulna maximum length (UMAXL): the maximum distance from the top of the olecranon to the most distal end of the styloid process.

Radius maximum length (RMAXL): the maximum distance from the proximal head to the most distal end of the styloid process.

Lumbar height (LUMBARH): the vertical distance measured from the most anterior-superior point on the body of L1 to the most anterior-inferior point on L5, when all five lumbar vertebrae are articulated in their assumed anatomical positions.

Femur maximum length (FMAXL): the maximum distance from the most proximal end of the femur head to the most distal end of the medial condyle.

Femur bicondylar length (FBICONDL): the maximum length from the most proximal end of the femur head to a line connecting medial and lateral condyles at the distal end.

Femur bicondylar width (FBICONDB): the maximum width of the distal condyles.

Femur head diameter (FHEADD): the maximum diameter of the femur head.

Femur subtrochanteric diameter anterio-posterior (FAPSUBTR): the minimum anterio-posterior diameter of the upper shaft, measured just below the lesser trochanter.

Femur subtrochanteric diameter, mediolateral (FMLSUBTR): the minimum diameter of the upper shaft measured just below the lesser trochanter and perpendicular to the anterio-posterior diameter.

Femur midshaft diameter, anterio-posterior (FAPMIDSH): the minimum anterio-posterior diameter of the midshaft.

Femur midshaft diameter, mediolateral (FMLMIDSH): the diameter of the midshaft measured perpendicular to the anterio-posterior diameter.

Tibia maximum length (TMAXL): the maximum length from the lateral condyle at the proximal end to the most distal end of the medial malleolus.

Fibula maximum length (FIMAXL): the maximum distance between the most proximal end and the most distal end of the fibula.

All measurements were taken on the left side, but right side measurements were used in cases where left side measurements were not possible. Metric and nonmetric data sets are available on the Smithsonian's DAM Server (Digital Access Management Server). The server also includes select groups of digital images both from the three field seasons and from the laboratory. Since our metric and nonmetric data are limited in the number of included variables, we have made allowance for accessing computed tomography images of each of the skulls so that researchers can download the CT information either as a series of individual 1.0 millimeter axial slices in the DICOM-3 format or as a three-dimensional model in the STL format. Dependent on the software, the STL formatted file can be used for defining and measuring variables and do rapid prototyping such as stereolithography if required. The STL model is also suitable for viewing certain well-defined and well-developed pathological lesions and other anomalies.

Bâb edh-Dhrâ' Metric Data

A total of 107 burials, including 47 males and 60 females, included elements that could be measured. The variable with the highest number of records is the minimum frontal breadth in both the male group (N = 34) and in the female group (N = 43; Tables 10.8, 10.9). The variable with the lowest number of records is the maximum length of the fibula in both the male and the female groups.

The relatively small sample sizes available for the metric variables make testing the sample for normality particularly relevant. Skewness, kurtosis, and the Shapiro-Wilkinson tests are included in Tables 10.8 and 10.9. A Shapiro-Wilkinson p-value of less than 0.05 indicates a nonnormal distribution. Two variables in the male group do not have a normal distribution (nasal height and femoral mediolateral subtrochanter diameter). Four variables in the female groups do not have a normal distribution (maximum length of the humerus, minimum frontal breadth, nasal breadth, and humeral epicondyle breadth; Tables 10.8 and 10.9). Missing data in some of our burials reduced the sample size to the point where the use of parametric statistics was questionable. Such cases either were removed or were subjected to nonparametric procedures if possible.

Estimation of Living Stature and Indices of Robusticity

The estimation of living stature has been calculated from the length of long bones (Table 10.10). Equa-

Table 10.8. Metric Variable, Males: Descriptive Statistics and Test for Normality

Variable	N	Mean	SD	Men	Max	Range	Skewness	Kurtosis	Shapiro-Wilk.	Prob
CRMAXL	34	186.6	7.2	175	204	29	0.5	−0.4	0.954	0.164
CRMAXW	32	133.2	5.3	124	148	24	0.7	0.7	0.968	0.435
BIAURIC	24	118.7	5.5	107	130	23	−0.2	−0.1	0.986	0.979
BASBREG	14	133.6	3.6	129	141	12	0.9	0.2	0.915	0.177
BASNAS	12	100.5	2.5	97	106	9	0.8	1.0	0.919	0.276
BASPROST	11	96.0	3.7	92	103	11	0.7	−0.8	0.904	0.207
CIRCUM	27	513.4	14.4	490	550	60	0.7	0.8	0.952	0.241
FRONTB	34	94.3	4.2	85	104	19	0.0	0.3	0.984	0.887
BIZYG	15	128.2	6.2	116	144	28	0.7	2.8	0.911	0.139
FACIALH	13	68.7	3.0	64	75	11	0.5	0.5	0.974	0.940
NASALH	13	50.5	1.1	49	52	3	−0.3	−1.4	0.822	0.013
NASALB	16	24.8	1.3	22	28	6	0.3	2.0	0.913	0.129
INTORBIT	15	23.3	2.3	20	28	8	0.5	0.0	0.941	0.398
ORBITB	16	38.3	1.8	35	41	6	−0.2	−0.8	0.953	0.533
ORBITH	15	33.3	2.3	28	37	9	−0.7	0.8	0.925	0.230
PALATEL	19	55.2	2.9	50	60	10	−0.2	−0.6	0.958	0.527
PALATEB	15	63.1	3.4	57	69	12	0.1	−0.7	0.978	0.957
MMAXL	19	102.4	4.3	92	110	18	−0.6	0.3	0.952	0.429
VERTICH	20	53.3	5.9	43	68	25	0.7	1.1	0.949	0.346
BODYL	21	74.4	5.3	62	87	25	−0.1	1.3	0.970	0.736
OBLIQUEH	20	59.4	4.5	51	69	18	0.6	0.3	0.940	0.245
GONIALA	22	32.5	6.6	15	43	28	−1.0	1.3	0.932	0.134
BIGONIAL	15	97.9	6.3	86	108	22	−0.3	−0.5	0.969	0.837
BICONDYL	8	121.3	8.9	105	130	25	−1.1	0.1	0.872	0.157
SYMPHYS	21	35.4	2.5	30	39	9	−0.6	0.3	0.918	0.077
RAMUSW	25	32.2	2.3	28	37	9	0.4	0.0	0.957	0.359
HMAXL	16	300.1	16.3	280	329	49	0.3	−1.4	0.900	0.081
HEPICOND	18	60.4	3.3	55	67	12	0.0	−0.7	0.944	0.338
UMAXL	11	255.5	11.9	236	272	36	−0.2	−1.0	0.942	0.547
RMAXL	12	235.1	9.2	214	247	33	−0.7	1.4	0.912	0.226
LUMBARH	10	129.4	5.2	121	138	17	−0.1	−0.5	0.981	0.970
FMAXL	22	431.6	18.4	404	472	68	0.8	0.1	0.924	0.092
FBICONDL	18	430.1	18.5	404	468	64	0.6	−0.1	0.939	0.273
FBICONDB	15	78.9	3.0	73	83	10	−0.5	−0.6	0.932	0.292
FHEADD	24	45.2	2.2	40	49	9	−0.4	0.1	0.956	0.371
FAPSUBTR	28	26.0	1.6	23	29	6	−0.2	−0.7	0.948	0.179
FMLSUBTR	28	33.9	3.0	30	43	13	1.0	1.7	0.909	0.019
FAPMIDSH	26	29.7	2.3	25	34	9	0.0	−0.5	0.967	0.551
FMLMIDSH	27	28.8	2.5	25	36	11	0.7	1.1	0.936	0.097
TMAXL	14	355.3	18.4	326	392	66	0.4	−0.1	0.981	0.977
FIMAXL	5	344.6	14.3	330	368	38	1.3	2.4	0.902	0.423

Note: Normal distribution is assumed when the Shapiro-Wilkinson P value is greater than 0.05.

tions derived from Caucasian skeletal elements with known living stature have been used for the male and female groups (Trotter and Glesser 1952). Table 10.10 shows the estimates for each selected long bone, and the average is given in both centimeters and inches. The estimated average male stature is 166 centimeters (65.25 inches), while the average female stature is 156 centimeters (61.5 inches). This compares well with data from Nippur (169 centimeters for males and 158 centimeters for females), Tepe Hissar (165 centimeters for males and 153 centimeters for females) and Armant I (168 centimeters for males and 159 centimeters for females). The Bahrain Bronze Age population and Neolithic populations from the United Arab Emirates produce taller estimates of 170 centimeters and 162 centimeters for Bahraini males and females, respectively, and 171 centimeters and 162 centimeters for Jebel Al-Buhars (UAE) males and females, respectively. In general, the Bâb edh-Dhrâ' population is slightly smaller in stature than other contemporary populations in the Middle East (Frohlich 1986; Kiesewetter 2006; Krogman 1940a, 1940b; Swindler 1956).

The following indices have been calculated: cranial index, platymeric index, pilasteric index, and femoral midshaft robusticity index. Cranial index is defined as the percentage ratio between the maximum length of the cranium (CRMAXL) and the maximum width of the cranium (CRMAXW). The platymeric index is

Table 10.9. Metric Variables, Females: Descriptive Statistics and Test for Normality

Variable	N	Mean	SD	Min	Max	Range	Skewness	Kurtosis	Shapiro-Wilk.	Prob
CRMAXL	41	178.4	5.9	165	190	25	0.3	−0.3	0.961	0.164
CRMAXW	39	130.6	5.2	120	140	20	0.1	−1.0	0.959	0.162
BIAURIC	27	115.1	4.9	106	124	18	0.1	−0.5	0.970	0.600
BASBREG	21	129.5	5.6	120	142	22	0.7	0.7	0.938	0.202
BASNAS	20	95.4	4.8	85	107	22	0.2	1.0	0.982	0.954
BASPROST	15	92.5	5.3	83	102	19	0.2	−0.2	0.972	0.881
CIRCUM	34	501.4	15.2	475	531	56	0.1	−1.0	0.971	0.498
FRONTB	43	91.9	4.2	85	104	19	0.8	0.6	0.938	0.022
BIZYG	20	123.5	6.5	112	140	28	0.6	1.2	0.969	0.725
FACIALH	20	64.6	2.3	61	68	7	−0.1	−1.2	0.938	0.218
NASALH	20	47.0	1.7	44	50	6	−0.5	0.0	0.921	0.104
NASALB	25	24.5	1.2	23	27	4	0.2	−0.7	0.905	0.024
INTORBIT	19	21.3	1.7	18	25	7	0.4	0.3	0.946	0.337
ORBITB	21	37.9	1.8	35	41	6	0.4	−0.4	0.925	0.107
ORBITH	19	31.4	2.2	26	35	9	−0.5	0.4	0.957	0.510
PALATEL	24	52.5	2.9	48	58	10	0.0	−1.1	0.950	0.273
PALATEB	18	59.1	2.7	55	64	9	0.3	−0.4	0.954	0.489
MMAXL	28	96.5	4.4	85	108	23	0.1	1.8	0.965	0.445
VERTICH	29	50.1	5.3	38	59	21	−0.3	−0.4	0.980	0.841
BODYL	29	70.0	3.1	64	76	12	−0.3	−0.2	0.956	0.264
OBLIQUEH	29	55.6	4.3	46	62	16	−0.7	−0.3	0.931	0.059
GONIALA	30	33.6	6.8	15	45	30	−0.9	0.9	0.933	0.059
BIGONIAL	20	92.7	7.8	81	113	32	0.7	1.0	0.955	0.454
BICONDYL	19	116.1	9.4	90	133	43	−0.9	2.2	0.939	0.256
SYMPHYS	25	31.2	2.4	26	36	10	−0.3	−0.3	0.966	0.543
RAMUSW	32	30.7	2.4	27	35	8	0.2	−1.1	0.938	0.067
HMAXL	20	282.2	10.9	258	296	38	−1.4	0.9	0.798	0.001
HEPICOND	24	54.0	3.2	50	62	12	0.9	0.3	0.912	0.038
UMAXL	14	237.3	13.3	213	258	45	0.2	−0.5	0.945	0.049
RMAXL	16	216.9	12.1	199	236	37	0.1	−1.2	0.950	0.483
LUMBARH	11	125.5	5.2	118	134	16	−0.1	−1.1	0.948	0.619
FMAXL	21	404.2	16.3	362	426	64	−0.9	0.8	0.932	0.147
FBICONDL	21	398.7	15.6	359	423	64	−0.8	0.7	0.952	0.360
FBICONDB	14	71.7	4.1	66	81	15	0.6	0.5	0.953	0.614
FHEADD	28	41.2	2.5	36	48	12	0.8	1.5	0.932	0.067
FAPSUBTR	30	24.2	1.7	20	28	8	−0.3	1.0	0.938	0.081
FMLSUBTR	30	30.0	2.0	26	34	8	−0.2	−0.3	0.962	0.350
FAPMIDSH	28	26.6	2.2	21	31	10	−0.5	1.1	0.944	0.137
FMLMIDSH	28	25.4	1.8	22	30	8	0.3	0.4	0.963	0.412
TMAXL	13	329.3	14.2	298	352	54	−0.5	0.7	0.963	0.799
FIMAXL	8	324.6	11.9	310	347	37	1.1	0.7	0.900	0.291

Note: Normal distribution is assumed when the Shapiro-Wilkinson P value is greater than 0.05.

defined as the ratio between the femoral subtrochanter anterio-posterior (FAPSUBTR) and mediolateral (FMLSUBTR) distances, and the pilasteric index as the ratio between the femoral midshaft anterio-posterior (FAPMIDSH) and mediolateral (FMLMIDSH) distances. The femoral midshaft robusticity index is calculated as the sum of the femoral midshaft anterio-posterior (FAPMIDSH) and mediolateral (FMLMIDSH) distances multiplied by 100 and divided by the femoral bicondylar length (FBICONDL; Cole 1994; Martin 1928). The indices are presented in Table 10.11.

The Bâb edh-Dhrâ' cranial index is 71.6 for males and 73.2 for females. This identifies them as elongated/long-headed, or dolichocephalic, and compares with earlier Neolithic populations in the United Arab Emirates and Bronze Age populations in Iran. However, the Bâb edh-Dhrâ' population is slightly more dolichocephalic than Predynastic populations in Egypt (Naqada and Badari) and significantly different from the Early Dynastic populations in Lower Egypt (Sedment). The two populations that deviate the most from Bâb edh-Dhrâ' are the Egyptian Sedment and the Greek Lerna populations. These results are partly substantiated by the multivariate comparisons between populations discussed later in this chapter.

The shape indices, platymeric and pilasteric, are calculated from femoral subtrochanter and midshaft

Table 10.10. Estimated Living Stature for Males and Females Based on Maximum Lengths of Long Bones

Element	Max. Length	Estimated Living Stature	Max. Length	Estimated Living Stature
Humerus	300	163	282	153
Ulnae	256	168	237	159
Radius	235	168	217	158
Femur	432	164	404	154
Tibia	355	168	329	157
Fibula	345	164	325	155
Femur + tibia	787	165	733	155
Average, centimeters		166		156
Average, inches		65.25		61.5

Table 10.11. Skeletal Indices Divided into Male and Female Groups and Year Excavated

	1977, 1979, 1981		1977, 1979		1981	
	Male	Female	Male	Female	Male	Female
Subtrochanter AP	26.0/28	24.2/30	25.4/20	24.2/19	26.3/8	24.2/11
Subtrochanter ML	33.9/28	30.0/30	34.5/20	30.4/19	32.4/8	29.4/11
Platymeric index	0.77	0.81	0.74	0.80	0.81	0.82
Midshaft AP	29.7/26	26.6/28	30.0/18	26.8/17	29.0/8	26.3/11
Midshaft ML	28.8/27	25.4/28	28.9/19	25.6/17	28.5/8	24.9/11
Pilasteric index	1.03	1.05	1.04	1.05	1.02	1.06
Bicondylar length	430.0/18	399.0/21	436.4/11	399.8/10	420.1/7	397.7/11
Midshaft robusticity index	13.6	13.0	13.5	13.1	13.7	12.9
Cranial index	71.7	73.2	71.4	73.0	73.0	73.8

diameters. These indices have been shown to reflect phenotypical changes caused by labor-induced stress on the lower extremities (Ruff 1987). For example, a flattening of the femoral shaft in the anterio-posterior direction may suggest resistance to increased stress in the mediolateral direction, a flattening in the mediolateral direction may suggest resistance to increased stress in the anterio-posterior direction, and the absence of flattening could suggest even resistance to increased stress or no increased stress from more than one direction (Cole 1994). Thus a platymeric index with the value of one reflects a round or circular shaft geometry, while a value less than one reflects a flattening of the subtrochanter area in the anterio-posterior direction. Similarly, a pilasteric index of one reflects a round or circular midshaft geometry, while a value greater than one reflects a flattening of the femoral midshaft in a mediolateral direction.

Various researchers have argued that a decline in the flattening of the femoral shaft both at the subtrochanter and midshaft levels is a temporal trend that can be traced back to the late Pleistocene and is most likely a product of less physical mobility as populations adapted to a more sedentary lifestyle (Cole 1994; Kiesewetter 2006; Larsen 1997; Ruff 1987). In fact, Kiesewetter (2006) has suggested that the platymeric and pilasteric indices should be labeled as "mobility indices." Cole (1994), Larsen (1997), and Ruff (1987, 1992, 1994) argue that changes in the platymeric and pilasteric indices correlate with variation in the amount of physical activity and that such variation can be linked to the transitions from a hunting and gathering behavior to an agricultural, and possibly concentrated bison hunting, behavior (Cole 1994). This approach has been applied successfully by Kiesewetter (2006) to two Neolithic populations in the United Arab Emirates in which the platymeric and pilasteric indices clearly showed a higher degree of shaft flattening in the highly mobile group living in the mountains when compared to the less mobile group living in the coastal area and presumably focusing on an increasingly sedentary and agriculturally based behavior pattern.

However, there are exceptions to these rules. Jantz (1994) observes that predicted morphometric differences between population groups with diverse subsistence behavior do not always follow this trend. There are many reasons for this. For example, although it is assumed that changes in the size and shape of shaft cross sections are caused by environmentally controlled phenotypic factors, they may be influenced by more genotypic factors than previously anticipated.

These genotypic factors would, therefore, place limits on an expected phenotypic change.

Another factor is the robusticity index. An increase in the femoral midshaft robusticity index is not always followed by an increase in the pilasteric index. The femoral midshaft robusticity index is calculated by dividing the sum of the two perpendicular midshaft diameters by the bicondylar length of the femur. Thus, any increase in either or both the anterio-posterior or mediolateral diameters and/or a decrease in the bicondylar length of the femur would automatically increase the value of the robusticity index. Table 10.11 shows the indices calculated for the Bâb edh-Dhrâ' population. The platymeric index is low for both males and females, 0.77 and 0.8, respectively, showing a femoral flattening in the anterio-posterior direction below the greater trochanter. The pilasteric index is, respectively, 1.03 for males and 1.05 for females and thus indicates circular-shaped midshafts with little to no flattening. This is contrary to what would be expected for a nomadic or seminomadic population and is more representative of a population with a long history of sedentism. However, the Bâb edh-Dhrâ' population also shows a high level of robusticity that most likely is a reflection of several factors, including a well-developed femoral linea aspera and a shorter femoral bicondylar length. The larger than normal values for the robusticity index are congruent with a circular configuration of the femoral midshaft geometry and may represent a mobility pattern that does not specifically result in a resistance to increased stress from one direction but responds to stress from more than one direction.

Cole (1994) and Ruff (1987) have argued that the increased division of labor found in nonsedentary/nomadic populations would develop different platymeric and pilasteric values in males and females, thus producing smaller platymeric index values and larger pilasteric index values in the more mobile male group. We found a lower average platymeric index value in the male group (0.77) than in the female group (0.81). However, there was no difference between the values for the male and female pilasteric indices (Table 10.11).

Earlier in this chapter we argue that burials excavated during the 1977 and 1979 seasons may represent an earlier phase of the EB IA period (one with little or less sedentary behavior), while the 1981 burials represent a later portion of the EB IA period (one with increased sedentism). This could be reflected in the higher platymeric index (more circular) and lower pilasteric index (more circular) in the 1981 groups. This is evidenced when comparing the male groups from 1977 and 1979 to that from 1981; there is an increase of the platymeric index from 0.74 to 0.80, and a slight decrease in the pilasteric index from 1.04 to 1.02. In a similar comparison, the female groups show a slight increase in the platymeric index, from 0.80 to 0.82, and a slight increase in the pilasteric index, from 1.05 to 1.06, between the two burial samples. Thus, there is a minimal trend between the male groups but little to no trend between the female groups. These differences are very minor and do not compare to significant larger differences found between Neolithic populations in the United Arab Emirates (Kiesewetter 2006). This is probably a function of the limited temporal distance between the Bâb edh-Dhrâ' groups within EB IA (less than 100 years) and significantly larger temporal distance between the UAE Neolithic groups (several hundred years).

As a final note, it is important to recognize that we and other researchers are limited in our quest for data and information by small sample sizes, a high number of missing data, uncertainties in determining correct temporal distances, and uneven methodologies followed by various researchers. The results should be evaluated with this in mind.

Sexual Dimorphism

The EB I people of Bâb edh-Dhrâ' exhibit considerable sexual dimorphism. Univariate statistics were selected to show the degree of dimorphism between males and females for each variable. Both parametric (student t-test) and nonparametric (Kruskal-Wallis one-way analysis of variance) methods were used because of the small sample size, high number of missing data, and presence of some variables that did not show a normal distribution (Tables 10.8, 10.9). The means, sample size, standard deviation for each sex, and probability values (P) for both the t-test and the Kruskal-Wallis test are seen in Table 10.12. Nineteen variables out of a total of 41 show significant group variation with $P \leq 0.001$. Of these, four variables yield $P \leq 0.0001$, nine variables $P \leq 0.00001$. The results found when using the Kruskal-Wallis test are very similar and do not suggest any major differences between parametric and nonparametric methods. As expected, the highest degree of dimorphism is seen in the extremities, especially the lower extremities, and the two most significant variables in the cranium are the cranial maximum length and the nasal height.

Missing data replacement procedures, using both Systat and SAS, were used to test for appropriate application of multivariate statistical procedures. However, our results indicated that missing data replacement significantly altered the descriptive statistics when compared to results from the same groups with no missing data replacements. We decided not to apply multivariate procedures to describe and analyze sexual dimorphism of the Bâb edh-Dhrâ' burials.

Genetic factors are important in sexual dimorphism, however, the degree of expression can be highly influenced by nongenetic factors. One hypothesis is that egalitarian societies have less sexual dimorphism than less egalitarian societies that may have a much stronger cultural division between males and females (Binford 1972; Frohlich and Ortner 2000). If this hypothesis is correct, we should find greater sexual dimorphism in sedentary populations when compared to nomadic populations. To explore this, we selected ten cranial variables in two samples: Bâb edh-Dhrâ', representing a nomadic or seminomadic behavior, and Tepe Hissar, representing a sedentary behavior (Schmidt 1937). Out of ten p-values for ten variables, five scored better (lower p-value) in the Hissar sample, three scored similar p-values, and one variable scored

Table 10.12. Sexual Dimorphism in Metric Variables

Variable	N	Males Mean	SD	N	Females Mean	SD	t-test P	Kruskal-Wallis P
CRMAXL	34	186.6	7.2	41	178.4	5.9	0.00000	0.00000
CRMAXW	32	133.2	5.3	39	130.6	5.2	0.04277	0.07169
BIAURIC	24	118.7	5.5	27	115.1	4.9	0.01707	0.01675
BASBREG	14	133.6	3.6	21	129.5	5.6	0.01362	0.00886
BASNAS	12	100.5	2.5	20	95.4	4.8	0.00047	0.00095
BASPROST	11	96.0	3.7	15	92.5	5.3	0.06172	0.05055
CIRCUM	27	513.4	14.4	34	501.4	15.2	0.00256	0.00686
FRONTB	34	94.3	4.2	43	91.9	4.2	0.01255	0.00764
BIZYG	15	128.2	6.2	20	123.5	6.5	0.03474	0.01678
FACIALH	13	68.7	3.0	20	64.6	2.3	0.00032	0.00031
NASALH	13	50.5	1.1	20	47.0	1.7	0.00000	0.00001
NASALB	16	24.8	1.3	25	24.5	1.2	0.57706	0.57917
INTORBIT	15	23.3	2.3	19	21.3	1.7	0.01168	0.01610
ORBITB	16	38.3	1.8	21	37.9	1.8	0.45324	0.40930
ORBITH	15	33.3	2.3	19	31.4	2.2	0.01996	0.01553
PALATEL	19	55.2	2.9	24	52.5	2.9	0.00590	0.00980
PALATEB	15	63.1	3.4	18	59.1	2.7	0.00091	0.00138
MMAXL	19	102.4	4.3	28	96.5	4.4	0.00006	0.00010
VERTICH	20	53.3	5.9	29	50.1	5.3	0.06252	0.10081
BODYL	21	74.4	5.3	29	70.0	3.1	0.00212	0.00085
OBLIQUEH	20	59.4	4.5	29	55.6	4.3	0.00469	0.01297
GONIALA	22	32.5	6.6	30	33.6	6.8	0.53115	0.46273
BIGONIAL	15	97.9	6.3	20	92.7	7.8	0.03804	0.02865
BICONDYL	8	121.3	8.9	19	116.1	9.4	0.19416	0.12914
SYMPHYS	21	35.4	2.5	25	31.2	2.4	0.00000	0.00001
RAMUSW	25	32.2	2.3	32	30.7	2.4	0.01922	0.02931
HMAXL	16	300.1	16.3	20	282.2	10.9	0.00093	0.01232
HEPICOND	18	60.4	3.3	24	54.0	3.2	0.00000	0.00000
UMAXL	11	255.5	11.9	14	237.3	13.3	0.00151	0.00564
RMAXL	12	235.1	9.2	16	216.9	12.1	0.00013	0.00069
LUMBARH	10	129.4	5.2	11	125.5	5.2	0.10730	0.12049
FMAXL	22	431.6	18.4	21	404.2	16.3	0.00001	0.00002
FBICONDL	18	430.1	18.5	21	398.7	15.6	0.00000	0.00001
FBICONDB	15	78.9	3.0	14	71.7	4.1	0.00002	0.00015
FHEADD	24	45.2	2.2	28	41.2	2.5	0.00000	0.00000
FAPSUBTR	28	26.0	1.6	30	24.2	1.7	0.00008	0.00020
FMLSUBTR	28	33.9	3.0	30	30.0	2.0	0.00000	0.00000
FAPMIDSH	26	29.7	2.3	28	26.6	2.2	0.00000	0.00001
FMLMIDSH	27	28.8	2.5	28	25.4	1.8	0.00000	0.00000
TMAXL	14	355.3	18.4	13	329.3	14.2	0.00038	0.00074
FIMAXL	5	344.6	14.3	8	324.6	11.9	0.03325	0.02789

Note: Tests for both normal distributed data and nonnormal distributed data are included (t-test and Kruskal-Wallis). Only P values are presented.

better (lower p-value) in the Bâb edh Dhrâʿ sample. Because of these p-values, we argue that the Tepe Hissar skeletal sample is more dimorphic than the EB IA sample from Bâb edh-Dhrâʿ. The results from the univariate t-tests were further confirmed by conducting similar tests using discriminant function analysis. Although it was not possible to compare the "between groups F-matrix" values because of different sample sizes, the reclassification scores for sex clearly showed better separation of sex in the sedentary Tepe Hissar sample (100% for males and 93% for females) when compared to the analogous scores in the nomadic or seminomadic Bâb edh-Dhrâʿ sample (94% for males and 95% for females).

A word of caution is necessary when evaluating these results, especially when using the Tepe Hissar sample, which has a disproportionate number of males (N = 111) over females (N = 43). Hemphill (1999) and Nowell (1978) express the opinion that, due to selective retention of burial remains, the available skeletal sample from Tepe Hissar is not representative of the human population of the site; this limits the usefulness of the osteological data for demographic statistics. Also, the increased sexual dimorphism seen in midshaft and subtrochanter dimensions in the femora of nonsedentary populations as reported by Cole (1994) and Ruff (1987) suggests that different kinds of physical activity may not be the only cause of phenotypic variation between groups with different activity levels. There may be other causes for the phenotypic variation when comparing nomads to agriculturalists or males to females within groups that have significant divisions of labor.

Nonmetric Recordings

Nonmetric traits, also called epigenetic or discrete traits, are scored as present or absent except for the observation of the mandibular tori. Size for this trait was recorded as 1 (small), 2 (medium), or 3 (large). Although genetic factors affect the presence of the torus (Suzuki and Sakai 1960), the size of the torus is influenced by other factors, such as the antemortem loss of teeth (AMTL), especially the premolars (Frohlich and Pedersen 1992). Genetic control over the expression of nonmetric traits is poorly understood and is generally thought to be fairly weak (Mays 1998). Nonmetric traits appear to be independent of the presence or absence of pathological conditions and may have no correlation with activity patterns (Kiesewetter 2006). The traits have been used to compare similarities and dissimilarities between populations, and since they are nonselective they could become superior to metric variables in such studies. Another advantage of nonmetric trait recording is that they can be observed in fragmentary bones, thus adding data for group comparison in cases where metric variables cannot be recorded. However, the unknown genetic control of the majority of the traits is still being evaluated. This evaluation includes discussion of the validity of the traits in group comparisons.

We have included 27 nonmetric traits in the skull, including the presence of ossicles or wormian bones at the following locations: coronal suture, sagittal suture, lambdoid suture, squamous-parietal suture, and the interfrontal suture. Ossicles are also recorded at the bregma, lambda, asterion, pterion, and parietal notch. The presence of the following tori has been recorded: maxillary torus, palatine torus, auditory torus, and mandibular torus. The following foramina have been recorded: supraorbital foramina, supraorbital notch, infraorbital foramina, zygomatic facial foramina, parietal foramina, mental foramina, and an incomplete foramina ovale. Additional nonmetric traits include the presence of the mylohyoid bridge, metopic suture, and keeling, or Inca bone (Table 10.13). For detailed description and definition of nonmetric traits, see Berry and Berry (1967) and Buikstra and Ubelaker (1994). A total of 98 crania, including 46 males, 43 females, and 9 unknowns, were observed. The frequencies for males and females are presented in Table 10.13. The highest frequency for any trait was for the zygomatic-facial foramina, expressed in 96% of the male group and 82% of the female group. Ossicles in the lambdoid suture and at the asterion yielded high frequencies in both the male and female groups. The Inca bone was present in only 4.4% of the males and 4.8% of the females, while only one male had a palatine torus. No other tori were found in either the male or the female group. Variations between the male and female groups were found in the following traits: ossicles at the asterion, supraorbital foramina, and supraorbital notch (Table 10.13). There are very few data available for group comparisons. Kiesewetter (2006) has recorded nonmetric frequencies from a Neolithic skeletal collection dating to 4700 BC from Jebel Al-Buhais in the United Arab Emirates. Although there are similarities between Bâb edh-Dhrâʿ and Jebel Al-Buhais, the small sample sizes and the potential of disparate recording criteria may prohibit a direct comparison between the two groups.

Table 10.13. Nonmetric Data

	Trait	Male			Female		
		Count	Total	Percent	Count	Total	Percent
Ossicles							
	coronal suture	11	27	40.7	14	33	42.4
	sagittal suture	10	20	50.0	12	30	40.0
	lambdoid suture	28	33	84.9	26	35	74.3
	squamous-parietal suture	3	17	17.6	2	13	15.4
	interfrontal suture	1	4	25.0	0	6	0.0
	at bregma	0	35	0.0	1	39	2.6
	at lambda	6	29	20.7	9	34	26.5
	at asterion	18	24	75.0	12	24	50.0
	at pterion	5	15	33.3	7	14	50.0
	at parietal notch	9	26	34.6	10	19	52.6
	Inca bone	2	45	4.4	2	42	4.8
Tori							
	maxillary torus	0	29	0.0	0	24	0.0
	palatine torus	1	37	2.7	0	32	0.0
	auditory torus	0	21	0.0	0	20	0.0
	mandibular torus	0	27	0.0	0	26	0.0
Foramina							
	supraorbital foramina	26	40	65.0	14	33	42.4
	supraorbital notch	8	34	23.5	2	31	6.5
	infraorbital foramina	1	16	6.3	1	16	6.3
	zygomatic facial foramina	22	23	95.7	13	16	81.3
	parietal foramina	21	36	58.3	21	39	53.9
	mental foramina	0	26	0.0	3	21	14.3
	incomplete foramina ovale	0	11	0.0	4	18	22.2
Other traits							
	mylohyoid bridge	1	21	4.8	1	18	5.6
	metopic suture	3	44	6.8	5	41	12.2
	keeling	0	46	0.0	0	42	0.0

Note: Percentage is calculated from total count of traits present and total samples available for each trait.

GROUP COMPARISONS

A major objective of our research is to learn about potential biological/genetic relationships between the Bâb edh-Dhrâ' population and other contemporary and noncontemporary populations in the Middle East. These populations include those found in Mesopotamia, Egypt, the Levant, and possibly as far away as Nubia or Persia. Unfortunately, no skeletal sample available for comparative studies includes enough information—both on the archaeological context of the sample and the data collection of the sample—to reliably assume that the material is as representative of its original population from demographic and statistical points of view as is the Bâb edh-Dhrâ' sample. Because of this limitation, we present such comparisons with a warning that conclusions drawn from such evaluations may not be valid due to the unknown factors associated with samples collected many years ago by excavators and other researchers. Although we will make some comparisons, they must be considered tentative. We do hope that our Bâb edh-Dhrâ' skeletal population will provide baseline data for future research when better analytical and statistical methods are possible and when there are more representative skeletal samples available. At such a time, the comparative analysis between populations will be meaningful.

At the moment, we have to rely on data of other researchers. In our initial statistical analysis, we calculated biological distances between published samples from the Middle East. The results of this research indicated that the differences between populations were more likely a measure of different methodologies used by various researchers. To minimize such problems, we have been selective in the kind of variables used in our analysis. For example, common variables such as maximum length and width of the cranium are most likely compatible between researchers, whereas other measurements, such as the dimensions of the left and right orbits, might be flawed because researchers may have used different landmarks and measuring methods.

Also, in selecting variables to be used for between-group analysis, we chose variables with the lowest number of missing data points. This becomes important when using multivariate procedures that require burials with complete data sets. The variables included in the analysis became a function of the frequency of missing numbers and the availability of data of a particular variable collected by different researchers. Variables utilized for this reason may not be the variables most responsible for significant group variation. Kiesewetter (2006) reports similar restrictions when comparing a Neolithic skeletal sample from the United Arab Emirates with other Middle Eastern populations. She succeeded in obtaining distance measurements between 26 different populations by lowering the number of used cranial variables to four and accepting relatively low sample sizes.

We have selected six sites for comparison with Bâb edh-Dhrâʻ (Table 10.14). These include a single sample from Greece (Lerna), dated to between 2500 BC and 1150 BC; two from Upper Egypt, both dated to the Predynastic Period (Badari and Naqada); and one from Lower Egypt dated to the First Dynasty (Sedment). The remaining two samples were from Iran and both dated to between 3000 BC and 2000 BC (Tepe Hissar and Share-i Sokhtar; Table 10.14).

The Badari sample is from Al Badari, located approximately 19 kilometers south of Asyut in the extreme northern part of Upper Egypt. The series was excavated by the British School of Archaeology in Egypt between 1924 and 1925 and is assumed to be dated to the Predynastic Period (Stoessiger 1927). The skeletons were studied first by Stoessiger (1927) and later by Morant (1935). The series consists of 58 crania (36 males and 22 females; Stoessiger 1927).

Naqada, also known as Nubt, or "The Gold Town," is located on the west side of the Nile River about 25 kilometers north of the modern town of Luxor. The Naqada sample includes three major stages: Naqada 1 from 4400 to 3500 BC, Naqada 2 from 3500 to 3200 BC, and Naqada 3 from 3200 to 3000 BC. Naqada 1 shows cultural similarities with the Badari complex. During Naqada 2 and continuing to the end of Naqada 3, the Naqada culture expanded both to the south and to the north, where it subsequently replaced part of the Lower Egyptian cultures. Interestingly, the pear-shaped mace heads found in the Bâb edh-Dhrâʻ tombs show up in the Naqada 2 graves, where they replaced the earlier disc-shaped mace heads. Naqada 3 corresponds to the late Predynastic Period and defines the emerging civilization of ancient Egypt. Graves are most common in the Naqada 2 and 3 periods. More than 3,000 graves were identified by Flinders Petrie. These graves became the cornerstone in identification of Egyptian Predynastic cultures, with relatively simple mortuary practices, especially when compared to the monumental mortuary architecture of later Dynastic periods.

The Sedment site is located in Lower Egypt, not far from Herakleopolis and approximately 110 kilometers south of Cairo. It represents the most southern distribution of the Maadi–Budo groups of sedentary and agricultural population. The Maadi–Budo culture slowly declined during the same periods when the Naqada 2 and especially Naqada 3 culture expanded from the south toward Lower Egypt in the north. At the end of the Predynastic Period, it was most likely that the Naqada culture had expanded to such a degree that this resulted in a unification of Upper and Lower Egypt and the initialization of the First Dynastic Period.

During the same period, the Naqada influence in Lower Egypt is associated with a significant expansion of trade with Palestine and possibly as far away as Mesopotamia. The peak of this development took place during the same period that the nomadic or seminomadic Bâb edh-Dhrâʻ population was moving toward a more sedentary society and developed the first settlements near the Bâb edh-Dhrâʻ burial ground.

Table 10.14. Available Skeletal Populations Used to Calculate Variation between Groups

Site	Country/Area	Period (BC)	Females		Males		Total
			n	F-matrix from BeD	n	F-matrix from BeD	
Bâb edh-Dhrâʻ	Levant	3000–4000	31		18		49
Badari	Upper Egypt	3000–4000, Predyn	32	1.4284	45	4.7173	77
Tepe Hissar	Iran	2000–3000	49	0.7712	113	2.4290	162
Lerna	Greece	1000–2000	19	4.4388	39	4.9563	58
Naqada	Upper Egypt	3000–4000, Predyn.	166	0.7883	139	3.4440	305
Sedment	Lower Egypt	2000–3000, 1st D.	30	3.3160	40	9.3586	70
Share-i Sokhtar	Iran	2000–3000	20	0.7548	24	6.8359	44
Total			347		418		765

Lerna is located at the head of the Argolikos Bay on the eastern part of the Peloponnesus in modern Greece. It was first excavated by the American School of Classical Studies at Athens, in 1952. The human skeletal remains, which derive from excavations from 1953 to 1957, have been extensively studied by Angel (1971). The Bronze Age sample consists of 58 individuals (19 males and 39 females) ranging from the Early Bronze Age (2500 BC) to the end of the Late Bronze Age (1150 BC; Table 10.14). All statistics have been based on the original data from Angel's publication (1971). The subadults assigned sex in Angel's publication have been excluded.

Tepe Hissar is located in the northeastern section of modern Iran, adjacent to the southeastern end of the Caspian Sea. The site was excavated in 1931 by a joint expedition of the University Museum of the University of Pennsylvania and the Pennsylvania Museum of Art (Schmidt 1937). The skeletal remains have been studied extensively by Krogman (1940a, 1940b) and more recently by Rathbun (1982).

Several strata were identified: Hissar I (4000 BC to 3500 BC), Hissar II (3500 BC to 3000 BC), Hissar III (3000 BC to 2000 BC), Sasanian Period (350 AD), and Islamic Period (800 AD). Only skeletal remains from Hissar II and III have been included in our analysis, resulting in a sample size of 162 (113 males and 49 females; Table 10.14). All measurements and sex assignment are based on data published by Krogman (1940a, 1940b). G. W. Nowell (1978) has suggested that the selective retention of burial remains has made it impossible for the sample to be considered representative of the original population; therefore, it cannot be used to generate meaningful demographic statistics. Nevertheless, the basic data do provide some indication of the cranial morphology that allows cautious comparison with the Bâb edh-Dhrâ' data.

Share-i Sokhta, or "The Burnt City," was a major urban settlement dating to the Bronze Age and located in the Iranian province of Sistan. It dates to around 3200 BC. Share-i Sokhta, possibly the location of the legendary land of Aratta, built its wealth on extensive trade with Mesopotamia, possibly Turkey to the West, and Afghanistan and possibly India to the East. It was famous for a blue semiprecious stone known as lapis lazuli that has been traced to Mesopotamia and Egypt. The Share-i Sokhta settlement ranks in importance with the contemporary sites of Tepe Yahya and Tepe Hissar, also located in Iran. However, it is believed that the skeletal sample from Share-i Sokhta may be a better representation of the original living people when compared to the biased distribution found in the Tepe Hissar sample.

In our statistical analysis comparing the Bâb edh-Dhrâ' sample with the six other samples, we limit our use of the data to the cranial and postcranial skeletal elements from secondary burials. The sample sizes available for comparison are limited (Table 10.14) and there is the further issue of the degree to which the sample actually represents the original population. However, the basic tests for normality as reflected in kurtosis and skewness values suggest that the distributions are normal or reasonably close to normal.

Data from archaeological skeletal samples from the Middle East and southwest Asia were made available from an extensive database of cranial metric data (Froment, pers. comm. 2006). We selected populations dated to between 4000 and 2000 BC. We added a sample from the site at Lerna in Greece and compared our EB I data from Bâb edh-Dhrâ' to all the samples. Significant factors in selecting the skeletal samples for comparison with our data were the need for adequate sample sizes and minimal missing data for all the variables we hoped to use in our analysis. This eliminated samples from Kish (Sumer), Tell Sultan (Jericho), Abydos (Egypt), Arment (Egypt), Al Amrah (Egypt), Hou (Egypt), Maadi (Egypt), Cyprus, and Natuf (Israel) that, had they been adequate for our analysis, would have been helpful. Instead, we used only the samples from the sites described above (Table 10.14).

Within the protocol of data we gathered on the Bâb edh-Dhrâ' sample, we selected variables based on the fewest numbers of missing data points, the availability of the same variables between the comparative samples, and the assumption that the same definitions for the variables had been used by the researchers supplying the data. In our multivariate analysis, the following seven cranial variables were used: CRMAXL, CRMAXW, BASNAS, BIZYG, FACIALH, NASALH, and NASALB. Only the Bâb edh-Dhrâ' sample has a relatively large amount of missing data. This is the result of the secondary burial funerary tradition in which the skeletal elements that were brought to the Bâb edh-Dhrâ' cemetery varied between burials; in some cases the secondary burial included only a single skeletal element, while in other cases the secondary burial was almost complete. To minimize problems in statistical analysis, we did not include cases in the multivariate analysis unless most of the data were available for the given burial.

Basic descriptive statistics were calculated, including the mean, sample size, maximum, minimum, range, standard deviation, kurtosis, and skewness for each of the selected seven variables. In no case did the kurtosis and skewness values exceed values suggesting a nonnormal distribution of the data. With a limited number of burials, we did use missing data replacement to ensure a reasonable size for the various samples. The Bâb edh-Dhrâ' male group had the lowest sample size (N = 18) and the Naqada female group had the largest (N = 166). To ensure that missing data replacement did not affect the basic quality of the sample, we compared means, standard deviations, maximum and minimum values, kurtosis values, and skewness values. The values for data sets with missing data replacement were the same or very similar to the values apparent in the data set before missing data replacement. This new data set made it possible to use multivariate procedures that require complete data sets for each case in the sample.

We used discriminant function analysis to evaluate the degree of relationship between the selected samples; males and females were analyzed separately. The between-group F-matrix was calculated between each set of samples. The F-matrix statistics are based on the Mahalanobis D^2 distance and reflects the equality of the sample means measured as the distance between the centroids of any two samples. In order to use the probability value associated with the F statistics, we make the inference that all samples are representative of the populations from which they derive. This may not be true. For example, the F values describing the linear distance between sample centroids require similar variances and sample sizes to be an accurate representation of the sample differences. This is especially true when multiple samples are analyzed. A more accurate picture of the relationship between samples is provided when both F values and the canonical scores plot, using the first and second canonical variables, are used to define the relationship. The plot has not been included in this volume because of the difficulties in displaying the results in a black-and-white mode.

The F-matrix values can be seen in Tables 10.15 for males and 10.16 for females. The distance between population centroids of the male groups is smallest between the Bâb edh-Dhrâ' sample and the two Iranian samples, followed closely by that from Naqada (Table 10.15). However, when the same samples are viewed in the canonical scores plot, the group difference becomes least significant between Bâb edh-Dhrâ' and Naqada. A similar relationship is seen between the female groups (Table 10.16).

Because of potential differences in the degree to which the various samples are representative populations, caution is needed in interpreting these results. The limitations of the samples are further amplified by the sample sizes in most of the groups and especially the uneven sample size between groups. The latter has the potential to significantly alter the F-matrix values. Thus, when the female groups show higher F-matrix values than the male groups, a direct comparison can be misleading because of the different sample sizes in the male and female groups (Table 10.14).

Table 10.15. F-Matrix Statistics for Males

	Bâb edh-Dhrâ'	Badari	Tepe Hissar	Lerna	Naqada	Sedment	Share-i Sokhtar
Bâb edh-Dhrâ'	0.0000						
Badari	1.4284	0.0000					
Tepe Hissar	0.7712	2.4334	0.0000				
Lerna	4.4388	6.7797	9.3550	0.0000			
Naqada	0.7883	2.0625	1.6421	9.3967	0.0000		
Sedment	3.3160	3.6990	8.0570	6.5865	7.4887	0.0000	
Share-i Sokhtar	0.7548	1.8079	1.1057	3.4910	2.0220	3.7160	0.0000

Note: Statistics are based on discriminatory analysis using seven cranial metric variables only.

Table 10.16. F-Matrix Statistics for Females

	Bâb edh-Dhrâ'	Badari	Tepe Hissar	Lerna	Naqada	Sedment	Share-i Sokhtar
Bâb edh-Dhrâ'	0.0000						
Badari	4.7173	0.0000					
Tepe Hissar	2.4290	2.9517	0.0000				
Lerna	4.9563	4.8758	3.9115	0.0000			
Naqada	3.4440	2.8434	2.1001	5.6293	0.0000		
Sedment	9.3586	3.5109	4.8716	4.8509	5.8980	0.0000	
Share-i Sokhtar	6.8359	2.7500	3.6056	5.3653	6.1200	3.3744	0.0000

Note: Statistics are based on discriminatory analysis using seven cranial metric variables only.

Assuming that the relationships suggested by Tables 10.14 and 10.15 represent the populations from which the samples are drawn, we find it interesting that the Bâb edh-Dhrâ' sample is closer to the Naqada sample than to the Badari and Sedment samples. Because the geographic distance between Bâb edh-Dhrâ' and the Sedment site is smaller than the distance between Bâb edh-Dhrâ' and the site providing the other samples, we thought the closest relationship would be between these two samples. However, on the basis of what is known about group movements during the later phase of the Predynastic Period, our results could be reasonable. For example, the Naqada 2 and 3 populations expanded rapidly north and south and most likely replaced the more sedentary Maadi-Budo populations to which the Sedment sample most likely belonged. During this phase the Naqada people were also responsible for an increase in trade between Lower Egypt and Palestine, the Levant, and possibly places as far as Mesopotamia, much farther to the east.

Comparison of samples available to show the relatedness of the Bâb edh-Dhrâ' people to other populations in the Middle East and Northeast Africa highlights some methodological issues that need clarification. For example, it is known that the Tepe Hissar sample is biased; it includes many more male skulls than female (Table 10.14). This creates the potential, if not the likelihood, that this sample is not representative of the population and may not be suitable for comparative tests between populations. What is not clear is how significant this bias may be in showing relationships between various Middle Eastern populations. For example, the bias may be reflected in the large distance between the two contemporary Iranian populations (Tepe Hissar and Share-i Sokhta). This distance is larger than the distance between the centroids representing the Share-i Sokhta and the Egyptian Badari and Sedment samples (Tables 10.15 and 10.16).

To use multivariate analysis in defining relationships between archaeological populations, there are several aspects of research methodology, including definition of variables by different researchers, consistency in the data protocol between researchers, and the problem of missing data, that are almost inevitable for any human archaeological skeletal sample. Lack of comparability between samples can result in spurious conclusions regarding multivariate analysis. The analysis we present in this chapter is suggestive but must remain tentative until some of the methodological issues are clarified.

CHAPTER 11

The Paleodemography of the EB IA People

BRUNO FROHLICH AND DONALD J. ORTNER

PALEODEMOGRAPHY ATTEMPTS to reconstruct demographic parameters from past populations and use such information to interpret the health and well-being of an archaeological human population (Hoppa and Vaupel 2002). Such reconstructions can be based on bioarchaeological data derived from human skeletal remains. However, several assumptions have to be made in paleodemographic reconstructions. These include the assumption that the skeletal population is representative of the living population and that age-at-death and sex estimates are accurate (Ubelaker 1989). Archaeological skeletal samples have inherent limitations under which these assumptions cannot be fully met. These limitations include inadequate sample size, difficulty in obtaining accurate age estimates (particularly in older individuals), inaccurate attribution of sex in some burials, movement of people from other societies into and out of the population being studied, difficulty in isolating the different generations interred in an archaeological cemetery, and sample bias due to incomplete retrieval of burials during either excavation or subsequent curation of the sample.

Related issues that affect paleodemography include factors such as the time span during which the cemetery was in use as well as the type of burial and the burial structure that may have resulted in a biased representation of those who died in the living population associated with the cemetery (Acsadi and Nemeskeri 1970; Hoppa and Vaupel 2002; Littleton 1998; Kiesewetter 2006; Smith and Horwitz 2007). Techniques and procedures developed by modern demographers based on living populations have been used as a baseline for research on ancient populations. Wood (1994) notes some problems using modern populations for comparison with bioarchaeological samples because of the change in the subadult mortality rates and changes in the reproductive patterns, especially a reduction in the number of live births per woman. However, research by demographers studying modern but less "developed" or more "primitive" and isolated populations has shown that such studies may be applicable to the interpretation of ancient demographic data (Howell 1976).

There are three categories of data used in the reconstruction of demographic profiles for ancient populations: (1) skeletal data from archaeological burials; (2) data from modern populations that are less influenced by modern advances in reproductive control and health care and with lifestyles assumed to be similar to those of ancient populations; and (3) indirect data derived from settlement patterns including house sizes, estimates of the average number of individuals associated with each room, and village or town sizes. Skeletal data and data from modern tribal populations traditionally have been favored over archaeological data such as settlement information (Ward and Weiss 1976).

In general, the adequacy of data collected from archaeological skeletal samples relative to the quality of the data obtained from modern populations is an important consideration when reconstructing and interpreting demographic profiles of ancient populations (Ward and Weiss 1976; Weiss 1973). In paleodemographic reconstructions, it is important to understand the limitations inherent in data from archaeological skeletal samples and utilize demographic methods that minimize the effect of these limitations in creating a demographic profile.

The Bâb edh-Dhrâʿ EB IA sample is important, based on several of the factors relevant in paleodemographic analysis, including: (1) it is represented by a large number of individuals, (2) both the age and sex distributions of the sample are congruent with the assumption that the sample is a good representation of those who died during the EB IA time period, (3) at least some skeletal elements were recovered for most of the individuals originally interred in the tomb chambers, (4) the archaeological context for the sample is carefully documented, and (5) the span of time (ca. 3300 to 3200 BC) associated with the burials is fairly well established. These factors address some of the problems identified by Smith and Horwitz (2007), Weiss (1973), Ward and Weiss (1976), and Buikstra and Konigsberg (1985).

The total number of burials recovered at Bâb edh-Dhrâ' is 578; 41% of these individuals are adults, while 59% are subadults. The gender ratio, based on all burials that could be assigned a sex, is 50.4% males to 49.6% females. This breakdown, combined with careful control of the excavations of tombs and relatively complete retrieval of most skeletal elements associated with the burials, creates a sample that is as representative of the EB IA living population as an archaeological skeletal sample can be. Therefore, we have increased confidence in the results of our demographic analysis reported below.

This chapter is divided into three sections: (1) data preparation and assignment of isolated skeletal elements to appropriate age and sex cohorts, (2) creation and evaluation of life tables and estimation of the living population size, and (3) speculation regarding the population's social composition.

DATA PREPARATION

Three adjustments have been made to the Bâb edh-Dhrâ' data. Age-at-death estimates, based on skeletal data, have been discussed in earlier chapters. In modern hunter and gatherer and nomadic societies the onset of female sexual maturity occurs at about 18 years of age and the first child is born around the age of 20 years (Howell 1976; Pennington 2001; Wood 1994). Because of this, we have defined the age of 18 as the dividing point between subadults and adults in the Bâb edh-Dhrâ' skeletal population. In addition, for the purpose of the construction of life tables, the population has been divided into five-year age cohorts starting with birth and ending with year 50. The upper end of the age scale has been selected because of the difficulties and range of error in assigning accurate age estimates to burials past 50 years of age. This creates a potential increase in the expected number of years to live after reaching 50+, with the last age cohort (50+) including burials represented by several age cohorts above year 50. We have not applied corrections such as the Gompertz model, which uses estimates of the 50+ cohorts by inferring a log-linear extrapolation of cohort data calculated from decrements in the mortality rates for cohorts younger than 50 years (Littleton 1998). For this reason, the calculated life expectancy for the last age cohort of 50+ may be unrealistically high.

The first stage in the data preparation involved assigning burials to five-year age cohorts. Any original age-at-death estimate including a range larger than one of the five-year categories has been assigned proportionately to two or more cohorts depending on the original estimate's age range (Table 11.1). In the second stage, we apportioned burials with the general age assignment of "adult" into specific adult five-year age cohorts. Eighty-three burials were initially assigned to the adult group (Table 11.1). In order to include these burials in our analysis, we calculated the percentage of burials in each age cohort in the subsample where an age estimate was possible and apportioned the 83 burials into the five-year adult age cohorts based on the percentage of burials that could be aged with reasonable accuracy. These adjustments were made independently for males and females (Tables 11.2, 11.3).

The third correction involved the allocation of an age category for subadult and some adult burials with no original assignment into a male or female group. We have seen in Chapter 10 that the number of male and female burials are equally distributed (Table 10.2). Because of this, we made the assumption, for the purpose of calculating life tables, that half the burials with no sex identified were assigned to each of the sex categories (Tables 11.2 and 11.3). This does not include five burials that were allocated an adult age at death but could not be assigned a sex (Table 11.1) because of the limited number of skeletal elements, which disallowed certainty about their adult status.

PROBABILITY OF DEATH AND LIFE EXPECTANCY

Fifty-nine burials (10.3%) have been aged as late prenatal and were not included in the calculation of life tables. The prenatal group will be included later in this chapter when we discuss the estimated time between births.

Table 11.1. Distribution of Males, Females, and Unknowns; Raw Data

Age	Males	Females	Unknowns	Total
PN	0.0	0.0	59.0	59.0
0–0.49	0.0	0.0	49.0	49.0
0.5–0.9	0.0	0.0	57.5	57.5
1–4.9	0.0	0.0	70.5	70.5
5–9.9	0.0	0.0	62.5	62.5
10–14.9	1.3	3.0	18.8	23.1
15–19.9	11.0	16.2	2.8	30.0
20–24.9	12.0	10.4	1.0	23.4
25–29.9	21.1	13.2	1.0	35.3
30–34.9	12.7	16.2	0.0	28.9
35–39.9	8.3	7.5	0.0	15.8
40–44.9	6.3	5.5	0.0	11.8
45–49.9	4.8	5.5	0.0	10.3
50+	7.8	7.5	0.0	15.3
adults	41.0	37.0	5.0	83.0
Total	126.3	122.0	327.1	575.4

Table 11.2. Distribution of Males, Corrected for Unknown Values

	n	Percent	Adjustment	Adjusted (1)	50% of Unknowns	Adjusted (2)
PN	0.0	0.0	0.0	0.0	29.5	29.5
0–0.49	0.0	0.0	0.0	0.0	24.5	24.5
0.5–0.9	0.0	0.0	0.0	0.0	28.8	28.8
1–4.9	0.0	0.0	0.0	0.0	35.3	35.3
5–9.9	0.0	0.0	0.0	0.0	31.3	31.3
10–14.9	1.3	1.5	0.6	1.9	9.4	11.3
15–19.9	11.0	12.9	5.3	16.3	1.4	17.7
20–24.9	12.0	14.1	5.8	17.8	0.5	18.3
25–29.9	21.1	24.7	10.1	31.2	0.5	31.7
30–34.9	12.7	14.9	6.1	18.8	0.0	18.8
35–39.9	8.3	9.7	4.0	12.3	0.0	12.3
40–44.9	6.3	7.4	3.0	9.3	0.0	9.3
45–49.9	4.8	5.6	2.3	7.1	0.0	7.1
50+	7.8	9.1	3.7	11.5	0.0	11.5
adults	41.0				2.5	
Total	126.3	99.9	40.9	126.2	163.7	287.4

Note: The total adjustment—Adjusted (2)—does not include 2.5 burials with unknown age distribution and unknown sex.

Table 11.3. Distribution of Females, Corrected for Unknown Values

	n	Percent	Adjustment	Adjusted	50% of Unknowns	Adjusted (2)
PN	0.0	0.0	0.0	0.0	29.5	29.5
0–0.49	0.0	0.0	0.0	0.0	24.5	24.5
0.5–0.9	0.0	0.0	0.0	0.0	28.8	28.8
1–4.9	0.0	0.0	0.0	0.0	35.3	35.3
5–9.9	0.0	0.0	0.0	0.0	31.3	31.3
10–14.9	3.0	3.5	1.3	4.3	9.4	13.7
15–19.9	16.2	19.1	7.1	23.3	1.4	24.7
20–24.9	10.4	12.2	4.5	14.9	0.5	15.4
25–29.9	13.2	15.5	5.7	18.9	0.5	19.4
30–34.9	16.2	19.1	7.1	23.3	0.0	23.3
35–39.9	7.5	8.8	3.3	10.8	0.0	10.8
40–44.9	5.5	6.5	2.4	7.9	0.0	7.9
45–49.9	5.5	6.5	2.4	7.9	0.0	7.9
50+	7.5	8.8	3.3	10.8	0.0	10.8
adults	37.0				2.5	
Total	122.0	100.0	37.1	122.1	163.7	283.3

Note: The total adjustment—Adjusted (2)—does not include 2.5 burials with unknown age distribution and unknown sex.

With 177.2 burials (31%) in the 0–5 age cohort, the expected mortality rates are similar to the rate reported by Buikstra and Mielke (1985). The mortality for the 15–20 year cohort is higher than that of the 10–15 year cohort; 264.8 burials out of a total 570.7, or 46.4% of burials, have age estimates of younger than 15 years at death (Table 11.4; Figures 11.1, 11.2, 11.3, 11.4). This figure does not include 59 prenatal burials.

The probability-of-death curve for the male and female groups and the combined male and female group is similar to the U-shaped curve reported in other paleodemographic research. In this curve, there is a high probability of death at the time of birth. The lowest probability of death occurs between 10 and 15 years and is followed by a steady increase in the death rate, from the 15–20 year cohort until the 40–45 year cohort; the death rate increases even more

Table 11.4. Distribution of Males and Females, Adjusted

	Adjusted (M)	Adjusted (F)	Total
PN	29.5	29.5	59.0
0–0.49	24.5	24.5	49.0
0.5–0.9	28.8	28.8	57.6
1–4.9	35.3	35.3	70.6
5–9.9	31.3	31.3	62.6
10–14.9	11.3	13.7	25.0
15–19.9	17.7	24.7	42.4
20–24.9	18.3	15.4	33.7
25–29.9	31.7	19.4	51.1
30–34.9	18.8	23.3	42.1
35–39.9	12.3	10.8	23.1
40–44.9	9.3	7.9	17.2
45–49.9	7.1	7.9	15.0
50+	11.5	10.8	22.3
Total	287.4	283.3	570.7

Note: The total does not include five burials with unknown sex and age distributions.

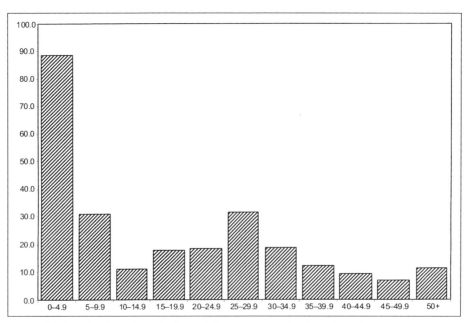

Figure 11.1. Number of deaths in each five-year age cohort, males only.

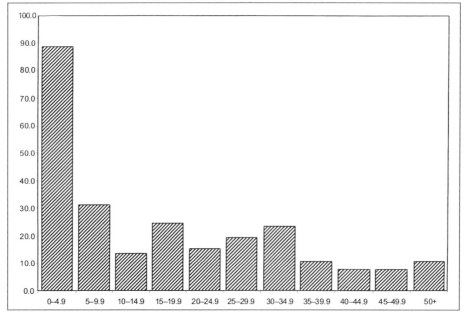

Figure 11.2. Number of deaths in each five-year age cohort, females only.

significantly for the older age cohorts (Figures 11.5, 11.6, 11.7.). There is a slight variation between male and female probabilities of death between 15 and 40 years. Females produce a higher value between 15 and 20 years and between 30 and 35 years, while males have an increased probability of death between 20 and 30 years (Figures 11.5, 11.6, 11.7; Tables 11.5, 11.6). Most likely, the increase in female probability of death between 15 and 35 years, with a slight drop around age 20 to 25, is related to higher female mortality caused by pregnancy and childbirth. The higher mortality in the male group between 20 and 30 years may be associated with increased labor-related risks for this group (Figure 11.7).

Life tables have been calculated separately for males and females as well as for the combined male and female group (Tables 11.5, 11.6, 11.7; Figure 11.8). The following codes for the table have been used (Littleton 1998; Ubelaker 1986): D_x is the number of deaths in each age cohorts as defined in Table 11.4. Note that the 0–4.9 year age cohorts for males, females, and combined males and females have been divided into three smaller age cohorts: 0–0.49, 0.5–0.9, and 1–4.9. d_x is the number of burials in each age cohort as a

254   CHAPTER 11

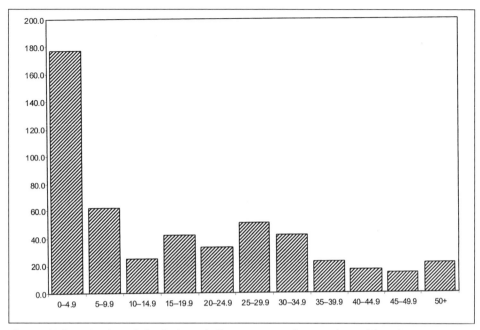

Figure 11.3. Number of deaths in each five-year age cohort, males and females.

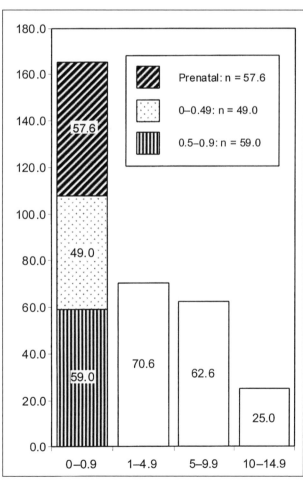

Figure 11.4. Number of deaths in subadult age cohorts. Infant column divided into three groups: (1) prenatals; (2) infants from 0 to 0.49 years; and (3) infants from 0.5 to 0.9 years.

ratio where $d_x = (D_x * 100)/(\text{sum of all } D_x)$. l_x is the number of individuals still alive in each category as a percentage of the original population: $l_x = l_{(x-1)} - d_{(x-1)}$. q_x is the probability of death where $q_x = d_x/l_x$. L_x is the total number of years lived by all burials in each age cohort: $L_x = n(l_x + l_{(x+1)})/2$, where $l_{(x+1)}$ is the number of survivors entering the following age cohort and n is the number of years included in each age cohort. T_x is the total number of years to be lived after reaching a specific age cohort where $T_x = L_x + L_{(x+1)} + L_{(x+2)} \ldots \ldots L_{(x+n)}$. e_x is the life expectancy or the number of years an individual can expect to live after reaching a specific age cohort. $e_x = T_x/l_x$.

The prenatal burials have not been included in the life table calculations (Tables 11.5, 11.6, 11.7). At the time of birth, an individual could have expected to live about 17.6 years (Table 11.7). If the individual survived the first five years, life expectancy after that time was 20.5 years; it was 14.3 years if the person survived the first 20 years (Table 11.7). The calculated life expectancies are remarkably similar for the males and the females, although males from birth until around 20 years of age appear to have had slightly higher life expectancy values than the females. This circumstance reverses between 20 and 30 years of age, when the females appear to have had a slight advantage relative to the males (Tables 11.5, 11.6; Figure 11.8).

In general, the differences between males and females as expressed in life expectancy values are less than expected and differ significantly from Littleton's results (1998) based on a sedentary Hellenistic sam-

Table 11.5. Life Table for Males

x	D_x	d_x	l_x	q_x	L_x	T_x	e_x
0–4.9	88.6	34.4	100.0	0.344	414.1	1783.4	17.8
5–9.9	31.3	12.1	65.6	0.185	297.9	1369.2	20.9
10–14.9	11.3	4.4	53.5	0.082	256.6	1071.4	20.0
15–19.9	17.7	6.9	49.1	0.140	228.5	814.8	16.6
20–24.9	18.3	7.1	42.3	0.168	193.6	586.3	13.9
25–29.9	31.7	12.3	35.2	0.350	145.1	392.7	11.2
30–34.9	18.8	7.3	22.9	0.319	96.2	247.6	10.8
35–39.9	12.3	4.8	15.6	0.306	66.0	151.4	9.7
40–44.9	9.3	3.6	10.8	0.333	45.1	85.4	7.9
45–49.9	7.1	2.8	7.2	0.382	29.2	40.3	5.6
50–54.9	11.5	4.5	4.5	1.000	11.2	11.2	2.5

Table 11.6. Life Table for Females

x	D_x	d_x	l_x	q_x	L_x	T_x	e_x
0–4.9	88.6	34.9	100.0	0.349	412.7	1725.8	17.3
5–9.9	31.3	12.3	65.1	0.189	294.6	1313.0	20.2
10–14.9	13.7	5.4	52.8	0.102	250.3	1018.4	19.3
15–19.9	24.7	9.7	47.4	0.205	212.5	768.1	16.2
20–24.9	15.4	6.1	37.6	0.161	173.0	555.7	14.8
25–29.9	19.4	7.6	31.6	0.242	138.7	382.7	12.1
30–34.9	23.3	9.2	23.9	0.384	96.6	244.0	10.2
35–39.9	10.8	4.3	14.7	0.289	63.0	147.4	10.0
40–44.9	7.9	3.1	10.5	0.297	44.6	84.3	8.1
45–49.9	7.9	3.1	7.4	0.422	29.1	39.7	5.4
50–54.9	10.8	4.3	4.3	1.000	10.6	10.6	2.5

Table 11.7. Life Table for Males and Females

x	D_x	d_x	l_x	q_x	L_x	T_x	e_x
0–4.9	177.2	34.6	100.0	0.346	413.4	1754.8	17.6
5–9.9	62.6	12.2	65.4	0.187	296.3	1341.4	20.5
10–14.9	25.0	4.9	53.1	0.092	253.5	1045.1	19.7
15–19.9	42.4	8.3	48.3	0.172	220.5	791.6	16.4
20–24.9	33.7	6.6	40.0	0.165	183.4	571.1	14.3
25–29.9	51.1	10.0	33.4	0.299	141.9	387.7	11.6
30–34.9	42.1	8.2	23.4	0.352	96.4	245.8	10.5
35–39.9	23.1	4.5	15.2	0.298	64.5	149.4	9.9
40–44.9	17.2	3.4	10.7	0.316	44.9	84.9	8.0
45–49.9	15.0	2.9	7.3	0.402	29.1	40.0	5.5
50–54.9	22.3	4.4	4.4	1.000	10.9	10.9	2.5

ple from Bahrain. In that sample, the life expectancies and probability-of-death values for females strongly suggest increased mortality rates in the female age cohorts between 15 and 30; these increased rates most likely are associated with childbirth. Why these differences are not observed in the Bâb edh-Dhrâ' population is not clear. Most likely there is no difference in the relative risk associated with childbirth between Bâb edh-Dhrâ' and the Hellenistic Bahrain populations. The difference might be the result of a higher risk associated with a nomadic or semi-nomadic lifestyle in the Bâb edh-Dhrâ' male group when compared with that of the sedentary behavior associated with the Hellenistic Bahrain male group. This is further supported by the higher probability-of-death value for the male group between 20 and 30 years of age (Figures 11.7, 11.8).

Factors affecting the high mortality during the first five years of life include poor nutrition and infectious diseases. Young children are particularly vulnerable to malnutrition when they are weaned from mother's milk and have a diet consisting largely of agricultural products. Such diets may be deficient in vitamins and minerals as well as the amino acids needed for protein synthesis. Weaning normally takes place between the ages of two and four years. In the Bâb edh-Dhrâ' EB IA sample, the earliest evidence of stress affecting dental development (enamel dental hypoplasia) occurs at about two years of age. This is discussed in greater detail in Chapter 12, but weaning, and the complications of malnutrition associ-

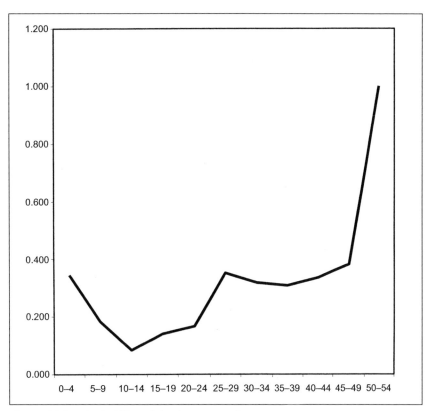

Figure 11.5. Probability-of-death curve, males.

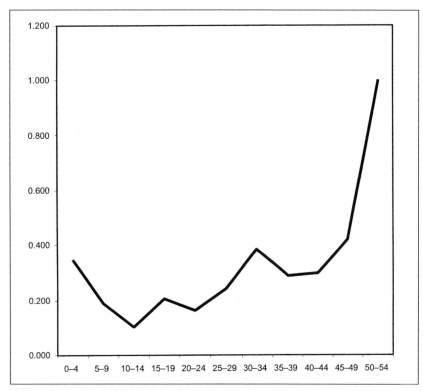

Figure 11.6. Probability-of-death curve, females.

ated with it, is among the most likely causes of this initial stress event. It is likely a major factor in the high mortality associated with the 1–5 year age cohort.

Despite the high subadult mortality and low life expectancy, the EB IA people survived and provided the base for developments in the EB IB period. During EB

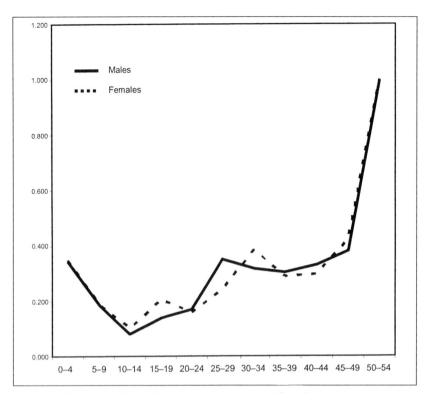

Figure 11.7. Probability-of-death curves, males and females.

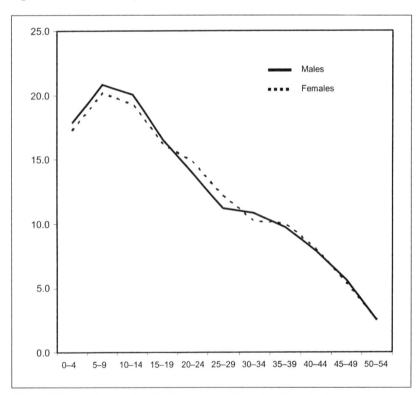

Figure 11.8. Life expectancies for males and females. Note the increase in life expectancies between time of birth and five to ten years.

IA, we assume a very low growth rate, probably in the range of 0.001 to 0.005 (0.1% to 0.5%) per year, that would not allow for any major increase of the population size within a hundred years. Therefore, there may have been no significant change in the population size between EB IA and EB IB. Between 55% and

64% of burials are less than 15 years old. Thus, more than half of the live births would have died before they reached the age of sexual maturity. To maintain a stable population, each woman would have had to bear five children in order for at least two of them to reach sexual maturity and contribute to the stability of the population size. With a life expectancy of 16.2 years for the females between 15 and 20 years of age, five children would have had to be produced within 16 years and with approximately 3.2 years (38 months) between each birth (Table 11.6).

The potential total of live births for a woman 35 years of age is about 20% lower than for a woman in her early twenties (Wood 1994). This means that the average time between live births in the lower age cohorts actually would have had to be shorter than the suggested time of 38 months. Other factors play important roles in determining time between births. Prolonged and intensive breast-feeding delays the resumption of ovulation and tends to increase the time between births. Because of the nutritional benefit inherent in mother's milk, it is likely that breast-feeding would have been prolonged whenever possible, giving the child a better chance of defense against infectious diseases.

Fetal and infant mortality also affect the time between births, since breast-feeding is terminated with infant death; the cessation of breast-feeding allows for ovulation to resume more regularly. Half of all the subadults who died before reaching sexual maturity died between birth and three years of age. How did this affect the time between births? Two-thirds of the subadults dying before they reached three years of age would have been less than one year old and, thus, died while the mother was still lactating. Considering a timeline that began with the birth of a living child, defined as the index child (Wood 1994), lactation would have been interrupted if the child died within the first year, and there would have been an early resumption of ovulation. This would have shortened the time between births and increased the fertility rate. If the index child died between one and two years, the effect would be similar but less significant. If the index child survived two years and the mother became pregnant before weaning the index child, then the chances of the index child dying would have increased. This would have been partially the result of an increase in the mother's steroid levels due to pregnancy, which would have been likely to interfere with lactation. Thus, death of the index child during the first year of its life may have shortened the time between births and increased the mother's fertility, while death of the index child during a new pregnancy or after the birth of a second child may have increased the odds of the child's death due to a decrease in the mother's milk production. In turn, there would have been increased competition between the two children to gain access to the mother's milk.

Howell (1976) characterized prehistoric populations as belonging to one of four types: (1) high fertility and low mortality (growing population), (2) high fertility and high mortality (stable population), (3) low fertility and low mortality (stable population), and (4) low fertility and high mortality (declining population). Subsequent research on an archaeological skeletal sample supports this observation (Littleton 1998). We argue that the Bâb edh-Dhrâ' EB IA population could be classified as the second group (high fertility and high mortality), which would have provided the population stability for the transition to the EB IB population. However, when the population's increased prenatal mortality is considered, it seems that there may have been times when the society would have been classified in the fourth group (low fertility and high mortality). This is partly the result of the expanded time between births as a result of prenatal mortality and would have allowed for a negative growth rate, or population decline.

One of the remarkable findings in the EB IA skeletal sample is the large number of fetal skeletons. Most of these individuals are in the late stage of fetal life, but a few are from the early phase of the third trimester in fetal development. The cause of death cannot be determined in any of these burials. The possible causes of death include: (1) maternal mortality from any of several causes, including complications of pregnancy, infection, or other disorders such as severe malnutrition; (2) abortion, either induced or from natural causes; or (3) genetic defects in the fetus that made it incompatible with life. In our study, we include all forms of intrauterine mortalities and group them under fetal loss (Wood 1994).

Clinical studies have shown that fetal loss is underreported because loss, particularly during early gestational periods, may be unreported. In bioarchaeological research it is rare to recover fetal remains and, in particular, it is rare to recover them in the numbers present in the EB IA sample. This is highlighted by the fact that no fetal remains and very few infant or young child skeletal elements were recovered in the EB IB sample. Our analysis of fetal loss in the Bâb edh-Dhrâ' EB IA sample is largely limited to skeletal material in the third trimester of fetal development. It is probable that there were fetal losses during earlier stages and that these deaths were not interred in the EB IA chamber; therefore, these losses are underrepresented in our sample.

All fetal losses result in an increase of the time between live births and a lowering of the fertility rate (Wood 1994). This is contrary to the potential decrease in time between births and the potentially higher fertility rate overall that result from infant loss or infant mortality. The added time is a product of several factors, including the incomplete gestation, a waiting period before the next ovulation, and a fecund waiting period before the next conception (Wood 1994). The percentage of pregnancies ending in fetal loss depends on factors such as the pregnancy order and the age of the mother. Thus, the estimated frequency of fetal loss in women between 15 and 18 years of age may be around 15%.

This frequency of fetal loss increases as the woman becomes older and has more pregnancies. There are exceptions to this; older women who experience their first pregnancies show higher fetal loss frequencies than similarly aged women who have gone through several earlier pregnancies. However, the more pregnancies that a woman, regardless of her age, experiences, the more likely it is that she will have a pregnancy resulting in fetal loss; in women between 15 and 25 years of age, more than 30% of pregnancies result in fetal loss if the individuals have undergone more than four or five previous pregnancies.

The increased time between births caused by fetal loss becomes more significant as the gestational age of the fetus increases. Thus, early fetal loss may increase time between births by only a few days, while fetal loss at the end of the gestational period, and just before a normal birth, could add two months to the time between births. The death of an infant results in an earlier resumption of ovulation and subsequent pregnancy, while the additional time between successful births increases independently as a result of fetal loss and possibly as a result of the loss of an infant or a child.

The number of prenatal burials in the Bâb edh-Dhrâʿ sample is 59, which is about half the total number of infant burials in our sample. Both prenatal and infant deaths increase the time between live births; infant death causes ovulation to resume earlier than would occur under normal circumstances and thus speeds up the possibility of subsequent pregnancy and potential successful birth. However, independent of fetal loss or infant and early childhood deaths, the increased time between live births results in added pressure on the female to reach the birth quota that is necessary for the maintenance of a stable population.

It is likely that the EB IA population lived a reproductive life resulting in a neutral growth rate. With a potential underrepresentation of early prenatal burials, the potential exists for a negative growth rate, in which case the population would not have been sustained. The adaptive pressures created by this situation could have been one of the contributing factors stimulating a change from a nomadic or seminomadic lifestyle to a more sedentary one.

Although it seems likely that gradual but significant cultural changes took place during the transition between the EB IA and EB IB societies, the cultural evidence suggests that this change developed within a continuing society and not as a result of the superimposition of a new society in the area. The factors that influenced this change are not known, but the evidence of improved health in the EB IB sample suggests better cultural adaptation through some mechanism. Perhaps a more settled agricultural community had a more stable food source that improved the community's nutrition.

ESTIMATION OF POPULATION SIZE

When life table data are combined with the number of burials and a time span expressed in years, we can estimate the average number of living people needed to produce a certain amount of burials in a certain number of years. This number is dependent on life expectancy at birth; with a fixed number of burials, a lower life expectancy will require a higher average number of people to be associated with the cemetery, and a higher life expectancy will require a smaller number of people. It does not take into consideration any fluctuation in mortality, changes in the male to female ratio, or increases or decreases in population size caused by external factors such as migration, emigration, or warfare. As limited as such numbers can be, they give us an idea about the likely size of the population within a given time span. Population size reconstructions become very powerful tools that allow us to define future research with a focus on clarifying some of the poorly understood demographic factors described above. Also, demographic reconstructions provide a basis for comparing the relative adaptive success of past human populations. However, these reconstructions will still be limited if samples are not representative of the living population.

The data obtained from the Bâb edh-Dhrâʿ cemetery are based upon a careful study of the landscape in association with known locations of several hundred tombs and burial chambers. Conductivity readings helped identify the potential density of tombs in different areas. Such data give us an estimate of the total size of the cemetery, and with that we can estimate the total number of people buried in the cemetery during the EB IA time period.

Within the cemetery there was a high tomb density zone clustered around area A in the eastern section of the cemetery. The other areas of the cemetery vary in tomb concentrations but generally are less dense. The high-density zone comprises an area that is approximately 15 hectares; the low-density area is about 18 hectares. Based on conductivity readings, the tomb density in the high-density zone is estimated to be 170 tombs per hectare. In the low-density zone we estimate that the average density is 17 tombs per hectare. This results in an estimate of 2,550 tombs in the high-density zone and 306 tombs in the low-density area. The combined estimate for the entire cemetery area is 2,856 tombs.

Our estimate of the total number of burials in the EB IA cemetery is based on a total of 512 burials and excludes the prenatal burials from the tomb chambers we excavated. These chambers contained an average of 8.1 individuals in each chamber and an average of 19.7 burials in each tomb. In accordance with the estimate of 2,856 tombs, there should be approximately 56,263 burials interred in the cemetery.

Given a life expectancy at birth of 17.6 years and about 100 years associated with the EB IA phase, the necessary size of the living population needed to provide 56,263 burials is about 9,902 individuals at any given time. Based on one person per square kilometer, which is the typical population density for nomadic and seminomadic populations, the people associated with the Bâb edh-Dhrâ' burial ground would have required about 10,000 square kilometers, or 100 kilometers by 100 kilometers, so ours is not an unrealistic estimate of the land area needed by the people of Bâb edh-Dhrâ' at this early stage in utilization of the site.

SOCIAL COMPOSITION

On the basis of current archaeological evidence, the EB IA people using the cemetery at Bâb edh-Dhrâ' probably were a nomadic or seminomadic society. However, remember that the evidence for this conclusion is limited and strong arguments can be made for an alternative economy for the EB IA society (Bentley 1987; see also Chapter 13). On the basis of research on modern nomads/pastoralists, the Bâb edh-Dhrâ' EB IA people would have required more than 10,000 square kilometers of land to support their way of life. We argue that the population had high mortality and fertility rates. This is reflected in low life expectancies (less than 20 years at birth), high numbers of prenatal and subadult mortality, and an even amount of males and females.

There are minimal differences between the probabilities of dying and life expectancies of males relative to females. There is a slight decrease in the female life expectancy between 15 and 20 and between 30 and 35 years, while males appear to have a lower life expectancy between 20 and 30 years. The lower life expectancies represented by the female group are most likely associated with childbirth. The minimal or limited difference between the males and females may be a reflection of males engaged in riskier activities, which made the likelihood of their deaths about equivalent to that of females during childbearing years.

The number of prenatals, infants, and younger children suggests a high level of subadult mortality and is similar to data from other preindustrial skeletal samples. What makes the Bâb edh-Dhrâ' sample somewhat different is the prevalence of burials within the infant and early childhood age cohorts. The larger number of infants in the 0.5–1 year cohort when compared to the 0.0–0.5 cohort likely reflects the loss of passive immunity that is present in newborn, breast-feeding infants and that lasts only about six months after birth. If an infant does not achieve adequate active immunity during the first six months of life, it will be at considerable risk from infectious pathogens. This risk will be even greater if the infant has insufficient nutrition.

High fertility rates combined with high mortality rates suggest a stable or stationary population with little or no growth rate. The possibility exists that, at least during some years within the EB IA period, lower fertility rates caused by an increase in fetal loss or prenatal mortality may have caused a temporary decline in the Bâb edh-Dhrâ' population. It may be that factors such as change in the environment, external population pressure, and high mortality rates, combined with seasonal lower fertility rates, may have been factors that compelled the EB IA people to adopt a more sedentary behavior and a heavier dependence on agricultural products.

We estimate that a total of 56,263 burials were interred in 2,856 tombs. If the span of EB IA was about 100 years, an average living population of about

10,000 people would have been required to supply this many burials. Considering that each tomb included an average of 19.7 burials, a visit to the cemetery once a month would have required the construction of approximately two tombs. A visit twice a year would have required the construction of 15 tombs every six months. The total disarticulation of most of the interred individuals argues for relatively extended periods between primary and secondary burials. This would have necessitated reasonable intervals between secondary burial ceremonies, which may have been held no more than one or two times each year. This break between funerary ceremonies may have diminished the transition between EB IA and EB IB; the articulation of some shaft tomb chamber burials was complete in EB IB and almost complete in what may have been late EB IA.

An average of 10,000 people would have been associated with the Bâb edh-Dhrâ' EB IA cemetery at any given time. These people would have been divided into many smaller groups, each following food resources for herds within the 10,000 square kilometer area required to sustain this many people. The development of a massive walled town starting in EB II as well as the evidence of interpersonal violence apparent in EB IB suggests increased pressure from hostile societies. Although the walled town probably did not contain more than 1,000–2,000 people most of the time, agriculturists living in surrounding villages probably would have depended on the walled town for protection during times of warfare.

Demographic reconstructions are restricted by many factors that have been described earlier in this chapter. Any deduction based on the results from the demographic analysis must be evaluated with these restrictions in mind. One of the greatest concerns is that burials from a cemetery represent the entire time period associated with the cemetery. In most archaeological contexts, it is impossible to allocate a specific time within the range of the cemetery to any given burial.

We argue that the population growth rate for the EB IA population was most likely very close to zero and may have been negative for part of the EB IA period. We have calculated an average of about 10,000 (9,870) people who must have lived in the area, utilizing the Bâb edh-Dhrâ' cemetery, in order to account for the estimated number of deceased people found in the cemetery. With a short time span of about 100 years for the EB IA period and a growth rate close to zero, it is unlikely that the size of the original population at the beginning of EB IA deviated much in size from the population at the end of EB IA and the beginning of EB IB.

A population size of almost 10,000 people would have needed an area of 10,000 square kilometers in order to support nomadic or seminomadic behavior. Although it is possible and not particularly unusual for a nomadic or seminomadic population to utilize such distances and areas, such a range would also mean that the people utilizing the Bâb edh-Dhrâ' cemetery would have extended their area of interest and control to the southern part of the Ghor; this would have included Numerira, Safi, Feifah, and Khanazir to the south and other areas several kilometers north of the Lisan Peninsula. Surveys in 1984 verified the presence of extensive cemeteries at Safi and Feifah, some of which may be associated with the EB I period. This was later confirmed by Schaub and Rast (pers. comm. 1990), when EB IA–type pottery was found at the Feifa cemetery, located 38.8 kilometers south of Bâb edh-Dhrâ'. We argue that cemeteries dated to the EB IA period within the 10,000 square kilometer area surrounding the Bâb edh-Dhrâ' cemetery to the north, east, and south suggest that the area controlled by people using the Bâb edh-Dhrâ' cemetery during the EB IA period may have been significantly smaller than the 10,000 square kilometers needed to support a population living in the region for a 100-year time span.

Since two of the three variables used to calculate population size are constants (number of deceased individuals and life expectancy), only the third variable (time span) can be altered to make the data conform to a more realistic scenario. This means that 100 years for the length of the EB IA time period is too short to produce the estimated number of burials located in the Bâb edh-Dhrâ' shaft tombs. For example, given that individuals from Bâb edh-Dhrâ' had a life expectancy at birth of about 17.6 years, a period of 200 years would have required a population size of about 5,000 people and a time span of 400 years would have required a population size of about 2,500 people to produce the expected 56,263 burials. Increasing the number of years to as many as 400 would allow for an increased growth rate and, therefore, supports the idea of a smaller population at the beginning of the EB IA period. At the transition to the EB IB period, this smaller population would have reached a size that is more consistent with the available data and the relatively short time span allotted to the EB IB phase.

CHAPTER 12

The Paleopathology of the EB IA and EB IB People

DONALD J. ORTNER, EVAN M. GAROFALO, AND BRUNO FROHLICH

IN ANTIQUITY DISEASE was undoubtedly a major factor within human societies as people in those societies related to each other and the other biological and social factors in their environment. The more restricted dietary variety associated with the development of agriculture increased the probability of malnutrition due to an inadequate intake of necessary trace elements, vitamins, and, in the event of an inadequate crop, protein and calories. This less varied diet tended to increase the prevalence of metabolic diseases, including those resulting from inadequate intake of vitamins C and D as well as trace elements such as iron.

By the Early Bronze Age, innovations in food production such as irrigation were well established in the Middle East. This, of course, increased the risk from waterborne pathogens. The importance of domestic animals for protein providing milk, eggs, and meat and the close contact that existed between people and their domestic animals increased exposure to zoonotic pathogens. The greater concentrations of people linked to urbanism raised the risk of infection, particularly by aerosol-borne pathogens. Higher population density also resulted in increased pathogen exposure from poor hygiene.

Despite these detrimental factors, it is clear that through the Holocene, human groups increased in size and somehow adapted through combinations of cultural innovation and biological adaptation. The net benefit of increased calories associated with agriculture provided the stimulus to population growth once this innovation was established. This, of course, leaves to be determined how human groups adjusted to the increased risk of disease that accompanied this innovation. One factor must have been the inherent and very substantial potential of human biology, particularly the immune system, to respond to the increased risk of disease. The immune system is inherited, creating the potential for the changing biological and cultural landscape to stimulate microevolutionary adjustments in those genes that affect the immune response.

The objective of this chapter is to present the evidence for skeletal diseases that we were able to identify in the burials. This evidence provides a direct indication of disease prevalence, but interpretation of this evidence relative to the health of the living population needs to be approached with considerable caution and a full awareness of the inherent limitations.

Biomechanical function is essential to human survival, so it is not surprising that the human skeleton is remarkably resistant to the effect of diseases that might compromise this function. It is often the last organ to be involved in a systemic disease process and many even serious diseases do not affect the skeleton at all. Of those disorders that have the potential to produce skeletal abnormalities, it is very uncommon for skeletal manifestations to occur in more than 10–20% of individuals who have the disorder. There are exceptions to this, such as carcinoma; about 90% of the patients with this general type of cancer will have skeletal involvement. It is also troublesome that there often is considerable overlap in the skeletal manifestations of many diseases. This, of course, makes specific diagnoses challenging and in some cases impossible.

Generally, diseases must be survived by a patient for a considerable length of time for skeletal involvement to occur. This means that it is chronic disease conditions that are most likely to be seen in archeological skeletal samples. As Ortner has noted in other publications (1998, 2003), skeletal disease usually occurs when the immune response is between an optimum response and an inadequate reaction at the other end of the response spectrum. At the optimal end, exposure to disease results in little if any morbidity and no skeletal disorder. At the other end of the spectrum are those individuals who die quickly from disease because of either the virulence of the disease or the individual's inability to react effectively to it. At both ends of the spectrum there is usually no skeletal evidence of disease. Somewhere between these extremes in the continuum of immune responses are the individuals with disorders that may have the potential to affect the skeleton. Furthermore, the prevalence of many diseases is linked to aging. Low life expectancy of the EB IA people of Bâb edh-Dhrâ' reduced the prevalence of many diseases because the people did not live long enough to acquire them.

Diseases that affect the skeleton range from those that are associated with slight morbidity and low mortality to diseases that have significant morbidity and relatively high mortality. Ideally, a researcher trying to reconstruct the health of a past human population will assign different weights to skeletal manifestations of disease depending on the expected morbidity and mortality associated with each disorder. Unfortunately, at this stage in the development of both medical research and paleopathology there are too many unknown and uncontrolled variables to permit this type of precision in the evaluation of health in past human groups. Nevertheless, even general information based on the prevalence of skeletal evidence of disease does provide useful information about the health of past populations. The recent research on the health of past Native American groups based on skeletal evidence of disorder (Steckel and Rose 2002) represents a promising strategy in reconstructing at least some aspects of health in past human groups.

An initial phase of research on the Bâb edh-Dhrâʿ EB IA burials associated with burials recovered in each of the field seasons (1977, 1979, 1981) concentrated on the basic osteology of the human remains. Some attention was paid to obvious evidence of skeletal disorder, and several publications have included some of this evidence (Frohlich and Ortner 1982, 2000; Ortner 1979, 1981, 1982; Ortner and Ribas 1997; Rashidi et al. 1991). At the time of this initial research, the authors were fully aware that a more detailed analysis of the skeletal remains for evidence of disease was necessary. The extended delay in conducting this analysis has had at least some benefit, since considerable research has been done in the past 25 years to clarify the diagnostic significance of several skeletal abnormalities encountered in archeological human burials. As part of the process of reanalysis of the EB I burials for evidence of disease, every bone and fragment of bone was evaluated for the presence of skeletal abnormality.

There are four basic abnormalities that occur in human skeletal remains: (1) abnormal bone formation, (2) abnormal bone destruction, (3) abnormal size, and (4) abnormal shape. In determining the diagnostic options for any skeletal abnormality or combination of skeletal abnormalities, the location within the skeleton is an important variable. If the abnormality involves multiple locations, that is, multifocal, an important diagnostic criterion is whether the abnormalities are bilateral and, if bilateral, whether they are symmetrical, that is, occurring at the same location on the contralateral bone and being approximately the same size.

All bones and bone fragments from the EB I burials were evaluated for the presence of these abnormalities. It was not possible in all cases to determine the diagnostic options and the most likely cause. The fragmentary and incomplete nature of some burials was a major barrier in this limitation. However, in the majority of cases it was possible to assign diagnostic options and at least the most probable general disease. While skeletal evidence of disease provides a very incomplete picture of the health of the EB I people of Bâb edh-Dhrâʿ, it does provide insight about the importance of disease in that population at that time. Furthermore, there are some interesting and potentially important differences between the two subphases of EB I.

There is no completely satisfactory classificatory system for skeletal disease. Ideally, any classification system should (1) contain categories that include all possible conditions that can affect the skeleton; (2) have mutually exclusive categories, that is, there is no ambiguity regarding which category a given abnormality belongs to; and (3) the criteria for assigning a category are well established and understood relative to the pathological processes that cause the abnormality. As new insight regarding pathogenesis is acquired, specific orthopedic diseases have been assigned to different categories. This, of course, reflects the healthy and dynamic development of medical knowledge, but it also means that those conducting research on paleopathology need to maintain a working knowledge of the current literature in orthopaedic pathology and radiology.

It is also important not to permit classification of skeletal abnormalities to obscure the pathological processes associated with a given disorder. A good rule to follow is to avoid diagnosis before carefully describing the type and distribution of the skeletal abnormalities. For example, the erosive arthropathies such as rheumatoid arthritis are classified with the general category of arthritis, which also includes osteoarthritis. However, it is now generally accepted that the triggering mechanism for the onset of at least most of the erosive arthropathies is an infectious agent. Infectious agents are generally not associated with osteoarthritis, and it is important not to obscure the different pathogenesis of osteoarthritis and erosive arthropathy by their association in a general category of skeletal disease. This is a fairly obvious example of the hazards of a classification system in paleopathology, but there are other disorders, such as Paget's disease, for which the cause is unknown, making classification difficult if not impossible at this stage of our knowledge. In some cases, a specific disease classification depends on its similarity to other disorders rather than the cause of the disease. For example, is iron deficiency anemia more appropriately classified as a metabolic disease than a hematopoietic disorder with other anemias since a dietary deficiency of iron intake is the underlying cause? This is further complicated by the possibility that chronic bleeding in the gastrointestinal tract can greatly increase the need for red blood cell formation, creating skeletal abnormalities identical to what occurs with anemia caused by malnutrition or genetics, even though infection may have been the cause of the bleeding.

In the following discussion of skeletal evidence of disease in the Bâb edh-Dhrâʿ EB I people, we have adopted the general classification (Table 12.1) used in Ortner's reference work (2003) on human skeletal paleopathology. Another very general classification for skeletal paleopathology is specific versus nonspecific disorders. The former are those skeletal or dental abnormalities that can be linked with reasonable certainty to at least one of the general categories of disease. Nonspecific disorders include skeletal or dental disorders that cannot be attributed to even a general category of disease. Examples of these disorders are dental hypoplasia and periostosis, both of which can be associated with several general abnormal conditions. Many, but not all, of the categories in Table 12.1 have been identified in at least one burial among the Bâb edh-Dhrâʿ remains, but some of the abnormalities can only be classified as nonspecific disease.

The major categories of specific disease in archaeological remains are trauma, infection, and arthritis, which are the most common disorders found in virtually all archeological skeletal samples. On the basis of the first author's experience in human skeletal paleopathology, in most archeological skeletal samples these disorders constitute 80–90% of the abnormalities encountered. As paleopathologists improve our knowledge of the less common skeletal manifestations of disease, diseases that have been underreported may become more prevalent, and this is likely to reduce the relative prevalence of the three major categories. For example, as we will see in greater detail later in this chapter, metabolic diseases are relatively common in the Bâb edh-Dhrâʿ sample. The diagnosis of some of these cases could not have been made twenty years ago.

Table 12.2 provides a summary of the skeletal disorders found in the human remains from the shaft tombs dated to EB IA. There was very little evidence of skeletal abnormality other than trauma found in the bones from the G1 Charnel House dated to the EB IB period or in the few burials found in the shaft tomb chambers (A100N and G2) also dated to this time.

TRAUMA

Although evidence of trauma is the most prevalent disorder identified in Table 12.2, this problem is not

Table 12.1. General Categories of Skeletal Disease

Trauma
Infection
Arthritis, including osteoarthritis and the erosive arthropathies
Circulatory diseases
Reticuloendothelial and hematopoietic disorders
Metabolic disorders
Endocrine disturbances
Congenital and neuromechanical abnormalities
Skeletal dysplasias
Tumors
Pathological conditions of the teeth and jaws
Miscellaneous skeletal diseases

Source: Ortner 2003.

Table 12.2. Human Skeletal Paleopathology Identified in the Human Remains of the EB IA People

Paleopathology	Prevalence
Trauma	
Depressed fracture of the skull	9
Spondylolysis	4
Slipped epiphysis	1
Compression fracture vertebra	1
Fracture of zygomatic bone	1
Femoral head with osteoarthritis	1
Fracture of innominate	1
Osteochondritis dissecans	8
Infection	
TB	2
Brucellosis	2
Osteomyelitis/ulcer	1
Osteoarthritis, severe	3
Metabolic	
Osteoporosis	
Female <35 years	2
Female >35 years	16
Male	2
Rickets	12
Scurvy	7
Tumor–benign	2
Congenital	
Craniostenosis	1
Biparietal thinning	1
Spondylolysis	4
Spina bifida occulta	2
Hip dysplasia	7
Cribra orbitalia–nonspecific	8

Note: Some skeletons had more than one pathological condition.

common in the EB IA people of Bâb edh-Dhrâʻ. One of the most interesting cases of trauma is a skull with unmistakable evidence of scalping (Ortner and Ribas 1997). This case is of an adult male burial (Burial 3, chamber A80SW) that was about 40 years of age at the time of death. Virtually the entire skull vault is characterized by destruction of the outer table (Figure 12.1a). This, of course, could be caused by postmortem degradation, and there is evidence that postmortem changes are superimposed on an antemortem disorder. The diagnostic evidence of scalping occurs at the margins of the destructive process involving the skull vault. At these margins, there is a zone of destructive remodeling (Figure 12.1b) that demonstrates the antemortem aspect of this disorder. The zone is characterized by an initial partial destructive remodeling of the outer table, leaving a depressed, compact-bone gutter between the unaffected outer area of the zone adjacent to normal cortex. The inner margin of the zone includes sharp edges that are the remnants of osteoclastic activity. Osteoclasts require elevated levels of vascular supply and highly oxygenated blood. They cannot function beyond about one centimeter from living tissue, and this leads to the very sharp border with the necrotic outer table of the large central area of bone abnormality.

An intriguing question is how this scalping occurred. Is the scalping indicative of interpersonal violence in which the scalp was intentionally removed by a combatant? Another option is that the scalp was lost as the result of accidental injury as occurs in recent times when hair gets caught in industrial machinery (Ortner 2003). We have not encountered any additional convincing evidence of scalping in the EB I hu-

Figure 12.1a. Skull vault with bone necrosis following scalping in Burial 3 from tomb chamber A80SW. This burial contains the skeleton of an adult male about 40 years of age.

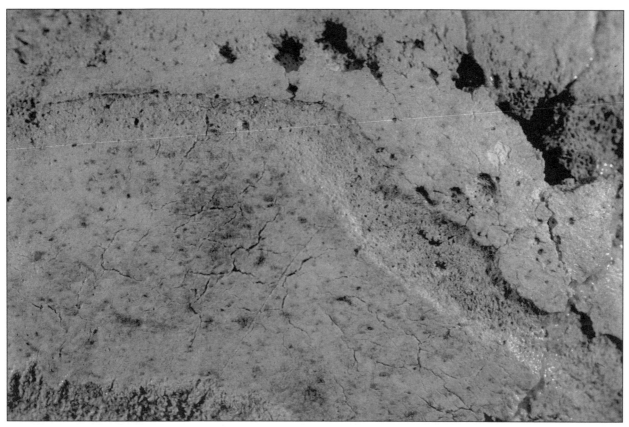

Figure 12.1b. Detail of the wide groove between normal bone and necrotic bone. The border with normal bone is rounded, in contrast with the sharp, undercut margin of the border with necrotic bone. The latter marks the attempt by osteoclasts to remove the necrotic bone.

man remains. This argues for an accidental cause, but intentional violence remains possible.

There are nine cases of depressed lesions of the skull in the EB IA burials, at least some of which probably were the result of interpersonal conflict. Depressed lesions of the skull can be caused by a blow to the skull, but can also be caused by pressure erosion associated with an overlying sebaceous cyst. The latter is a fairly common finding following exposure of the skull during surgery or autopsy. In general, skull lesions caused by sebaceous cysts do not involve the inner table, so if there is inner table involvement associated with a depressed lesion, the almost certain cause is a blow to the head with recovery and remodeling. In some cases, the lesions in the Bâb edh-Dhrâ' EB I skulls are elongated (Figure 12.2), and this contrasts with depressions caused by sebaceous cysts that are circular. An elongated, depressed lesion is probably caused by a glancing blow, perhaps from a sword or axe. Almost all of the lesions have evidence of remodeling, indicating that the individual survived the trauma.

Burials from the EB IB phase also show evidence of skull trauma. One of the tomb chambers dated to this phase (G2) was badly silted, which was further complicated by extensive ceiling collapse. The human remains were in very poor condition, limiting the observations that could be made. In the bone fragments that were recovered, there was no evidence of any type of bone disorder. In the other shaft tomb chamber dated to the EB IB phase (chamber A100N), the skull of the adult male burial (Burial 1) has two depressed lesions (see Figure 6.53) that probably are due to trauma. In the skeletal remains recovered from the G1 Charnel House, which is also dated to the EB IB phase, there are two fragmentary skulls that show unmistakable evidence of healed axe wounds (see Figures 7.22a,b, 7.23). In both cases, the trauma penetrated both tables of the skull but probably did not penetrate the connective tissue layer protecting the brain (dura), thus minimizing the risk of infection. Both injuries are well remodeled, with compact bone sealing off the diploë, indicating long-term survival after the traumatic event.

In the EB IA postcranial skeleton fractures are uncommon. There are five cases of compression fracture of one of the vertebrae. The most likely cause of this

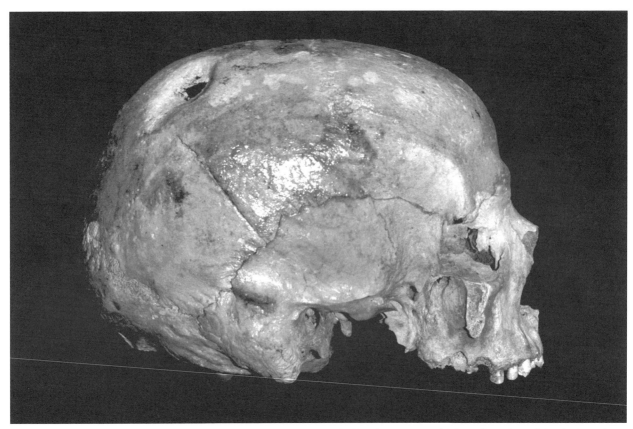

Figure 12.2. Healed trauma to the skull vault of Burial 5 from tomb chamber A111W. This burial consists of the skeleton from an adult male about 35 years of age. The lesion is long and relatively narrow, suggestive of an axe or sword wound.

trauma is a fall from a height, as from a wall or rocky escarpment. These fractures also occur in osteoporosis from what would normally be minor trauma, but because of the loss of both compact and trabecular bone in osteoporosis, the vertebrae are less able to withstand biomechanical stress and fracture more easily. Other evidence of spinal trauma is apparent in two cases of spondylolysis. In this condition, the posterior arch breaks away as the result of chronic stress affecting the lower spine over a fairly long period of time (Figure 12.3). There is one case of healed fracture of the innominate (os coxae) in Burial 4, chamber A111N.

There is remarkably little evidence of long bone fracture in the EB IA burials. There is one case of slipped epiphysis of the proximal humerus (Burial 50, chamber A80N; see Figure 6.24). A more common site for this type of trauma is the femoral head, but no example of this disorder occurs in the EB I burials. Slipped epiphysis is usually the result of trauma during development in which the epiphysis of a bone separates from the metaphysis at the cartilaginous growth plate. Commonly, the epiphysis will reattach, but often in an abnormal position. Disruption of the growth plate may result in abnormal shortening of the bone.

In individuals in whom growth has stopped, the joint end of a long bone may be separated by fracture. A common site for this general type of fracture is the neck of the femur. This type of fracture is particularly associated with cases of osteoporosis, but in this case the fracture, which had healed in abnormal relationship with the neck, was probably due to trauma in early adulthood. The abnormal angle created a defective relationship between the joint components, resulting in osteoarthritis of the femoral head.

There is one case of dislocation of the femoral head (see Figures 6.60a, b). This example of trauma occurs in the right hip of an adult male Burial 1 from chamber A100E. This condition can occur as a result of defective development of the acetabulum (hip dysplasia), but it can also occur in an otherwise normal hip because of severe trauma. In this case, the acetabulum is abnormally shallow and the diameter enlarged, so it seems likely that this is a congenital condition. Chronic subluxation of the femoral head resulted in pressure erosion where the head was in contact with the rim of the acetabulum. There is also a very irregular groove, probably the result of pressure erosion caused by the ligamentum teres, which would

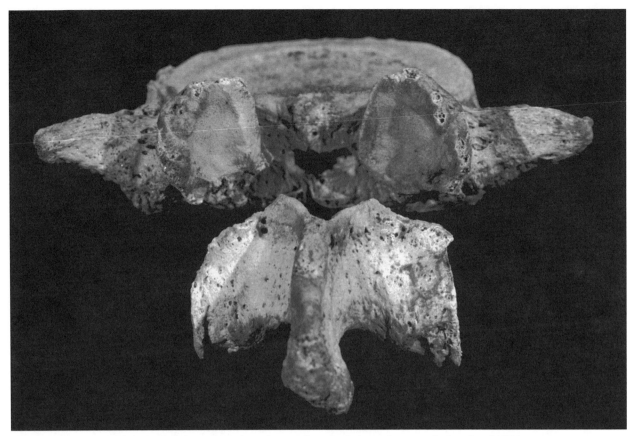

Figure 12.3. Spondylolysis of L5 from the spine of an adult male skeleton from tomb chamber A100E.

have been abnormally stretched from the chronic subluxation of the femoral head relative to the acetabulum.

In the EB I human remains, we have seen eight cases of focal destruction of the subchondral bone of a joint; this condition is called osteochondritis dissecans. The most common site for this abnormality is the bones of the knee, but it can occur in other joints as well. There are several examples of this abnormality in the EB I human remains (see Figure 6.38). The pathogenesis of this condition remains to be resolved, but trauma to the joint is the most likely cause.

Comparison of evidence of trauma associated with interpersonal violence between the EB IA and EB IB burials does not provide convincing evidence of any significant difference in prevalence. Both have cases of skull injuries probably caused by interpersonal violence. However, the prevalence relative to the total sample from each time period is small.

INFECTION

Although the immune system is invoked in most of the disorders affecting the skeleton, infectious diseases are the most important disorder in relation to the human immune response. This is due to the fact that infection is one of the most common disorders encountered in archaeological human remains and almost certainly was the most serious health problem experienced by ancient societies. It is also the case that the gene pool related to the immune response within a human group needed to adjust genetically to the challenges posed by new infectious pathogens that resulted from the changes that occurred in human groups during the Holocene. These changes undoubtedly increased the risk of infectious disease as exposure to existing infectious pathogens increased, but also as new pathogens were encountered through closer contact with animal vectors, through man-made changes in the human landscape, and through the evolution of pathogens.

Most of the infectious diseases that probably were serious potential problems for human societies during the Holocene did not affect the skeleton. However, a few did in some cases, including tuberculosis, brucellosis, and osteomyelitis. Variation in the prevalence of these diseases has the potential to provide an index regarding the effect of infection on earlier human societies.

Tuberculosis

Most cases of human tuberculosis are caused by either of two species of *Mycobacterium* bacteria, *M. tuberculosis* and *M. bovis*. The coevolutionary history of these two pathogens and humans is a matter of significant scientific interest today. Because of the link between *M. bovis* and cattle, an early hypothesis was that this bacterium occurred in early domesticates of aurochs, the early ancestor of today's cattle, and coevolved into *M. tuberculosis* as humans utilized cattle as a source of both meat and milk. However, recent research based on DNA analysis suggests that *M. bovis* may be a more recent development of the genus (Brosch et al. 2002 in Roberts and Buikstra 2003), which, if confirmed, would make it unlikely that *M. tuberculosis* evolved from *M. bovis*.

There are several summaries of the many attempts to establish the antiquity of tuberculosis on the basis of evidence in ancient human remains and how far back in time this disease occurred in human groups (e.g., Aufderheide and Rodríguez-Martín 1998; Ortner 2003; Roberts and Buikstra 2003). Aufderheide and Rodríguez-Martín (1998) note the lack of any skeletal evidence for tuberculosis in any of the reports of human remains dating to the Paleolithic period. An early report of spinal lesions possibly attributable to tuberculosis has this disease in human remains from the European Neolithic period (Bartels 1907), but this finding has been questioned. The earliest convincing evidence of skeletal lesions (Pott's disease) possibly caused by tuberculosis is from sites in Italy dated between 4000 and 3500 BC (Formicola et al. 1987; Canci et al. 1996).

We have identified plausible evidence of tuberculosis (Figures 12.4a, 12.4b) in lower lumbar vertebrae in the burial of a male about 18 years of age at the time of death (Burial 73, A100E). In this case, the inferior portion of the L4 vertebral body was destroyed and reactive bone formation is apparent on the anterior bodies of L3 and L5. There is also a cloaca penetrating the reactive bone formed on the left lateral side of the body of L4 that is probably associated with drainage

Figure 12.4a. Lumbar vertebrae from Burial 73, tomb chamber A100E, showing destruction of the lower vertebral body of L4 and reactive bone formation apparent in L3 and L5.

for chronic infection. Another probable case of tuberculosis is apparent on an unassociated vertebra from A89SE.

Another possible case occurs in the skeleton of a four-year-old child (Burial 2, A100N) that is probably dated to the EB IB phase at Bâb edh-Dhrâ' (Ortner 1979). The destructive lesion occurs in the cranial base, which is an uncommon location for skeletal involvement in tuberculosis (Ortner 2003). The lesion in this case has more marginal sclerosis than is typical of cranial base lesions in tuberculosis. However, with the presence of one relatively convincing case in the EB IA skeletal sample, tuberculosis remains at least a possibility. This case is the only example of possible infection of any kind identified in the EB IB human remains.

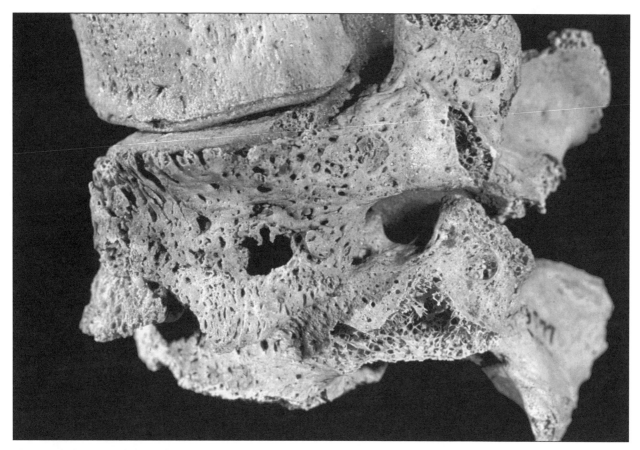

Figure 12.4b. Lateral view of partially destroyed L4 vertebra with reactive bone formation and an opening (cloaca) for draining pus from the chronic infectious focus.

Brucellosis

There are many animals, both wild and domestic, that are potential vectors for brucellosis. Because of this, it seems likely that the disease has been present in human groups for several thousand years. Unfortunately, little attention has been paid to possible evidence of this disease in archeological human skeletal samples. A complicating factor is that the skeletal manifestations of brucellosis overlap with other infectious diseases, such as echinococcosis; both diseases can produce multifocal destructive lesions of the spine and can exhibit considerable sclerosis indicative of a very chronic disease. However, both radiologically and anatomically, the typical lesion of the spine in brucellosis is the development of osteophytes on the anterior superior rim of a vertebral body known as a parrot's beak lesion. These osteophytes look very much like the osteophytes associated with osteoarthrosis or osteophytosis of the spine. There does tend to be an extensive thickening of trabeculae associated with the parrot's beak osteophyte that is very uncommon in osteoarthritis, but certain diagnosis remains somewhat elusive. The important aspect of the pathogenesis of brucellosis is that in most cases it is a very chronic disease with mild morbidity and very low mortality. The destructive lesions that are associated with brucellosis predilect the vertebrae but also occur in the sacroiliac joint and other areas of the skeleton. Because of the chronic nature of the disease, marginal sclerosis of destructive lesions is a typical finding. The important distinction between tuberculosis and brucellosis is that bone lesions in the latter disorder tend to be multifocal, whereas destructive lesions of tuberculosis are usually unifocal, although contiguous vertebral bodies may be affected.

Despite the probability of a long relationship between brucellosis pathogens and humans, little is known about the antiquity of brucellosis. In the Old World, a possible case dated to the Early Bronze Age phase at Jericho is one of the earliest possible cases (Brothwell 1965:690, 692–693). Capasso (1999) reports sixteen possible cases in a skeletal sample from the Roman site of Herculaneum in Italy (79 AD). Capasso provides supporting historical evidence of Roman use of untreated milk of sheep and goats in their diet.

Ortner (2003) provides brief descriptions of possible cases of brucellosis in archaeological human skeletons, two of which were from sites in Norway associated with Lapp herders. One of the cases involves multiple destructive lesions, including end plate vertebral body lesions of the inferior ninth and tenth thoracic vertebrae, suggesting an initial focus for the disease in the intervertebral disc. This is an unlikely focus for tuberculosis. Ortner (2003) also reports a possible case of New World brucellosis in human remains from a historic archeological site in Merida, Mexico. The pathology is multifocal, including a destructive lesion of the right sacroiliac joint.

Rashidi et al. (2001) conducted a survey of the EB IA vertebrae for evidence of brucellosis and identified vertebral body lesions (miscellaneous vertebrae from A108NW, A108SW, A111N, and G4) that could have been caused by brucellosis (Figure 12.5). In the burial from A108NW, there is a well-developed osteophyte projecting from the anterior superior margin of the vertebral body. There is also a considerable increase in the density of the adjacent trabeculae, which does not normally occur in association with osteophytes stimulated by arthrosis of the spine. These osteophytes develop as a reaction to a chronic focus of infection of the vertebral body and may be accompanied by a destructive lesion affecting the superior anterior end plate of the body.

Osteomyelitis/Periostitis

Osteomyelitis differs from both tuberculosis and brucellosis in that it does not describe a disease caused by a single organism, although today *Staphylococcus aureus* is the most common pathogen causing this disorder, and it is likely that this pathogen has a long history with humans. However, various species of *Streptococcus* cause osteomyelitis, and indeed any infectious pathogen that has a primary site in the marrow can be classified as osteomyelitis. Today the use of osteomyelitis is usually limited to infections by either *Staphylococcus* or *Streptococcus* organisms.

Correct usage of the term requires that osteomyelitis be limited to infections where the primary site is the marrow, but there is a tendency to use the term in conditions where staph or strep organisms are the pathogen and marrow is involved, whether or not it is the primary site. In archaeological human remains, it may not be possible to establish the primary site of infection. This is not true of the only case of probable staph infection that occurs in the EB I burials.

The case of a bone lesion associated with an overlying skin ulcer occurs in the skeleton of a female, Burial 2 from tomb chamber A100E, with an estimated age at death of over 50 years (see Figure 6.61a). Here the primary site is almost certainly from an injury to the anterior tibia that introduced pathogens, probably staphylococcal, into the tissue underlying the skin. Staphylococcal organisms commonly exist on the skin surface and are easily carried into the underlying tissues in the case of injury. There is evidence of extensive periostitis involving both the entire diaphysis of the left tibia and also the fibula, resulting in bony fusion of the two long bones. There is also evidence of periostitis of the contralateral tibia. However, the latter manifestation of periostitis is much less severe on the right tibia. It is possible that this condition results from hematogenous dissemination of the pathogens from the left lower leg. However, the slight degree of involvement suggests that this may be a neurovascular response to the stimulus created by chronic infection apparent in the left tibia and fibula and not a direct response to pathogens disseminated to the site.

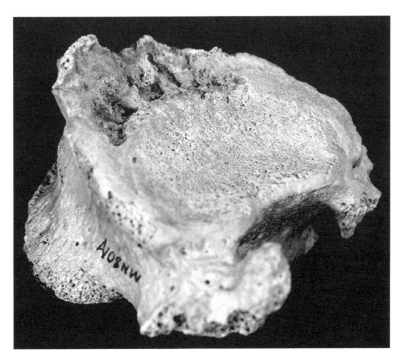

Figure 12.5. Superior, oblique view of an unassociated lower lumbar vertebra from tomb chamber A108NW. Notice the abnormal bony ridge projecting from the anterior margin of the body and the destruction of the anterior end plate.

ARTHRITIS, INCLUDING OSTEOARTHRITIS AND THE EROSIVE ARTHROPATHIES

Osteoarthritis

Unlike in Brothwell's report (1965:687) on the Early Bronze Age human skeletal sample from Jericho, significant osteoarthritis, the most common type of arthritis, in the human remains from Bâb edh-Dhrâʿ occurs in only two burials (Table 12.2). This difference, of course, may reflect no more than the criteria used to define significant manifestations of osteoarthritis, although a major difference in the prevalence of this disorder probably would remain even if the criteria were identical.

A partial explanation for the low prevalence of osteoarthritis is the high mortality associated with low life expectancy present at Bâb edh-Dhrâʿ in which few people lived long enough to acquire skeletal changes associated with osteoarthritis. However, there were at least a few people who did attain old age with minimal evidence of any type of arthritis, including osteoarthritis. This, in combination with the above average robusticity of the people, suggests that whatever the basis of the Bâb edh-Dhrâʿ EB IA economy, the population more likely experienced fairly constant but moderate physical activity rather than intermittent, high-intensity physical stress to the joints and associated tissues.

Destructive joint disease, or erosive arthropathy, includes the various forms of rheumatoid arthritis, the erosive arthropathies that affect the spine, spondyloarthropathy, and other destructive joint diseases such as gout. Differential diagnosis between the various types of erosive arthropathy can be challenging, and indeed there is evidence that osteoarthritis has an erosive variant that can easily be mistaken for the erosive arthropathies (Dieppe and Lim 1998:3, 12). Osteoarthritis is primarily an expression of biomechanical action of the joints, although there is likely a genetic component to its manifestation in a given patient or skeleton. Most, if not all, the erosive arthropathies have a genetic component, but the disorder is triggered by one or more infectious pathogens.

Erosive arthropathy can occur in any age category, although there is a tendency for its prevalence to increase with age. Differential diagnosis of multifocal skeletal disorders in archaeological human remains generally depends on careful analysis of the type of lesion and the distribution of lesions in the skeleton. This is particularly the case in the identification and diagnosis of the erosive arthropathy in archaeological human skeletons. Both the relatively early age at death and the common absence of one or more bones militates against finding examples of erosive arthropathy and even more so against specific diagnosis of these disorders when evidence of erosive arthropathy is present.

There is no convincing evidence of any type of erosive arthropathy in any of the EB I burials from Bâb edh-Dhrâʿ. There is some evidence that rheumatoid arthritis is a relatively recent disorder (Aufderheide and Rodríguez-Martín 1998:101) but at least some of the erosive arthropathies have been identified in other archaeological skeletal samples (Ortner 2003). Given modern prevalence rates for the erosive arthropathies that are on the order of magnitude of 5–10%, one would expect to find some evidence of the disorders if they were present in the population. The lack of evidence is not conclusive but certainly argues for a probability that these disorders were not present in the Bâb edh-Dhrâʿ EB I people.

CIRCULATORY DISEASES

Circulatory diseases are caused by a disruption, either temporary or permanent, of blood supply to bone. Usually this is limited to part of the bone, and the most common site is the joint. Circulatory disorders tend to occur most often in male subadults (Ortner 2003) and are usually activity related.

During the growth period the epiphysis is attached to the metaphysis through a cartilage growth plate. This growth plate is vulnerable to biomechanical stress, particularly where shearing stress occurs. The femoral head is a common site for the epiphysis to be detached from the femoral neck in a condition known as slipped epiphysis. The growth plate for the femoral head remains with the head when it is detached and continues to grow because vascular supply continues through the artery that accompanies the ligamentum teres and attaches to the fovea of the femoral head. Growth of the femoral neck ceases when the head is detached. Typically, the femoral head eventually reunites with the neck, but usually in an abnormal, inferior position. One case of slipped humeral head epiphysis occurs in the Bâb edh-Dhrâʿ sample, in Burial 50 from chamber A80N (see Figure 6.24).

RETICULOENDOTHELIAL AND HEMATOPOIETIC DISORDERS

The reticuloendothelial system is part of the immune response. The primary function of the cells of the reticuloendothelial system is to remove dead cells and tissues. In this role, reticuloendothelial cells break down dead or abnormal cells and tissues prior to their removal and, for this reason, have considerable potential for destructive or lytic activity. When the mechanisms controlling these cells are defective, destruction of normal cells and tissues occurs, and disease results. This abnormality is rare, although there are some cases reported in the paleopathological literature (e.g., Barnes and Ortner 1997). There is no evidence of this type of disease in the Bâb edh-Dhrâ' EB I human remains.

Hematopoietic disorders include the various manifestations of anemia, including anemia caused by genetic abnormalities such as sickle cell anemia and thalassemia. However, any condition that causes an abnormally increased demand for red cell formation can be classified as a hematopoietic disorder. Iron deficiency anemia is included in this category of disease, although it is also a metabolic disease because of its link to inadequate nutrition. Intestinal bleeding caused by various infectious pathogens also causes anemia.

Diagnosis of anemia in archaeological skeletal remains depends on evidence of abnormally increased space for hematopoietic marrow. This is most likely to occur in infants and young children, since virtually all available marrow space is utilized for red blood cell formation in a healthy infant or child. Any increased demand for red blood cell formation occurs at the expense of normal cortical bone. In long bones, this results in larger marrow cavities and thinner cortical bone. In the skull, the diploë enlarges and replaces the outer table with spongy bone. If the condition is severe, as can occur particularly in thalassemia, the inner table may be replaced with spongy bone as well. Other areas of the axial skeleton may also be affected. Angel (1966) applied the descriptive term *porotic hyperostosis* to the porous hypertrophic lesions he identified with possible thalassemia in his research on archaeological human skeletal samples in the eastern Mediterranean. This is a very useful descriptive term that has, unfortunately, become almost synonymous with anemia in the scientific literature on paleopathology when, in fact, several conditions result in porous hypertrophic lesions of bone, including metabolic diseases and infection. When porotic hyperostosis is associated with marrow hyperplasia, anemia becomes a plausible diagnostic option, but the crucial factor is the presence of marrow hyperplasia indicating increased demand for hematopoietic marrow.

In the Bâb edh-Dhrâ' EB I sample there is one example of porotic hyperostosis with marrow hyperplasia. The case is Burial 8 from tomb chamber A107W, the incomplete skeleton of a child between three and four years of age at the time of death. Porotic hyperostosis occurs on the right parietal (Figure 12.6a). Marrow hyperplasia is apparent in the CT coronal image of the parietal (Figure 12.6b). Note that a remnant of

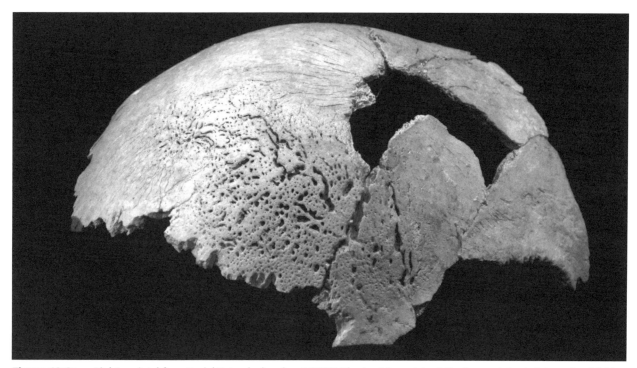

Figure 12.6a. Right parietal from Burial 8, tomb chamber A107W. The burial consists of the incomplete skeleton of a child between three and four years of age. Porous hypertrophic bone formation is present in a large lesion affecting the parietal.

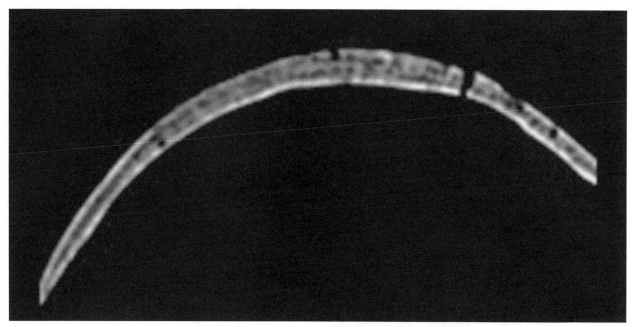

Figure 12.6b. Coronal CT slice through the porotic bone lesion of Burial 8 from A107W showing enlargement of diploë. Evidence of marrow hyperplasia is strongly suggestive of anemia.

the original outer table remains as a line of increased density within what has become an abnormally enlarged diploë.

In some cases of anemia, enlargement of marrow space occurs in the orbital roof (cribra orbitalia), usually in conjunction with increased space for the diploë. Abnormal porosity of the orbital roof, including porotic hyperostosis, also occurs in scurvy and probably chronic infection of the eye, so a diagnosis of anemia solely based on porotic hyperostosis of the orbital roof is inappropriate. Nevertheless, the presence of one case of anemia in the Bâb edh-Dhrâ' sample raises the possibility that some of the nonspecific cases of cribra orbitalia recorded in Table 12.2 could be associated with anemia. What type of anemia remains problematic. However, the strong evidence for malnutrition in both children and adults argues for iron deficiency anemia perhaps complicated by intestinal bleeding from some infectious pathogen. Even allowing for the possibility that some of the examples of cribra orbitalia are caused by anemia, the prevalence of this disorder in the Bâb edh-Dhrâ' sample is low and far less than the prevalence of scurvy and rickets.

METABOLIC DISORDERS

Metabolic disorders include a rather disparate group of diseases. Some of these disorders are linked to a specific nutritional deficiency such as rickets and scurvy. Anemia caused by a nutritional deficiency of iron could be included in this category, but we have discussed this condition in the section on hematopoietic disorders. In addition to diseases that result from a specific dietary deficiency, there are other conditions such as osteoporosis or its milder expression, osteopenia, that can be linked to more than one cause.

In modern Western society, osteoporosis is most often linked to postmenopausal hormone changes in older women. This manifestation does occur in antiquity but is less common than today in Western societies because most women in antiquity did not live long enough to acquire the disorder. However, osteoporosis simply means an abnormal loss of bone mass, and other disorders, including chronic infection, cancer, and severe malnutrition, can result in this abnormality.

In archaeological human burials, diagnosis of subnormal bone density is complicated by taphonomic processes that affect bone density in the burial environment. Clinically, osteoporosis is defined as bone density that is 30% less than normal based on comparisons with normal patients of the same size and sex. In archaeological remains, comparison of bone weight between individuals of similar size and sex is a useful initial variable. Further clarification of cortical thickness is possible with plain film and CT radiography.

As indicated in Table 12.2, there are 16 cases of subnormal bone density in female skeletons with age estimates in excess of 35 years. We have unequivocal evidence of osteoporosis is some of these cases, such as the female skeleton Burial 2 from tomb chamber A100E. The estimated age at death of this burial is in excess of 50 years. Both spongy bone and cortical bone are greatly reduced (see Figure 6.61b). In cases of this type, a diagnosis of postmenopausal osteoporosis is certainly plausible, although other causes, such as protein/calorie malnutrition, cannot be ruled out. Another probable case of postmenopausal osteoporosis occurs in the skeleton of another woman over age 50 at the time of death (Burial 61, chamber A102E).

In Table 12.2, there are two cases of osteoporosis in female skeletons with an estimated age at death of less than 35 years. Hormonal changes related to aging are unlikely to be the cause of subnormal bone density in these cases. Furthermore, there are two cases of low bone density in male skeletons. One of the female skeletons is Burial 3 from chamber A110NE. The estimated age at death of this burial is between 25 and 35 years. In the CT images of the femur (Figure 12.7), there has been a dramatic loss of cortical thickness in the diaphysis, but the spongy bone in the femoral neck is also greatly reduced. In this case, protein/calorie malnutrition is probably the most likely diagnostic option. This is particularly the case in the context of other evidence of malnutrition discussed below.

Scurvy

Unlike most other mammals, higher primates are unable to synthesize vitamin C. This vitamin is critical to several physiological processes, including the synthesis of collagen, the major protein component of bone but also an important constituent of blood vessel walls. In most primates, the diet is heavily weighted with fresh plant foods, which contain significant amounts of vitamin C. However, with the advent of agriculture, human diets tended to become less variable, and deficiencies in one or more essential elements such as iron or vitamins such as C and D became more likely.

Although a deficiency in vitamin C can affect the skeleton of adults (Maat 1982), the effect of vitamin C deficiency is much greater in infants and young children because the greatest relative growth is taking place during this stage in development and the synthesis of collagen is a vital component of the growth process. Several papers have been published on the manifestations of scurvy in the subadult skull (Ortner and Ericksen 1997; Ortner et al. 1999; Ortner et al. 2001). All of the skeletal lesions are formed in response to chronic bleeding resulting from minor trauma to defective blood vessels. In the skull, the primary lesion is porosity of the cortical bone in contact with a chronic site of bleeding. Several of the lesions are associated with muscles linked to chewing.

The youngest example of probable scurvy is the skeleton of an infant about six months of age (Burial 1, A111E). Bilateral abnormal cortical porosity is present in the greater wing of the sphenoid (Figure 12.8). In the long bones, abnormal spicules of bone formed and cortical porosity developed at sites where bleeding from defective blood vessels occurred.

Rickets

Evidence of childhood vitamin D deficiency (rickets) is present in 12 subadult burials. This condition was relatively mild in all cases and was diagnosed on the basis of abnormal curvature of the tibia. On the tibia, the interosseous line normally is straight in both the anterio-posterior and medio-lateral axis. Deviation from a straight line is indicative of abnormal bending,

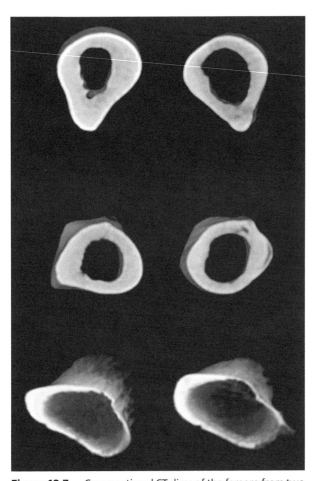

Figure 12.7. Cross-sectional CT slices of the femora from two female skeletons at midshaft (top), subtrochanteric (middle), and femoral neck (bottom). CT slices from the normal skeleton (Burial 1 from A114N; female 25–30 years) are on the left, and osteoporotic bone is on the right (Burial 3 from A110 NE; female 25–35 years). In addition to the loss of cortical and trabecular bone apparent in the CT slices of Burial 3, quantitative CT measurements indicate a 30% reduction in bone density.

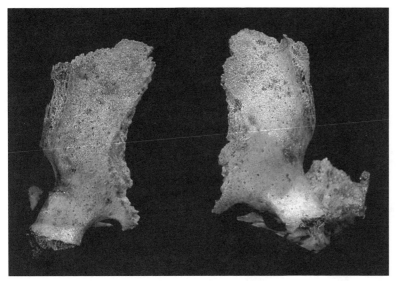

Figure 12.8. The left and right greater wing of the sphenoid from an infant burial about six months of age, showing abnormal porosity but not bone hypertrophy. This case is Burial 1 from tomb chamber A111E. There are other abnormal lesions in this case, which, along with the abnormal sphenoids, suggest a diagnosis of scurvy.

and by far the most plausible cause of this is rickets. Bending may occur from fracture but, with the exception of green-stick partial fractures (infractions), most healed fractures are easily identified and distinguished from bowing caused by rickets.

An example of abnormal bowing is apparent in Burial 61 from chamber A100S. The burial is of a child about three years of age. Both tibias are present, and the abnormal bowing is bilateral (Figure 12.9).

ENDOCRINE DISORDERS

Endocrine disorders include conditions created by a malfunction of one or more glands, such as the pituitary, thyroid, and parathyroid glands. Some of the endocrine disorders cause abnormalities of the skeleton, for example, the dwarfism resulting from a deficiency of pituitary hormone during development. There are also complex interactions between some glands. Endocrine disorders that affect the skeleton are not common, so it is not surprising that there is no evidence of these disorders in the Bâb edh-Dhrâʿ EB I skeletal sample.

CONGENITAL AND NEUROMECHANICAL ABNORMALITIES

In the living individual, the skeleton can be affected by abnormalities in vascular supply, nerve disorders, and defects in muscular function. Abnormalities in any combination of these systems tend to be more severe if they occur early in life, during development, but can result in skeletal defects in the adult as well. Evidence of these types of disorders in the Bâb edh-Dhrâʿ EB I skeletal sample is minimal and includes conditions such as spondylolysis, which is the result of abnormal biomechanical stress on the pedicles of the lower vertebrae, and spina bifida occulta, a developmental failure in which one or more of the posterior arches of the vertebrae and sacrum fail to develop, leaving an open neural canal.

Spondylolysis can also be classified as trauma, since it is a stress fracture through the pedicals of the lower vertebrae brought on by biomechanical stress. There may be an underlying genetic predilection for this disorder, but hard physical activity that puts stress on the lower spine is the major factor. The examples of this condition are present in the Bâb edh-Dhrâʿ sample, including a fifth lumbar vertebra from chamber A100E that is from an adult male skeleton (Burial 1; see Figure 12.3). This finding supports other evidence indicating a relatively vigorous existence for the people of Bâb edh-Dhrâʿ.

Developmental failure of one or more arches of the spine can involve severe disability with partial to complete paralysis of the lower extremities. Severe forms are known as spina bifida and would likely have prevented long-term survival. There is no skeletal evidence that this disorder is present in the Bâb edh-Dhrâʿ sample. A less severe manifestation of this disorder is typically confined to the sacrum and is not associated with any dysfunction. This condition is known as spina bifida occulta (chamber A100E, Burial 73) and is present in the Bâb edh-Dhrâʿ material (Figure 12.10).

Another congenital disorder that is well known in the scientific literature on human skeletal paleopathology is a failure of the acetabulum to develop properly. This may be unilateral or bilateral. The skeletal manifestation of this disorder is a shallow acetabulum that does not support the weight-bearing biomechanical function of the femoral head. A common result is dislocation of the hip joint with formation of a secondary pseudojoint on the lateral surface of the ilium. A less severe manifestation is chronic subluxation of the joint, in which the femoral head slides out of the acetabulum during use of the lower limb, but usually returns to a relatively normal position when the limb is not bearing weight. This type of hip dysplasia tends to result in osteoarthritis and may exhibit pressure erosion of the femoral head. This condition is known as hip dysplasia and is usually the result of congenital defect, but can also be caused by trauma in childhood. An important example of the disorder occurs in the right hip of an adult male (chamber A100E, Burial 1; see Figure 6.60a, 6.60b). Pressure erosion is apparent on the femoral head, resulting from both the rim of the acetabulum as it contacts the head in the subluxed position and also from the ligamentum teres, which was undoubtedly stretched during subluxation of the joint and pressed on the femoral head when the head was in a more normal position.

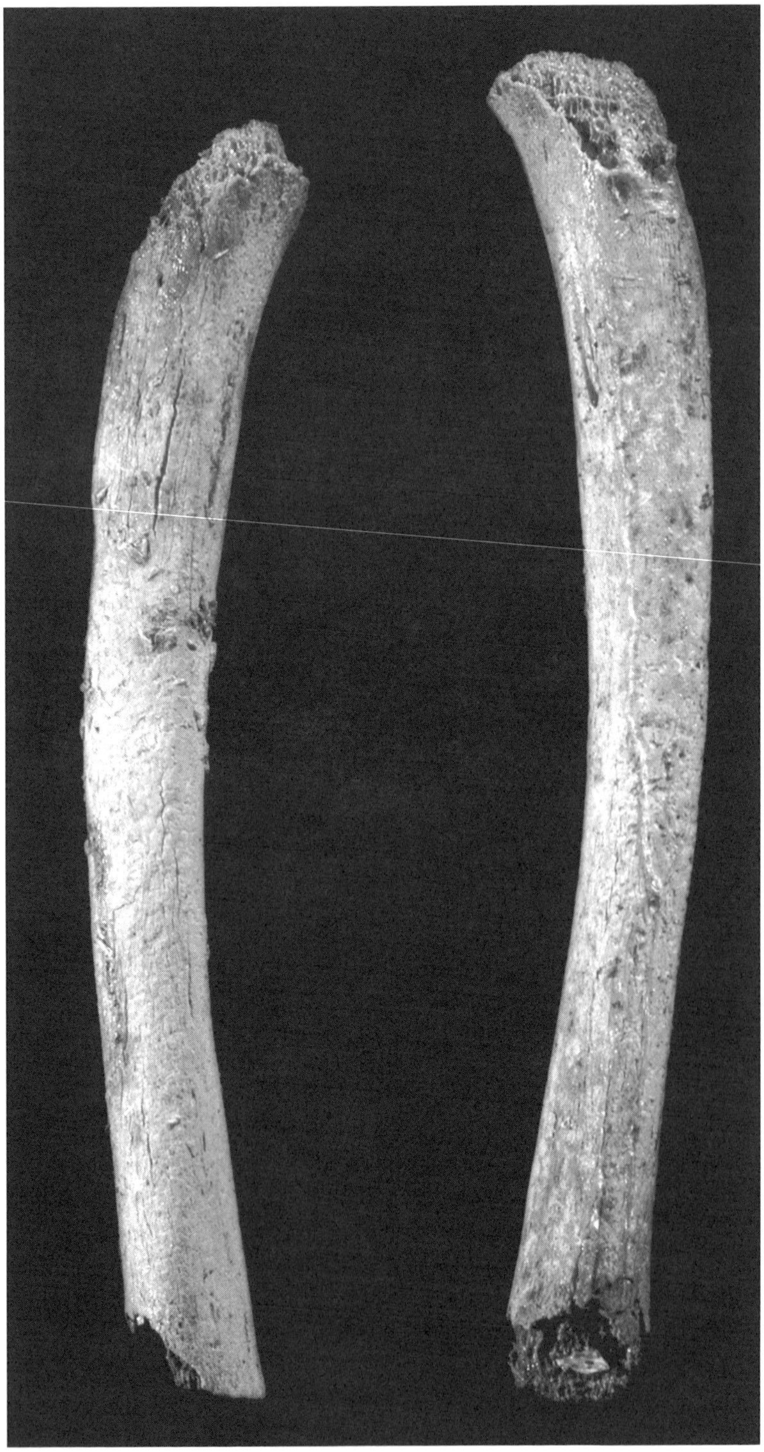

SKELETAL DYSPLASIAS

Like congenital disorders presented in the previous section, skeletal dysplasias are usually manifest at birth. Unlike the abnormalities reviewed in the previous section, skeletal dysplasias usually have a general effect on development, and many if not most bones are affected. The abnormalities tend to be bilateral. Included in this category of bone disorders is the most common manifestation, achondroplasia, but there is a very long list of very rare skeletal dysplasias, most of which have not been reported in the literature on human skeletal paleopathology. There is no evidence of any of the skeletal dysplasias being present in the EB I skeletal sample from Bâb edh-Dhrâ'.

TUMORS

Tumors can be classified in several ways, but the basic distinction is between benign tumors and those that are malignant. Benign tumors do not have the potential to metastasize from the primary site to other tissues or organs in the body, although some benign tumors do, in some cases, undergo a malignant transformation; but this is rare. Both bone-forming and bone-destroying tumors may be benign, and benign tumors can be detrimental to the health of the patient and may even be the cause of death. Malignant tumors have the potential to metastasize from the primary site to other tissues and organs. Morbidity

Figure 12.9. Lateral view of the left and right tibias from Burial 61, tomb chamber A100S. This burial consists of the incomplete skeleton of a child about three years of age. Abnormal bowing is apparent bilaterally and probably was caused by rickets.

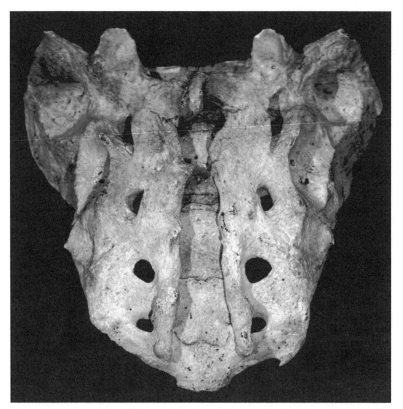

Figure 12.10. Spina bifida occulta in Burial 73, tomb chamber A100E. Unlike more serious manifestations of spina bifida, this anatomical variant is not associated with any dysfunction.

and mortality vary depending on the type of malignant tumor, but both morbidity and mortality tend to be significant.

There is no evidence of malignant tumors (carcinoma or sarcoma) in the Bâb edh-Dhrâʿ burials. There are only two cases of probable benign tumors, and a more specific diagnosis is not possible in either case. One of these tumors occurs on the parietal of Burial 2 from chamber A102S. The lesion is characterized by the formation of poorly organized bone (see Figure 8.7). It is unlikely that significant morbidity was associated with this tumor.

PATHOLOGICAL CONDITIONS OF THE TEETH AND JAWS

Disorders of the teeth and jaws are common in all human archaeological skeletal samples. These disorders include caries, abscess, enamel hypoplasia, and periodontal disease. Tooth wear from a coarse diet is common in the Bâb edh-Dhrâʿ EB I sample. This condition tends to reduce caries and abscess formation, since the surfaces in the crown of the tooth that are most susceptible to caries are worn away during mastication of food. Nevertheless, both caries and abscesses are present in the human remains.

However, the most striking dental problem encountered is enamel hypoplasia, in which enamel development is disrupted, resulting in one or more horizontal grooves in the enamel. An example of a case with multiple linear enamel hypoplasias is Burial 5, a female skeleton around 15 years of age from chamber A114N (Figure 12.11). Less severe manifestations of enamel hypoplasia are much more common, however, and occur in about 25% of the burials from Bâb edh-Dhrâʿ. Dental hypoplasia is caused by a stress event during infancy and childhood that is of sufficient severity to temporarily affect the cells that form enamel (ameloblasts) and diminish or prevent enamel formation while the stress is severe. Generally, the effect on ameloblasts is of relatively short duration, so the hypoplastic lines tend to be about one millimeter in width. However, teeth in some burials may have multiple hypoplastic lines, indicative of multiple stress events.

NONSPECIFIC ABNORMALITIES OF THE SKELETON

The human skeleton is affected in relatively few cases of disease. Furthermore, there may be considerable overlap in the skeletal manifestations of different diseases. Nevertheless, even nonspecific evidence of skeletal disorders provides useful data on the health of archaeological human populations. In the discussion of infectious diseases, we have emphasized that reactive bone formation on the outer cortical surface is often caused by infection, either local or systemic, that stimulates the periosteum. We have also noted that pathological conditions other than infection can stimulate the periosteum, and that it may not be possible to determine the cause of periosteal reactive bone formation, in which case the term *periostosis* is appropriate. There are cases in the Bâb edh-Dhrâʿ sample where the cause of periosteal reactive bone formation cannot be determined and the diagnosis must remain nonspecific periostosis.

Similarly, dental enamel hypoplasia, discussed in the preceding section, does indicate that a stress event occurred, but greater specificity is not possible. However, both periostosis and enamel hypoplasia provide evidence that a stress event of some type took place,

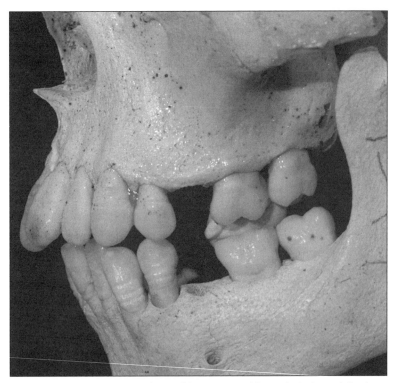

Figure 12.11. Multiple defects of linear enamel hypoplasia in Burial 5 from tomb chamber A114N. This burial consists of an almost complete skeleton of a female between 12 and 18 years of age.

and the presence of either condition is evidence that the individual survived the stimulating event.

A potentially important question that emerges from data on the prevalence of various diseases in a human archaeological skeletal sample is the impact of these diseases on the human population represented by the skeletal sample. Are the people with skeletal disease better or worse off than those who do not have skeletal disease? Another question is, do some manifestations of skeletal disease present a more serious problem than others? The answer to this question is certainly "yes," but given the chronic nature of most skeletal diseases, it is certainly not clear that the presence of skeletal disease is associated with shortened life expectancy relative to burials that do not show evidence of skeletal disease. The question of differential frailty within a human population is one that has been addressed as an important issue in interpolating the relative health of a living population on the basis of skeletal evidence of disease (Wood et al. 1992).

We have not explored this issue very far in our analysis of the Bâb edh-Dhrâʿ EB I sample. However, we did evaluate whether or not there was any difference in the mean age at death of adults with dental hypoplasia and those without this disorder. The mean age in these two subgroups is not significantly different. Admittedly, dental hypoplasia may be a very poor index of frailty but at the very least our finding raises a cautionary note about the significance of dental hypoplasia as an indicator of frailty and thus its utility in any research that attempts to compare the health of archaeological skeletal samples.

In our discussion of metabolic disease, we noted the presence of scurvy, rickets, osteoporosis related to post-menopausal hormone changes in women, and osteoporosis probably caused by protein/calorie malnutrition. It is highly probable that the actual prevalence of these metabolic diseases in the living population of Bâb edh-Dhrâʿ was much higher than the skeletal evidence for these disorders shows. It seems likely that malnutrition was a fairly common problem in the EB IA phase.

We explored this possibility by comparing the subadult age estimate based on dental development with age estimates based on long bone length. Long bone estimated age is consistently younger than dental age estimates. Dental development is much less affected by general health conditions, including malnutrition, than long bone growth. There may be some methodological issues that need to be considered relative to the formulae used for each of the age estimates. However, we argue that a case can be made that this difference indicates that growth was at least slightly delayed due to protein/calorie malnutrition.

CHAPTER 13

Dental Analyses of the Bâb edh-Dhrâ' Human Remains

GILLIAN R. BENTLEY AND VICTORIA J. PERRY

THIS CHAPTER CONCERNS the dentitions excavated from the EB IA shaft tombs during the 1977, 1979, and 1981 archaeological seasons at Bâb edh-Dhrâ'. The tombs themselves are described in more detail in Chapters 6, 8, and 9. There are two parts to this chapter. The first deals with the nonmetric, morphological dental traits of almost 300 individuals, excavated from 25 of these tombs, for which relevant teeth were available, originating from a total of 578 (often incomplete) skeletal burials. The second part of the chapter constitutes an analysis of 140 relatively complete EB IA skulls for evidence of linear enamel hypoplasia (LEH) and other dental pathology in order to extend our knowledge of the general health of the Bâb edh-Dhrâ' EB IA population.

As outlined earlier, almost all of these individuals represent secondary burials, a major factor—together with the condition of the tombs at excavation—contributing to the incomplete preservation of the human remains. The initial purpose of the nonmetric analysis was an attempt to reconstruct the kinship and social relationships of the EB IA period with the hypothesis (first suggested by Walter Rast and Thomas Schaub) that individuals buried in individual shaft tomb chambers were closely related and represent family groups. The inclusion of both sexes and a full range of age groups within the tombs, as well as the very formal and consistent layout of the skeletal remains and grave goods within each tomb chamber, together with the varying periods of time of primary burial for different individuals, also support this reconstruction. An essential part of this hypothesis dealt with the social structure of the EB IA Bâb edh-Dhrâ' population, whom Rast and Schaub believed to be nomadic pastoralists (Rast 1999; Rast and Schaub 2003; Schaub and Rast 1984:35), but who Bentley (1987, 1991) has argued were sedentary agriculturalists.

The chapter is reconstructed both from research conducted by the first author while a doctoral student at the University of Chicago and a predoctoral fellow at the National Museum of Natural History (NMNH), Smithsonian Institution (1984–1986; Bentley 1987, 1991), and from research undertaken more recently by the second author at NMNH while a master's student at University College London (Perry 2005).

METHODS AND MATERIALS FOR ANALYZING THE DENTITIONS

Dental Nonmetrics as a Method of Estimating Biodistance

Dental nonmetrics have a long history of utility in the quest for analyzing biological relationships between individuals and family groups, whether in an archaeological or forensic context. They have been used to estimate biological distance at various levels, including within and between populations (e.g., Berry 1978; Coppa and Macchiarelli 1982; Coppa et al. 1998; Hammond et al. 1975; Hanna 1962; Prowse and Lovell 1996; Rosenzweig 1970; Sofaer et al. 1972; Townsend and Brown 1979; Turner 1971; Turner and Scott 1977; Ullinger et al 2005). The development of a series of stoneware dental plaques in which grades of expression of particular traits have been molded has led to the standardization of techniques for dental morphological analyses and considerably reduced interobserver error (Turner et al. 1991).

The dentitions analyzed for morphological traits comprise those recovered from 25 of the 26 EB IA shaft tombs excavated during the 1977, 1979, and 1981 seasons at Bâb edh-Dhrâ'. These tombs, the history of their excavation, and the condition in which they were found when excavated are outlined in more detail in Chapters 6, 8, and 9. Due to the secondary nature of the burials in these tombs, many loose teeth were discovered in the tomb chambers. Where possible, these were matched together by the first author of this chapter on the basis of morphological structure, metric characteristics, wear patterns, and coloring. This exercise was easier to accomplish in younger individuals suffering from less attrition. Very often loose teeth could be matched to a corresponding mandible or maxilla. Table 13.1 provides information on the sample frequencies of the maxillary and mandibular dentitions based on this matching, including an indication of whether the teeth are loose or associated with other skeletal parts.

Of the dentitions used for nonmetric analyses, 98 represent fairly complete samples comprising both maxillary and mandibular elements, 49 represent partial dentitions associated with gnathic fragments (26 with maxillary and 23 with mandibular fragments), and 330 loose teeth (176 maxillary and 154 mandibular) which could not be matched to other skeletal parts. Matching of the teeth was designed to provide a conservative and yet realistic profile of the skeletal population to complement the figures available from other osteological evidence. Older individuals, however, may in fact be somewhat overrepresented by loose and excessively worn teeth. This was the case undoubtedly in tombs that were heavily silted and where the skeletal remains were damaged or incomplete (e.g., Tombs A78, A89, A112, G4, and C10). Taking all the tombs together, and including counts of loose teeth, the average number of individuals per tomb chamber (using just dentitions) is 11. When the sample is restricted to those tombs where remains are significantly more complete, the average number is five.

Age estimates for individuals were made on the basis of dental attrition using Brothwell's chart (1981), while the actual scoring of attrition followed the scale devised by Molnar (1971). Where dentitions could be associated with other skeletal parts, these age estimates were then compared with those made by Ortner and Frohlich and adjusted where necessary, since age estimation using these other indicators was considered to be more reliable. Sexing of individuals was restricted to those cases where the dentitions could be associated with other skeletal parts useful for gender assignment, and these assignments were also made by Ortner and Frohlich.

Dental nonmetric traits were scored using three sets of stoneware casts or plaques of particular teeth referred to above that exhibit grades of expression of specific dental traits, including the Arizona State University Dental Anthropology System (Turner et al. 1991) and plaques created by Dahlberg (1956) and Hanihara (1960). In general, the Turner plaques represented a refinement in scaling criteria over the earlier Dahlberg plaques and were used more often in assessing the Bâb edh-Dhrâ' dentitions, but where more qualitative rather than quantitative assessments were required the Dahlberg series were used and seemed better adapted, in some cases, to the Bâb edh-Dhrâ' materials. The Hanihara dental plaques were produced specifically for the deciduous dentitions. Table 13.2 lists the maxillary and mandibular nonmetric traits scored in the Bâb edh-Dhrâ' dentitions as well as the choice of plaque series used in assessing traits.

Dental Pathology

The dentitions analyzed for LEH and other pathology were restricted to 140 complete or partial skulls from the EB IA shaft tombs, except for remains from tomb chamber A100N, which dates to the EB IB. All of the dentitions originate from the A Cemetery except for five crania from the C Cemetery. Due to the large proportion of missing teeth from postmortem disturbance and some antemortem tooth loss, analyses are restricted to 23% of teeth (1,156 in number) that were present and could be measured. This reduced the total sample size to 93 individuals. Of these, 43 were of indeterminate sex, while of the remaining 52 individuals, 29 were male and 23 female. The American Dentists' Universal Numbering System (permanent teeth = 1–32, deciduous teeth = 51–70) was used to number teeth; those few loose teeth that were analyzed were identified for tooth type using indicators described by Hillson (1996). Patterns of dental eruption were modified from Buikstra and Ubelaker (1994) and were used to identify teeth of infants and juveniles. Ages were calculated using the formulae (Table 13.3) from Lovell and Whyte (1999).

Teeth were examined for LEH and other pathology using a hand-held magnifier/hand lens under a moveable lamp that was positioned to give the best surface contrast view (Reid and Dean 2000). Defects and pathologies investigated were LEH, caries cavities, calculus deposits, and abscesses. Periodontal disease was not assessed. Sex of individuals was noted if this information was available. The ages of individuals were noted as adult (only permanent dentition), juvenile (age two to adulthood), or infant (two years and under). Presence or absence of teeth, as well as caries and abscess sites, were coded according to Buikstra and Ubelaker (1994). Those very small caries that could be confused with noncarious pits were investigated using a fine needle or wire to determine the internal dimensions and therefore ascertain the nature of the feature. Caries size was taken as small (≤ 1mm), medium (>1–3 mm), and large (>3 mm, including those that had obliterated most or all the tooth). Calculus was scored according to Brothwell (1981).

LEH were scored individually on each tooth, and the position of each hypoplastic event was noted and measured from the cemento-enamel junction (CEJ), that

Table 13.1. Number of Complete and Fragmentary Dentitions Used for Morphological Analyses

Tomb Chamber	No. Complete Dentitions[1]	No. Partial Dentitions[2]		No. Loose Teeth	
		Maxillary	Mandibular	Maxillary	Mandibular
A78NW				21	11
A78NE		2			
A78SW				7	10
A79N				2	6
A79S				4	1
A79W		1		3	
A80N	1	1		2	1
A80S	2		1	2	2
A80W	4		1	1	
A80E	1		1	5	1
A86NE		1			
A86SW	1			1	1
A86SE	1				
A88		3	1	10	11
A89NW		2			1
A89NE	1				
A89SE				5	4
A89?			3	2	1
A100S	2		1	3	2
A100W	2	1			
A100E	4				2
A101N	1				
A102S	6	2	1	10	7
A102W	3	2	1	4	4
A102E	5	1			
A102NE		2		6	7
A103S	1				
A105NW	2	1		7	9
A105NE	2	2	1	2	3
A105SW	1		1		
A105SE	2		1		1
A106S	1		1	3	
A107N	2		1	4	2
A107S	6			3	3
A107W	2	3		3	3
A107E	5		1	2	1
A108N				3	4
A108NW		1		3	8
A108NE	2			4	10
A108SW	1	1			1
A108SE				2	2
A108?				1	1
A109N	1		2		
A109S				1	
A109E				1	1
A110NW	3				
A110NE	3				
A110SE	6				1
A111N	4		1	2	
A111W	5				
A111E	3				
A112S				4	4
A112E				1	1
A114N	5		1		
A120S					2
A121N				4	4
G4	3		1	3	1
C9	1		1	13	7
C10E				22	13
C11	3		1		
Total	98	26	23	176	154

1. Associated with both mandible and maxilla.
2. Associated with gnathic parts.
? Indicates unknown chamber attribution within the tomb.

Table 13.2. Morphological Traits and Associated Plaques (Where Available) Used to Analyze the EB IA Dentitions from Bâb edh-Dhrâ'

Tooth	Maxillary Trait	Plaque	Mandibular Trait	Plaque
Permanent Teeth				
Central incisor	Crown shoveling	Arizona1		
	Cingular tubercle	None available		
	Grooved cingulum	None available		
	Finger-like projections	None available		
Lateral incisor	Crown shoveling	Arizona		
	Cingular tubercle	None available		
	Grooved cingulum	None available		
Canine	Marginal ridging	None available		
	Cingular tubercle	None available		
Premolars	Extra labial cusp	None available	Extra labial cusp	None available
	Extra lingual cusp	None available	Extra lingual cusp	None available
	Labial marginal ridging	None available	Labial marginal ridging	None available
	Lingual marginal ridging	None available	Lingual marginal ridging	None available
Molars	Cusp number	Dahlberg2	Occlusal surface pattern	Dahlberg
	Carabelli's trait	Dahlberg	Protostylid	Dahlberg
	Size of hypocone	Arizona	Fifth cusp size	Arizona
			Sixth cusp size	Arizona
			Seventh cusp size	Arizona
			Anterior fovea	Arizona
Deciduous Teeth				
First molar	Occlusal surface pattern	Hanihara3		
	Stylar ridging	None available	Stylar ridging	None available
Second molar	Cusp number	Dahlberg	Occlusal surface pattern	Dahlberg
	Carabelli's trait	Dahlberg	Protostylid	Dahlberg
	Size of hypocone	Arizona	Fifth cusp size	Arizona
			Sixth cusp size	Arizona
			Seventh cusp size	Arizona
			Anterior fovea	Arizona

1. Turner et al. 1991.
2. Dahlberg 1956.
3. Hanihara 1960.

Table 13.3. Formulae for Calculating Age of Disturbance Leading to Visible Hypoplasia

Tooth	Maxillary Teeth	Mandibular Teeth
Canine	Age = –0.609375 × Ht + 6.0	Age = –0.58823529 × Ht + 6.5
First molar	Age = –0.45384615 × Ht + 3.57	Age = –0.45769231 × Ht + 3.56

being the furthest extent of the formation of enamel before the root of the tooth to the occlusal edge of the feature. This is in accordance with Hillson and Bond (1997), who state that the occlusal wall is indicative of the actual hypoplastic event, and that the floor of the groove and the cervical wall represent the recovery period from the stress event. Events were measured to tenths of one millimeter and taken uniformly with the same pair of thin-lipped calipers. Hypoplastic events were noted on the scoring sheets at the same measurements that they occurred on the tooth to allow for future analysis of possible age at disturbance. LEH were scored according to Buikstra and Ubelaker (1994), with modifications for instances when calculus deposits were so large that they obscured some or all of the surface features of the tooth.

Statistical Analyses
Statistical tests were performed using SPSS and Excel. For purposes of analyzing the dentitions for dental

morphological traits, as many teeth as possible were included in the statistical analyses. Due to the problem of heavy attrition on the Bâb edh-Dhrâʿ teeth, nonmetric traits could not always be scored, and a "missing value" code was assigned in such cases to the database. Such codes were also assigned where individual teeth were missing. All teeth present were scored for each individual as were all the loose teeth in the sample.

For purposes of analyzing the dental pathologies, only teeth that were coded as present in the jaw, fully developed, and in occlusion were included for statistical analyses. The average number of LEH per tooth for each individual was calculated by summing the total number of hypoplasias found per individual and dividing by the number of teeth present in that individual.

Descriptive statistics were used to test for frequencies and distribution of traits for both the dental pathologies and nonmetric traits. For purposes of nonmetric trait evaluation and statistical analysis, additive frequencies followed the "individual count" method outlined by Scott (1980), representing best estimates of both trait and individual expressions (Scott 1977; Turner and Scott 1977; Turner 1967). This method assumes a single genotype for each trait, and therefore scoring on one side only (usually the left) is counted. However, where bilateral asymmetry occurs, the trait is scored according to the highest expression recorded, whether the left or right antimere. Where the left side is missing, the corresponding antimere is substituted. This counting procedure is representative of the dental remains and, with an archaeological and often fragmentary sample, maximizes potential frequency counts without compromising the reliability of trait scoring.

Given the fragmentary and incomplete nature of the data set, with several missing values, univariate tests were used for all analyses of the nonmetric morphological traits. Kruskal-Wallis one-way analysis of variance was used to assess possible sexual dimorphism in trait expression, which may have confounded the analysis of biodistance between shaft tombs. Kruskal-Wallis was also used to test for significant differences in trait occurrence and expression between groups of individuals buried in each shaft tomb, as well as clusters of tombs either located near each other or randomly ordered in groups. Rank order correlations (Kendall's tau) were used to test for possible intertrait associations that could lead to consideration of a number of redundant traits for assessment of biological distance and an inflation of the importance of potential results. Fisher's Combined Test was used to assess trends of significance across the range of p-values derived from the Kruskal-Wallis one-way analyses of variance for both the individual tombs and groups of tombs. Mann-Whitney U tests were used to test for significant differences between A and C cemeteries, where tomb ceramics appear to be qualitatively different (Chapter 4), and where trait frequencies exceeded five in number. Where Kruskal-Wallis analyses had yielded statistically significant, or close to significant, results for intertomb differences, Mann-Whitney U tests were applied on a tomb-by-tomb basis to ascertain between which tombs the greatest differences might lie. The same test was also applied between the groups of randomly clustered tombs referred to above.

For the dental pathologies, general linear models (GLM) were used to test for significant differences in distribution of traits between sexes, between ages, and between tombs. Analysis into the possible age of disturbance shown by hypoplastic lines was conducted by taking the earliest and latest seen hypoplasias on the canine and first molar teeth. Significance level was set at $p = 0.05$ for all tests.

RESULTS OF THE STATISTICAL ANALYSES

Dental Morphological Traits

In general, there is a high degree of homogeneity in the dental traits of the Bâb edh-Dhrâʿ population. No significant differences were observed in trait expression between the sexes that might have confounded analyses of biodistance between tombs. A number of traits was significantly correlated (Table 13.4).

Using Kruskal-Wallis one-way analysis of variance to test for significant differences in trait occurrence and expression between groups of individuals buried in each shaft tomb, only one trait (the hypocone on the maxillary first molar) was significant ($p < 0.05$, Table 13.5). None of the mandibular traits exhibited any significant intertomb differences (Table 13.5). A second series of tests were run on clusters of tombs grouped together on the basis of geographical proximity, since this might reflect the practice of burying kin in nearby tombs (Table 13.6). Again, only one trait (the occlusal surface pattern of the maxillary third molar) was significant ($p < 0.01$). No mandibular traits exhibited significant differences (Table 13.6). Kruskal-Wallis tests run on two sets of randomly grouped tombs yielded significant differences ($p = 0.02$) in traits

Table 13.4. Significant Rank Order Correlations (Kendall's Tau) between Dental Morphological Traits

Teeth	Traits	Correlated with Tooth	Trait	n	r
Maxillary					
Central incisor	Crown shoveling	Second molar	Occlusal surface pattern	15	−0.44*
	Crown variation	Lateral incisor	Crown shoveling	18	0.47†
		First molar	Carabelli's cusp	10	−0.58*
		Second molar	Occlusal surface pattern	15	−0.49*
			Hypocone	14	−0.43*
		Third molar	Occlusal surface pattern	9	−0.5*
			Carabelli's cusp	7	−0.87†
			Hypocone	8	−0.52*
Lateral incisor	Crown shoveling	Second molar	Carabelli's cusp	13	−0.46*
		Third molar	Occlusal surface pattern	8	−0.59*
			Hypocone	7	0.63*
Canine	Crown variation	Second premolar	Crown variation	5	0.93*
		First molar	Hypocone	19	0.41*
First premolar	Crown variation	Third molar	Hypocone	6	0.75*
First molar	Carabelli's cusp	Second molar	Occlusal surface pattern	21	0.37*
		Second molar	Carabelli's cusp	19	0.41*
			Hypocone	21	0.45†
Second molar	Occlusal surface pattern	Second molar	Hypocone	57	0.6**
Third molar	Occlusal surface pattern	Third molar	Carabelli's cusp	38	0.41†
			Hypocone	46	0.85**
	Carabelli's cusp	Third molar	Hypocone	35	0.44**
Mandibular					
First premolar	Crown variation	Second premolar	Crown variation	12	0.59†
Second premolar	Crown variation	First molar	Occlusal surface pattern	10	0.55*
		Second molar	Occlusal surface pattern	10	−0.49*
		Third molar	Occlusal surface pattern	7	0.58*
First molar	Occlusal surface pattern	Third molar	Occlusal surface pattern	18	0.35*
First molar	Protostylid	Second molar	Protostylid	32	0.55**
Second molar	Occlusal surface pattern	Third molar	Protostylid	24	0.43†

* p ≤0.05; † p ≤0.01; ** p ≤0.001

for the third maxillary molar in just one of these sets (Table 13.7), but no significant differences in mandibular traits. Fisher's Combined Test used to assess trends of significance across the range of p-values derived from the Kruskal-Wallis tests yielded no significant results. Mann-Whitney U tests used to test for significant differences in traits between tombs in areas A and C could only be used in the case of the first maxillary molar. Expression of the hypocone differed significantly (p <0.01) between tombs in these different areas (Table 13.8).

Mann-Whitney U tests applied on a tomb-by-tomb basis also showed that Tomb A102 consistently illustrated greater differences in comparison with others in area A in expression of the maxillary first hypocone and the maxillary third molar occlusal surface pattern, while Tomb A105 appeared consistently more different in expression of the second molar hypocone. Mann-Whitney U tests applied between each group of randomly clustered tombs where significant results were found in expression of the maxillary third molar traits showed no consistent pattern of differences in particular tomb groups.

Dental Pathologies

CARIES. Caries were rare in the population, with a total of only 4% of teeth investigated showing any type of carious lesion (Figure 13.1). Of this proportion, interproximal caries accounted for 65% of cases (2% of the total population), while occlusal caries accounted for 18% (<1% of the total population). Large caries, where so much of the tooth had been destroyed that the original site of decay could not be ascertained, accounted for 10% of the total caries, with the remainder made up of noncarious pulp cavity exposures, most likely the result of rapid enamel attrition. Caries size was relatively evenly distributed, with small and medium caries having frequencies of 13 occurrences each, accounting for 70% of cases (35% each). Large caries, those with a dimension of

Table 13.5. Kruskal-Wallis One-Way Analysis of Variance for Dental Traits between Tombs

Tooth	Trait	Tombs	n	x^2
Maxillary Permanent				
Central incisor	Crown shoveling	78, 102, 107, 111	23	0.81
	Crown variation	78, 102, 107, 111	25	4.21
Lateral incisor	Crown shoveling	102, 111	14	0.67
	Crown variation	102, 111	15	0.16
Canine	Crown variation	102, 111	11	2.78
First molar	Carabelli's trait I*	(C9,10,11)[a], 80, 102, 107	25	5.18
	Carabelli's trait II**	(C9,10,11)[a], 80, 102, 107	25	0.87
	Size of hypocone	(C9,10,11)[a] 80, 100, 102, 107, 108, 110, 111	55	16.38*
Second molar	Occlusal surface pattern	100, 102, 105, 107, 108, 111	38	6.93
	Size of hypocone	100, 105, 107, 108, 110	27	8.45
Third molar	Occlusal surface pattern	79, 88, 102, 107, 110	99	6.14
	Size of hypocone	80, 100, 107, 110	17	2.56
Mandibular Permanent				
First premolar	Crown variation	(C9,10,11)[a], 102, 107, 108	18	1.05
First molar	Occlusal surface pattern	(C9, 10,11)[a], 80, 100, 102, 107, 108, 110, 111	58	10.13
	Protostylid	(C9,10,11)[a], 80, 100, 107,108	29	2.9
Second molar	Occlusal surface pattern	(C9,10,11)[a], 78, 100, 102, 107, 108, 110, 111	52	6.49
	Protostylid	(C9,10,11)[a], 100, 107, 108, 110	31	6.11
Third molar	Protostylid	(C9,10,11)[a], 107, 110	12	0.98

* Combines first three grades on the Dahlberg plaques: (a) smooth surface, (b) pit, (c) groove.
** Combines last three grades on the Dahlberg plaques: (a) small cusp, (b) medium cusp, (c) large cusp.

Table 13.6. Kruskal-Wallis One-Way Analysis of Variance between Tomb Clusters

Tooth	Trait	Tomb Groups	n	x^2
Deciduous				
First molar	Occlusal surface pattern	1, 3, 5	28	5.1
Maxillary Permanent				
Central incisor	Crown variation	1, 3, 5	40	0.75
	Crown shoveling	1, 3, 5	39	4.23
Lateral incisor	Crown variation	3, 5	29	0.66
	Crown shoveling	3, 5	30	2.99
Canine	Crown variation	1, 3, 5	40	0.95
First molar	Carabelli's trait I*	1, 2, 3, 5, 6	67	3.04
	Carabelli's trait II**	1, 2, 3, 5, 6	67	4.14
	Size of hypocone	1, 2, 3, 4, 5, 6	90	4.41
Second molar	Occlusal surface pattern	1, 2, 3, 4, 5	64	4
	Carabelli's trait I*	3, 5	36	0.96
	Size of hypocone	1, 3, 4, 5	54	0.49
Third molar	Occlusal surface pattern	1, 2, 3, 4, 5	53	14.35
	Carabelli's trait I*	1, 2, 3	67	3.04
	Carabelli's trait II**	1, 2, 3, 5	67	4.14
	Size of hypocone	1, 2, 3, 4, 5	51	3.99
Mandibular Deciduous				
Second molar	Occlusal surface pattern	1, 3, 4, 5	32	2.68
	Protostylid	1, 3, 4, 5	26	2.68
	Fifth Cusp	1, 3, 4, 5	28	6.3

Table 13.6. continued

Tooth	Trait	Tomb Groups	n	x^2
Mandibular Permanent				
First premolar	Crown variation	1, 3, 4, 5	33	1.19
First molar	Occlusal surface pattern	1, 2, 3, 4, 5, 6	70	1.74
	Protostylid	1, 3, 4, 5, 6	48	1.68
	Fifth cusp	1, 3, 5, 6	22	1.82
Second molar	Occlusal surface pattern	1, 2, 3, 4, 5, 6	69	8.39

Note: Tomb Groups: A78, A80, A100 = 1; A86, A88, A89 = 2; A101, A102, A107 = 3; A120, A121, A108 = 4; A110, A111, A114 = 5; C9, C10, C11 = 6.

* Combines first three grades on the Dahlberg plaques: (a) smooth surface, (b) pit, (c) groove
** Combines last three grades on the Dahlberg plaques: (a) small cusp, (b) medium cusp, (c) large cusp.

Table 13.7. Kruskal-Wallis One-Way Analysis of Variance for Maxillary Dental Traits between Randomly Clustered Tombs

Tooth	Trait	Tomb Groups	n	x^2
Deciduous				
First molar	Occlusal surface pattern	1, 3, 4	19	4.24
Maxillary Permanent				
Central incisor	Crown shoveling	1, 2, 3, 4	34	2.14
	Crown variation	1, 2, 3, 4	33	5.85
Lateral incisor	Crown shoveling	1, 3, 4	23	0.38
	Crown variation	1, 3, 4	24	0.65
Canine	Crown variation	1, 2, 3, 4	29	2.21
First molar	Occlusal surface pattern	1, 2, 3, 4	63	1.52
	Carabelli's trait	1, 2, 3, 4	46	7.58
	Size of hypocone	1, 2, 3, 4	72	2.72
Second molar	Occlusal surface pattern	1, 2, 3, 4	56	4.55
	Carabelli's trait	1, 2, 3, 4	28	2.31
	Size of hypocone	1, 2, 3, 4	46	3.26
Third molar	Occlusal surface pattern	1, 2, 3, 4	50	10.15*
	Carabelli's trait	1, 2, 3, 4	35	10.01*
	Size of hypocone	1, 2, 3, 4	43	10*

Note: Tomb Groups: A78, A107, A121, C11 = 1; A88, A100, A105, A120, C9 = 2; A86, A102, A114 = 3; A80, A108, A110, C10 = 4.
* p = 0.02

Table 13.8. Results of Mann-Whitney U Tests between Areas A and C

Tooth	Trait	No. in A Tombs	No. in C Tombs	z
Permanent Maxillary				
First molar	Occlusal surface pattern	68	6	0
	Carabelli's trait	50	5	-0.84
	Size of hypocone	81	5	-2.97*

*p <0.01

over three millimeters, were only slightly less prevalent than smaller ones, with a frequency of 11, and accounted for 30% of cases.

ABSCESSES. Abscesses were also rarely encountered. A total number of 23 sites (<1%) were found in the entire population, with buccal/labial abscess drainage holes outnumbering lingual perforations with a frequency of 15 to 6 (65% to 26%). Two sites where drainage holes were found on both sides were reported. The presence of abscesses is frequently correlated with tooth loss. When only those teeth in occlusion were analyzed, the frequency of abscess

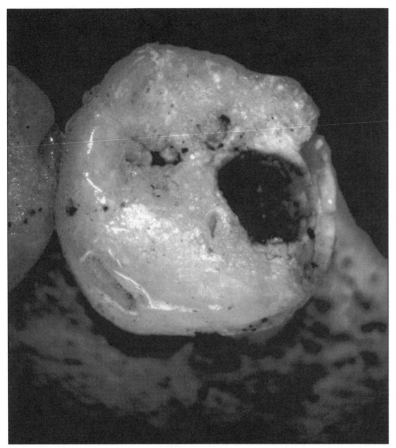

Figure 13.1. Caries cavity on mandibular third molar showing destruction of enamel and deep hole in dentine layer.

perforations declines to four cases out of 1,156 teeth studied (<1%).

CALCULUS. Calculus deposits were found on 709 teeth, or 61% of the population (Figure 13.2). There were slight deposits on 600 of these teeth (85% of deposits, 52% of the population), medium deposits on 104 teeth (15% of deposits, 9% of the population), and 5 cases (0.7% of deposits, 0.4% of the population) with considerable calculus deposits.

LINEAR ENAMEL HYPOPLASIA. Analysis of the total number of hypoplasias per tooth against the tooth number (American Dentists Universal Numbering System) showed that hypoplastic lesions were so infrequent on deciduous teeth in the sample that they were excluded from further analysis, and frequencies for pathologies were only undertaken for the permanent dentition. The most frequent and highly hypoplastic teeth were the canines (both maxillary and mandibular; Figure 13.3), the central and lateral incisors (maxillary and mandibular), and the first molar (again, maxillary and mandibular). For this reason, canines and first molars were chosen for calculations of first and last age at disturbance using formulae from Lovell and Whyte (1999).

Hypoplasia numbers were calculated as the frequency of the total number of hypoplasias found on each tooth as shown in Table 13.9 and Figure 13.4. The proportion of the population with hypoplastic teeth was high, at 87%, with the number of hypoplasias per tooth ranging from one to nine. The highest frequencies were those teeth with between one and four hypoplasias (Figure 13.4). When teeth were assessed visually for the position of hypoplastic events, it was found that in the majority of cases, hypoplastic lines occurred in the cervical half, and most often the cervical third, of the tooth, indicating that disturbances occurred after the age of two. When the canine and first molar were analyzed to assess average age at disturbance, results showed a range from approximately two years in the youngest instance to six in the oldest (Table 13.10). Results from GLM testing for significant differences in the frequency of hypoplasias based on sex and location of individuals within shaft tombs were not significant (Table 13.11).

ASSESSING KINSHIP RELATIONS AT BÂB EDH-DHRÂ': EVALUATION OF THE STATISTICAL RESULTS

The extensive homogeneity evident in the Bâb edh-Dhrâ' dentitions could indicate a biologically inbreeding or endogamous group at the site (cf. Coppa and Macchiarelli 1982; Bondioli et al., 1986). The similarity in traits between the sexes would also support this possibility.

Kruskal-Wallis one-way analysis of variance, testing for between-tomb differences in morphological traits, yielded only one statistically significant result in the cases of both inter-tomb group differences and the clustered tomb groups. These values are low for a positive evaluation of the hypothesis that family groups are buried together. Furthermore, the significant differences in third molar traits between randomly clustered tombs are unexpected. However, the third molars in general display the most marked variability in expression and are likely to contribute most to observed differences when using observations based on pooled data. Alternatively, shaft tombs located near each other might not represent wider

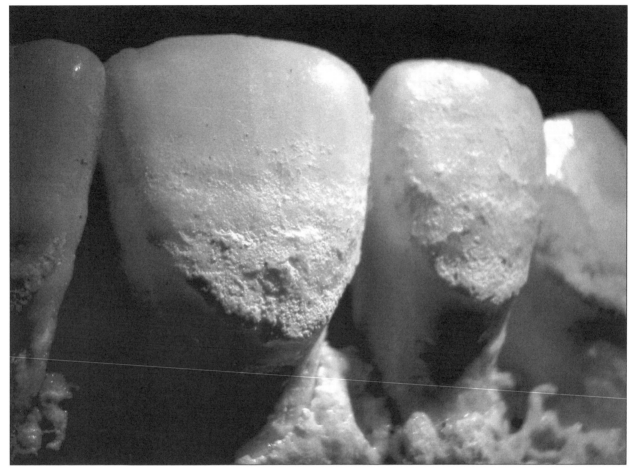

Figure 13.2. Maxillary incisors of a Bâb edh-Dhrâ' specimen showing medium calculus deposits.

extended families or clan groups choosing to locate burials adjacent to each other, so that differences between individual tombs might be expected wherever located in the cemetery.

Butler (1939) was the first to discuss what is commonly referred to as the "field concept" where each tooth type (i.e., incisors, canines, and molars) is developmentally determined in size and form by their relative position within the series. Those teeth placed more anteriorly within their field (e.g., the first premolar or molar) are more genotypically stable and exhibit less variation than the more posteriorly placed teeth (e.g., the lateral incisor and third molar). Moreover, as confirmed by the high number of traits that are correlated with each other (Table 13.4), traits within the same or different morphogenetic fields are clearly not genotypically independent in expression (Garn et al. 1966; Keene 1965, 1968; Scott 1975, 1977:461–66). The existence of a high number of intertrait associations in the Bâb edh-Dhrâ' sample would suggest that caution should be used in the use of possible redundant traits for the assessment of biological distance.

Lack of significant results might also originate from the very small sample sizes within each tomb that could be used in the statistical analyses, given the high degree of dental attrition that eroded observable traits, possibly leading to type II error. In fact, significant p-values tend to occur in those cases where sample sizes reach a maximum available for the individual tombs examined. Larger samples would also be helpful for evaluating further potential differences between cemetery areas A and C, but even here one morphological trait out of the three that could be examined was significantly different.

When individual tombs are examined for the distribution of specific traits to see if these might cluster together, there is greater support for family groups to be buried in the same tombs. For example, while 84% of the Bâb edh-Dhrâ' population shares the same degree of expression of the hypocone, Tombs A102 and C9 contribute most of the variation in this trait. Specifically, most individuals at Bâb edh-Dhrâ' scored grade "3" in degrees of expression using the Arizona State University system, but individuals in Tombs dividuals. Finally, stylar ridging on the deciduous first molars—a particularly unusual trait—occurs in eight individuals within five tombs (Figure 13.6). Again, Tomb A111 has three out of the eight children with this trait. Similarly, Tombs A102 and A105 contribute a number of individuals with this rare trait. Tomb A102 was also noted as having individuals with the greatest degree of expression of the maxillary first molar hypocone.

ASSESSING THE HEALTH OF THE BÂB EDH-DHRÂ' POPULATION: EVIDENCE FROM DENTAL PATHOLOGIES

The frequency of caries at Bâb edh-Dhrâ' is very low, at 4%. This is not surprising for the EB IA period, since caries frequencies are known to be low in premedieval populations (Brothwell 1959). The transition during mediaeval times in Europe to higher rates of caries is attributed to the increase in reliance on carbohydrates and consumption of sugared foods that occurred during this period. The paleobotanical data from Bâb edh-Dhrâ' studied by McCreery (1980, 2003) would also lend support to a low probability of dietary items that would contribute heavily to caries

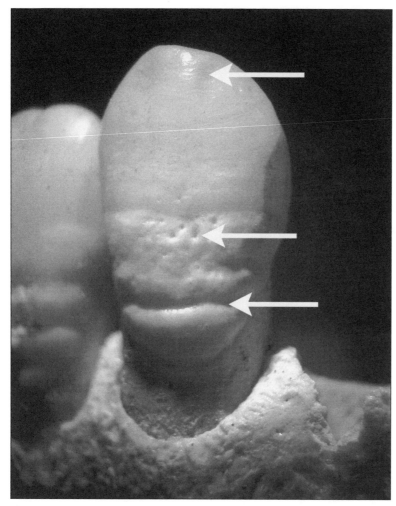

Figure 13.3. Mandibular canine from a Bâb edh-Dhrâ' specimen. The top arrow shows perikymatic grooves, the middle arrow shows hypoplastic pitting, and the bottom arrow shows linear hypoplasia.

A102 and C9 also scored in Grades "4" and "5". Moreover, when some of the more rarely expressed traits are examined for their distribution across tombs, they are found to be clustered in particular tombs. For example, the sixth cusp, which is expressed in only 14 individuals, occurs in just six tombs. Again, third molar agenesis (confirmed in each case by radiographic analysis) is characteristic in 16% of the population and appears in only five tombs associated with 11 individuals (Figure 13.5a, b). Tomb A111 alone houses five out of these 11 in-

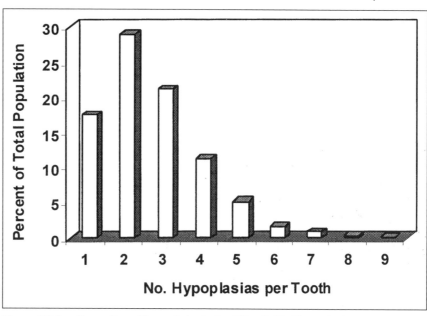

Figure 13.4. Proportion of hypoplasias in the Bâb edh-Dhrâ' sample.

Table 13.9. Frequencies and Percentages of Hypoplasias Found in Restricted Sample of Present and Occluded Teeth

Total Number of Hypoplasias/Tooth	Frequency	Percent of Total Population	Percent of Hypoplasias Seen
1	203	17.6	20.2
2	335	29.0	33.3
3	245	21.2	24.4
4	129	11.2	12.8
5	60	5.2	6.0
6	19	1.6	1.9
7	10	0.9	1.0
8	3	0.3	0.3
9	1	0.1	0.1
Total	1005 (of 1156)	86.9	100

Table 13.10. Estimates for Youngest and Oldest Ages of Hypoplastic Disturbances on Canine and First Molar Teeth

Tooth Analyzed	n	Age at First Event (years)	Age at Last Event (years)
Right maxillary canine	6	2.58	4.84
Left maxillary canine	11	2.87	4.85
Left mandibular canine	22	2.88	5.72
Right mandibular canine	27	2.69	5.67
Right maxillary first molar	3	2.11	2.80
Left maxillary first molar	14	2.06	2.91
Left mandibular first molar	19	2.18	2.98
Right mandibular first molar	30	2.28	2.94

Table 13.11. Results from General Linear Model Testing for Significance between Sex, Shaft, and Tomb

Variable	df	F	R^2	p-value
Sex	2	1.405	0.440	0.254
Shaft	16	1.303	0.440	0.410
Tomb (shaft and chamber)	40	1.135	0.466	0.331

formation at EB IA Bâb edh-Dhrâʿ. Similar low caries rates have been found by Greene (1972) for populations in Egypt and Nubia during the EBA. Some exceptions in the Middle East are seen for populations consuming a high proportion of fruits such as dates and figs (possibly in dried form) where the incidence of caries and antemortem tooth loss is high due to the high sugar content in these fruits (Littleton and Frohlich 1993).

The low prevalence of occlusal caries at Bâb edh-Dhrâʿ and the correspondingly high percentage of interproximal caries (65%) is also a pattern frequently seen in archaeological samples, especially those with a great deal of dental attrition (Hillson 2001). Given the high level of attrition on the dentitions of the Bâb edh-Dhrâʿ population, such wear would have obliterated the pits and fissures (along with any developing caries) present on the molar teeth that are usually the sites for caries initiation. The level of interproximal caries is therefore high in such situations, since spaces between the teeth allow for accumulation of food and plaque particles. There is, however, a lack of root caries (although the 10% of caries seen that were so large that their sites of origin were unidentifiable may have been of root caries origin). High levels of wear are known to cause a compensatory migration of the teeth to maintain functional occlusion (sometimes referred to as "continuous eruption," it is a modification of the dento-alveolar complex) therefore exposing the root surface to conditions that could instigate a carious lesion (Hillson 2001; Kaifu 2000).

A feature that may have added to the low level of caries at Bâb edh-Dhrâʿ is the fact that only well-developed carious lesions were identified and noted. Hillson (1996, 2001) states that, as caries form by initiation of a lesion below the enamel surface while the surface still appears intact, in many cases they are only readily apparent when radiolucency analysis of the tooth in question is undertaken. Therefore, without such analyses, it is possible that a number of early caries lesions would not have been observed. The low level of caries (especially occlusal) also helps to explain

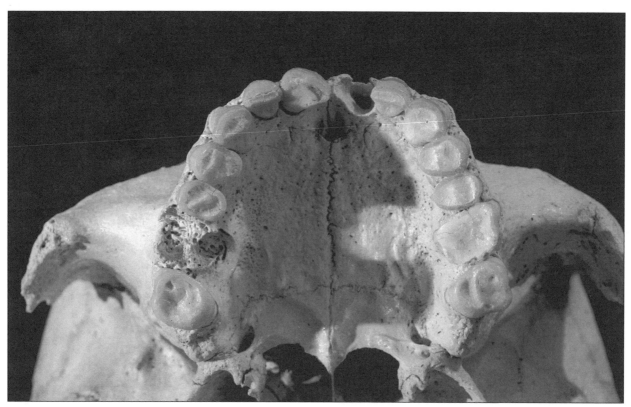

Figure 13.5a. Maxillary dentition showing third molar agenesis in Burial 5, an adult male about 45 years of age from tomb chamber A111W.

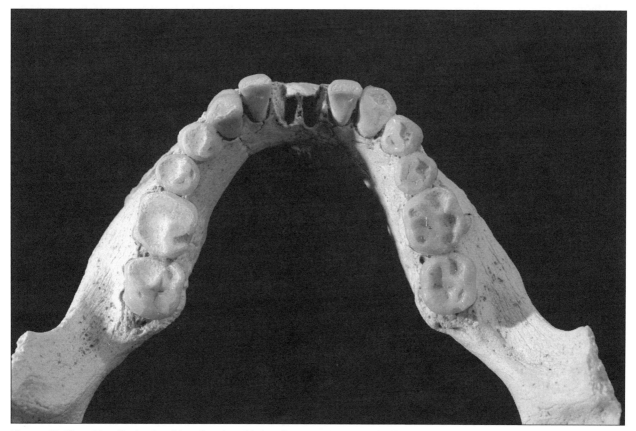

Figure 13.5b. Mandibular dentition showing third molar agenesis in Burial 5 from tomb chamber A111W.

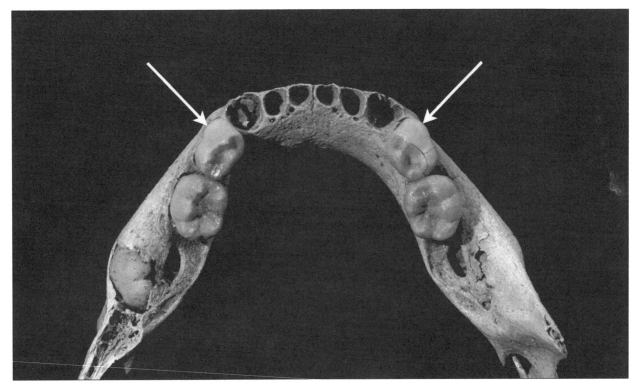

Figure 13.6. Mandibular stylar ridging in Burial 1, a two-year-old child from tomb chamber A111N.

the correspondingly low level of abscesses found in the population.

It has been noted in some publications that, due to the weakening of enamel at the site of a significant hypoplastic event, the tooth may be predisposed to carious lesions at that site (Mellanby 1927; Cook and Buikstra 1979; Hillson 2001). This effect was not found, however, in the sample studied from Bâb edh-Dhrâ', and could possibly indicate that hypoplastic lesions in this population did not reach the level of severity needed to induce this effect. Conversely, since caries were low in number overall, the etiology for formation of caries at hypoplastic sites was not produced.

The high level of calculus seen on the teeth (61% of those in occlusion) and the lack of apparent dental hygiene could have led to periodontal disease but also protected against caries formation. Indeed, Figure 13.3 shows calculus deposits below the cemento-enamel junction, and what appears to be alveolar bone recession indicative of periodontal disease. Manji et al. (1989) show that, in populations with high calculus deposit rates, caries rates are likely to be correspondingly low. This is because of different processes that lead to the formation of the two pathologies: mineralization in plaque deposits leading to calculus, and demineralization of the tooth enamel leading to caries. It is possible, however, that active caries may be residing below calculus deposits on teeth (Hillson 2001). Despite the possibility of periodontal disease, rates for antemortem tooth loss that could be related to this pathology are generally low at Bâb edh-Dhrâ'.

The evidence in the Bâb edh-Dhrâ' dentitions for extensive hypoplasias that developed during childhood, together with the paleopathological data for metabolic and nutritional disorders such as scurvy and rickets during the EB IA, suggests that the diet during this period was inadequate for optimal growth and development. Although McCreery's paleobotanical data (1980, 2003) show a range of domesticated and wild plants being available at the site during the EB I, including wheat, barley, chickpea, lentil, pea, fig, and grape, it is likely that the Bâb edh-Dhrâ' population relied heavily on just a few staples such as wheat and barley for their sustenance, thus explaining the poor diet for children. This is borne out by the low proportion of fruits relative to cereals (McCreery 2003).

However, the low level of hypoplasias on the deciduous teeth is taken to indicate little or no prenatal stress events. A number of scholars have reported that LEH on the deciduous teeth are rare in most human populations, except for those groups with particularly poor nutritional status (e.g., Cook and Buikstra 1979; Goodman et al. 1984) or where mothers suffer from particular problems such as hypocalcemia (Nikiforuk and Fraser 1979; Purvis et al. 1973), hypoparathyroidism (Myllarniemi and Perheentupa 1978), or renal conditions (Oliver et al. 1963). Any defects found on deciduous teeth are therefore likely to be used as a proxy for maternal health during the time of pregnancy.

Using microscopic analysis of modern anterior teeth, Reid and Dean (2000) found that little or no surface enamel forms in the first year of life. Any stresses, therefore, that might occur during this time would be unidentifiable. The lack of LEH in children between the ages of one and two years at Bâb edh-Dhrâ' would indicate that they were protected from malnutrition at least until weaning. Katzenberg et al. (1996) state that the process of weaning is associated with an increase in infant morbidity and mortality due to three factors: compromised nutritional status due to insufficiently nutritious weaning foods, the loss of passive immunity previously gained through the consumption of breast milk, and increased exposure to a host of new infective pathogens depending on the cleanliness and preparation of weaning foods and water.

It is logical to assume that weaning foods at Bâb edh-Dhrâ' would be based on wheat and barley cereals that, if not supplemented with foods containing higher levels of protein, vitamins, and minerals, would be nutritionally insufficient to meet the demands of growing children. After a child experiences an event resulting in a hypoplastic lesion, Goodman and Armelagos (1988) suggest the teeth may be rendered weaker and more susceptible to events in the future. This could account for the repeated number of hypoplasias seen in the majority of individuals. The age range for hypoplastic events from two until five years of age and beyond in the Bâb edh-Dhrâ' sample correlates with known ages at which children are most susceptible to infection and malnutrition (Goodman and Armelagos 1988).

What is apparent, though, is that these events did not cause mortality, since the individuals survived the stress and died mostly during adulthood. However, using heterogeneity models of frailty on the large skeletal series from the medieval village of Tirup in Denmark, Boldsen (2007) has recently demonstrated a clear negative effect on mortality for those individuals with LEH compared to those without (log rank test statistic for males = 9.96, p = 0.002; F-test for equal variances for females = 2.66, df = (17, 21), p = 0.0034). It can therefore be conjectured that those individuals at Bâb edh-Dhrâ' with higher scores for LEH were at higher risk for mortality (and possibly other morbidities) throughout their lifetimes.

Several publications refer to a decrease in health and nutrition accompanying the advent of agriculture in the Near East and other regions, with increasing evidence from the paleopathological data for health problems (e.g., Cohen and Armelagos 1984). Comparative data from archaeological sites near Bâb edh-Dhrâ' from the Natufian period onward show increasing rates of LEH. The Natufian site of Kebara has LEH frequencies of 23%, the Neolithic site of Abu Ghosh has frequencies of 47%, and the Chalcolithic site of Azor has frequencies of 80% (Smith et al. 1984). This could well indicate an increase in both nutritional disturbances and infectious diseases as societies became more agricultural in their subsistence base. The paleopathological dilemma referred to by Wood et al. (1992), however, means that the incidence of hypoplasias seen in the Bâb edh-Dhrâ' skeletal materials reflects an increase in survivorship of chronic conditions during childhood.

CONCLUSIONS

Reconstructing Kinship at Bâb edh-Dhrâ'

The use of dental nonmetric traits for a biodistance study of the Bâb edh-Dhrâ' EB IA shaft tomb population has illustrated the problems associated with morphological analyses where small samples and fragmentary bioarchaeological data are used. However, it has confirmed the possibility of a closer examination at what amounts to the forensic level for particular characteristics within individual dentitions, and confirmed the utility of dental morphological traits for such analyses.

The indications of close kinship at the intratomb level that the clustering of rarer traits would support offers credence to, but not definitive proof for, the hypothesis that family groups are buried together within the Bâb edh-Dhrâ' shaft tombs. The high degree of homogeneity within the entire population again argues for an inbreeding population and what Bentley (1987) referred to as a "corporate group structure" (cf. Chesson 1999, Rast 1999). The finding from the dentitions also bolsters the more general theory advanced earlier that the presence of a formal cemetery argues not for a nomadic pastoralist population but instead for a corporate group structure that was both sedentary and territorial, and where lineage claims to resources were legitimized by the presence of ancestors in nearby tombs (Bentley 1987, 1991).

Reconstructing Health at Bâb edh-Dhrâ'

Evidence from the dentitions at Bâb edh-Dhrâ' supports earlier paleopathological data presented in Chapters 6, 7, and 12 pointing to chronic nutritional insufficiencies in the EB IA population. The high proportion of hypoplasias in individuals, evidence of growth disruption from weaning onward, suggests a diet that was poor in macro- and micronutrients typical perhaps of agriculturalists relying heavily on a narrow range of cereal crops. Lack of differences in the proportion of dental pathologies between tombs suggests that there are few health inequalities between groups buried in the EB IA tombs.

The low rates of caries among the population as a whole points to a diet that did not contain foods conducive to tooth decay, but caries prevalence was also mitigated by the high degree of attrition resulting from the coarse diet eaten by the Bâb edh-Dhrâ' population that abraded tooth surfaces.

From the increase in frequencies of LEH, Smith et al. (1984) conclude that health in the Levant from the Natufian to the Bronze Age declined as agriculture and permanent settlements increased. Declining health is attributed to the higher levels of endemic diseases that the new settlements faced rather than to food shortages, although poor dietary quality may have been responsible for much of the growth disruption observed from the dental pathology. If the rationale of Wood et al. (1992) is correct, EB IA individuals rallied from nutritional disruptions and diseases more effectively than at previous times in the history of the area and outlived the chronic problems they faced as children. That such early growth disruptions, however, would have compromised overall survival is suggested by longitudinal frailty models of individuals with LEH, albeit from a population removed in time and space (Boldsen 2007).

Further comparative data from the EB IB–IV phases at Bâb edh-Dhrâ' that might emerge from future excavations would be invaluable for analyses of whether the prevalence of hypoplastic lesions and other pathologies increases or decreases during later periods of urban development at the site, and whether the general health of the population improved or deteriorated over time.

CHAPTER 14

Summary of Findings and Conclusions regarding the EB I People of Bâb edh-Dhrâ', Jordan

WITH OVER 700 BURIALS in the Bâb edh-Dhrâ' EB I skeletal sample, it is the largest archaeologically documented human skeletal sample from the Near East. Preservation and burial completeness range from a single skeletal element to the entire skeleton. Because we made a major effort to recover every skeletal fragment originally placed in the tombs, there is a high probability that almost every burial originally interred in the EB IA and EB IB shaft tomb chambers is represented in the burials included in our analysis. This means that at least the EB IA shaft tomb burials are probably a reasonably representative sample of the people associated with these tombs. We are less confident about the relationship of the G1 Charnel House burials to the people associated with this funerary tradition, because of the clear age bias in the sample and the vandalism that had occurred in the charnel house in antiquity and more recently.

Variation in completeness of burials creates some important limitations on the generalizations that can be made about the people represented by the interments, and these will be reviewed briefly later in this chapter. Nevertheless, there is much that can be said, and the objective of this chapter is to provide a review of the findings identified in earlier chapters and explore the broadest possible implications of the data. Admittedly, some of our conclusions will be speculative, but we hope that in presenting even tentative observations we will stimulate discussion and additional research as new burial sites in the Middle East are excavated in the future.

THE EB I SOCIETY OF BÂB EDH-DHRÂ'

As we have noted in previous chapters, particularly Chapter 5, the societies in the southern Levant were relatively independent, at least politically. Undoubtedly, some cultural traits were transmitted between societies, as Schaub notes in Chapter 4, but the archaeological differences apparent in excavations of various sites thus far indicate considerable cultural variability between societies. For obvious reasons, the focus of this book is the information regarding the society that is revealed by the funerary traditions. Funerary traditions, of course, are a very limited component of any society. Nevertheless they do provide insight about the society and enrich what can be determined on the basis of other cultural evidence.

There is little evidence for permanent settlement at Bâb edh-Dhrâ' during EB IA on the basis of excavations conducted at the site thus far. Some evidence of a permanent EB IA house structure was identified near the town site just to the north of the cemetery (Rast and Schaub 2003:71), but this appears to be late in the EB IA period. However, there is only very limited evidence for this type of structure. The current paucity of evidence for permanent structures is the basis for the hypothesis that during EB IA the Bâb edh-Dhrâ' people used the site as a stop on their circuit of a nomadic-pastoralist economy. Clearly, the presence of elaborate funerary traditions at the site argues for it being of considerable cultural significance and the likely location for frequent and perhaps extended stays at the site by some people.

Binford argues that funerary traditions reflect important aspects of the social and economic structure of a human society (Binford 1972:222, 227–231). Although we agree with this generalization, our experience with the Bâb edh-Dhrâ' tombs and burials from the EB IA period indicates that the linkage is complex, and inferences about social structure based on these traditions may not be direct or transferable between societies. For example, Binford (1972:230) makes the observation that in less sedentary societies, sex and age are more likely to be significant factors than in more settled communities. If this is true, during the EB IA phase at Bâb edh-Dhrâ' we would expect to see differences in adult versus subadult inhumations and between male and female burials. Our data indicate that during this period neither age nor sex were factors affecting burial. Male and female skeletons were commonly found in the same tomb chamber. In addition, at least some skeletal elements from all ages between fetal to old age were recovered from the primary burial site and returned for interment in the tomb chambers.

If, as Binford argues, burial tradition reflects social structure, the burial of all individuals in shaft tomb chambers at the time of death argues for a relatively egalitarian society in which all individuals were valuable and treated the same, at least in the funerary tradition. However, the evidence for an egalitarian society is not entirely consistent. The higher prevalence of early onset osteoporosis in some female skeletons suggests that adult women may not have had equal access to food. This is a fairly common situation in some third world countries today in which the hierarchy for access to food is adult and subadult males followed by subadult and adult females (Ortner 2003). In situations where food is adequate for all members of the society, this food access hierarchy may not have been a problem. However, if food was in short supply, adult women likely would have suffered.

Furthermore, as Schaub discusses in greater detail in Chapter 4, there is no evidence of a stratified social hierarchy exhibited in the cultural artifacts that were placed in the shaft tomb chambers. There is variation in the number of artifacts placed in the chambers, but this variation is not associated with variation in any of the biological characteristics of the burials in the chamber. Mace heads may very well be an indication of the relative importance of one of the individuals represented by the chamber burials. Nineteen of these cultural artifacts were found in the chambers, but there is no association of these artifacts with male or female burials. It may be that the presence of a mace head signifies that one of the burials represents an individual of special social significance, but we have no way of determining which burial in a chamber was associated with the mace head. The fact that multiple individuals were included in the same chamber argues for the conclusion that whatever social differences may be symbolized by the presence of a mace head, it probably does not represent more than the head of an extended family.

All the other evidence from the burials and the cultural materials recovered from the shaft tombs suggests a society that was relatively egalitarian, at least during the EB IA phase. There is no evidence of any variation in cultural contents of the tombs that can be linked to either the sex or age of the individuals buried in the tombs (see Chapter 4).

The funerary tradition appears to change somewhat with the introduction of charnel house tombs in EB IB. At least some of the burials continue to be secondary, but fetal, infant, and young children's bones were absent or rare in the G1 Charnel House.

Unfortunately, the disturbance associated with the G1 Charnel House burials precludes any plausible insight regarding a shift in the egalitarian nature of the EB I people. However, the change in both the tomb architecture and who is interred in the charnel house does suggest a change in culture that might have had a significant impact on social structure.

The investment made in the construction of both shaft tombs and charnel houses as well as the value of the objects placed within the tombs argues for the importance of the funerary traditions throughout the EB I period. Placement of food offerings, jewelry, and other personal items implies a belief in the continuation of life beyond normal human existence. The presence of 12 unfired clay female figurines in the chambers seems likely to have some symbolism, perhaps linked to fertility or rebirth but, again, there is no identifiable association with male or female burials or with burials of a certain age.

This egalitarian social structure may have changed in EB IB, at least with respect to the age of individuals accorded full funerary rites. If the evidence from the G1 Charnel House is broadly representative of the funerary tradition in EB IB, burial was usually limited to individuals past the age of two years, unlike universal burial in EB IA. This may signal a somewhat more stratified human society in EB IB.

An important component of the EB I society is the economy of the community. The very limited evidence of permanent settlement of the site in EB IA does support the possibility of at least a seminomadic-pastoralist society in early EB I gradually making a transition to a more sedentary agriculturist economy beginning in the EB IB period, when evidence of permanent settlement becomes more widespread. Nevertheless, nomadic pastoralists usually represent a subspecialization of an agricultural economy that takes advantage of marginal land resources to produce animal products that are important to the agricultural economy.

Bentley (1987) argues that a more plausible hypothesis is that the EB IA people were relatively sedentary and that nomadic pastoralism was not an important component of the Bâb edh-Dhrâ' economy until EB IV, a period she argues is associated with declining agriculture caused by increasing salinity of the soil. As in many attempts to reconstruct important aspects of past human societies, the evidence is simply insufficient to resolve all the questions that beg to be answered, although the current weight of archaeological evidence favors a transient use of the site until at least late EB IA, with more permanent settlement being established in EB IB. However, the relative abundance of water via what today is called Wadi Kerak, as well as the large area of flat land in the Ghor, seem to be resources that likely were exploited with some intensity. At the very least, Bâb edh-Dhrâ' would have been a site of importance to the EB I people living in proximity to if not at the site itself.

FUNERARY TRADITIONS AT BÂB EDH-DHRÂ' DURING EB I

The funerary traditions apparent during the EB I phases argue for a society with a rich cultural heritage. Considerable time and effort was invested in creating the shaft tombs and charnel houses. On the basis of the time needed to excavate the shaft tombs during the 1977 field season, Ortner (1981) estimates the time needed to create a four-chambered tomb at about 15 man days. The cultural artifacts associated with the burials consist mostly of fine ware pottery that required both expertise in making the pottery but also the time to invest in an activity that was not directly related to the preparation and consumption of food. Other cultural artifacts, such as the basalt vessels, required considerable time and energy resources to create. The time and effort needed to bury a deceased member of the society at the primary burial site and subsequently recover in some cases even the very small bones of fetuses for secondary burial also represent a substantial investment of resources by the society. All of the above factors make a strong case for the social significance of the funerary tradition during EB I.

Clearly embedded in the funerary tradition of the EB IA people is the culturally defined sense of afterlife. The food offerings in the EB IA chambers certainly provide an argument for this conclusion. Given this cultural emphasis on afterlife, ceremonies to mark the transition between life and afterlife are probably inevitable. A tradition that buries the dead temporarily but recovers the remains for final burial in the society's cemetery represents a good cultural strategy for nomadic pastoralists. What remains to be determined is why secondary burial continued to be practiced as the society became more sedentary. One possibility is simply a continuation of a cultural tradition in the context of a different settlement pattern. Another possibility is that nomadic pastoralism remained a significant component of the EB IB society so that some, if not many, members of the society would die some distance from Bâb edh-Dhrâ' and their bodies would be buried where they died until the soft tissue had decayed, facilitating transportation to the society's ceremonial burial ground.

Burial Type

Both primary and secondary burials occur in the EB I tombs of Bâb edh-Dhrâ', although secondary burial is far more common. In primary burial, there is only one interment event; in secondary burial there is an initial, temporary burial of some type, with later recovery of the remains, final burial in the tombs, and cultural funerary artifacts included with the human remains. We know very little about the primary burial environment for the EB I people. There is evidence that at least some of the burials were below ground. However, one burial in A88L has two millimeter holes most likely made by dermestid larvae (Ortner 1981). These larvae do not penetrate soil beyond a few centimeters, so this burial may have been on the surface or just below it in the primary burial site.

One of the frequent questions asked of those conducting research on secondary burials is why a society would develop a burial tradition that adds substantial costs to the burial process. Undoubtedly, the answers to this question vary between human societies, and in archaeological populations any answer is speculation at best. However, if the EB IA people of Bâb edh-Dhrâ' were nomadic pastoralists, secondary burial would be a useful strategy in dealing with the problem of what to do with a body as a mobile society moves from place to place with the animal herd.

Variation in bone preservation of different burials from same tomb suggests that in some cases at least the deceased were initially buried in different primary burial environments. Different preservation of different burials is apparent is several tombs, an example of which is Tomb A89 described in Chapter 6. This taphonomic variation in preservation is further complicated by evidence of breakage during excavation at the primary burial site and transportation to the shaft tomb chambers. Evidence of breakage between primary and secondary burial is particularly apparent in the burials from A80S but is a rather common feature in other tomb chambers.

The type of burial in the EB IB G1 Charnel House certainly includes secondary burials, as we have discussed in Chapter 7. However, we do have limited evidence of primary burial during EB IB in shaft tomb chambers (A100N and A88 upper), which are among the few shaft tomb chambers associated with that time period. This raises the possibility that some of the burials in G1 were primary. Evidence of other changes in the funerary tradition, including the lack of fetal bones and the presence of very few infants and young children, suggest that some changes did occur in the transition between EB IA and EB IB, perhaps as the result of the society becoming increasingly sedentary, as is suggested by other archaeological evidence. However, the utilization of both shaft tombs and charnel houses in the EB IB funerary tradition argues for gradual change of the tradition with considerable overlap between the two types of funerary structure.

Secondary burials pose multiple problems for the bioarchaeologist. Among the most significant of these are: (1) multiple burials typically occur in a tomb and skeletal elements from different burials are mixed together, (2) it is not always possible to associate cranial with postcranial bones, (3) breakage of skeletal elements during recovery from the primary burial site and transportation to the secondary site, and (4) skeletal elements are often not recovered from the primary burial site or are lost during movement to the secondary site. These problems mean incomplete data for some individual burials, and this causes difficulty during statistical analysis, as we have emphasized in Chapter 10.

Tomb Type

SHAFT TOMB. In Chapter 5, we have provided a general summary of shaft tomb architecture, and in Chapters 6, 8, and 9 we have described in detail each of the tombs excavated during field seasons in 1977, 1979, and 1981. In the A Cemetery the spacing between tombs was very tight. The density of the tombs creates the likelihood that some method was used to identify the location of shaft tombs, at least during the EB IA period. The evidence for this is less convincing in EB IB, since some of the EB IB tombs did serious damage to EB IA tombs. However, the reuse of tomb chamber A100N in EB IB (Chapter 6) suggests at least some long-term knowledge of tomb location.

Within shaft tomb chambers, the organization of the remains supports the opinion that the contents were from a single burial event. An exception to this seems likely in A89NW described in Chapter 6; this is clearly the case in the EB IB chamber of A100N.

Siltation, which was common in the shaft tomb chambers, could have an adverse effect on preservation and recovery of burials during our excavation of chamber contents. Usually the silt was relatively soft and could be removed carefully to expose the burials and artifacts. In some cases, such as in A86NE, the silt became very hard and not all skeletal elements placed in the chamber in antiquity could be recovered.

Some silt gradually flowed into the chambers over fairly long periods of time through cracks and holes, long after the burials and artifacts had been placed in the chambers, the chambers had been sealed, and the shaft filled. This type of siltation created relatively fine layers of silt, indicating multiple silting events, probably on an annual cycle related to the rainy season. This type of siltation is independent of the funerary tradition.

The presence of silt that entered through the entryways of the chambers offers insight into the shaft tomb burial tradition. We see evidence of this type of siltation in Tombs A79, A80, and A86. This silt must have entered the chamber through the open shaft, since the silt is most concentrated at the entryway and diminishes in thickness toward the center and rear of the chambers. This type of siltation implies a gap between the creation and initial use of the shaft tomb and the silting event, which must have occurred during one or more rainy seasons. This suggests that the shaft of a tomb would have been left open for at least several months after initial use of one or more chambers.

There is evidence suggesting a family relationship of individuals interred in a tomb chamber (Bentley 1987; Chapter 13). If this is the case, a close genetic relationship between many of the individuals buried in all the chambers of a given tomb is likely as well. With few exceptions, all burials within a chamber were interred at the same time. The variability in the number of chambers per tomb and the presence of some unfilled chambers in some tombs suggests that different chambers of a tomb were used in different burial events. This would have required knowledge of the tomb location, and, given the evidence of silt flowing in through the entryway of some chambers, one source of information about shaft tomb location would be shafts that remained opened for some time after initial excavation and use.

The association of specific tombs with a specific extended family group is also supported by the differences between the artifacts included in the tomb chambers of the A and C cemetery areas, as discussed by Schaub in Chapter 4. This difference is apparent in the pottery and mace heads that are more common in the C Cemetery tombs. This variation in tomb chamber contents may reflect differences between kin groups regarding the artifacts placed in the chambers.

The possibility of separate interment events for different chambers within a shaft tomb would fit well with an extended family association with a specific tomb. The number of deaths per year within an extended family might easily be insufficient to require all the potential chambers of a shaft tomb during a single secondary burial event. Keeping the shafts open for subsequent use would keep the location of the tomb well known to the family and limit the time and energy needed to create a chamber when additional deaths within the family required burial space.

Charnel House

The charnel house burial tradition beginning in EB IB provides an interesting mix of cultural information indicating both some continuity with the EB IA burial tradition and significant differences; the former hints at the stability of the society, the latter points toward the dynamic nature of the culture of the Bâb edh-Dhrâ' people. The tension between the old and new burial traditions is highlighted by the continued, although limited, use of shaft tombs in EB IB and the evidence of at least some secondary burials occurring in the G1 Charnel House.

Charnel houses represent, in many ways, a different funerary tradition. These tombs are above ground, predominantly mudbrick structures that contained, in most tombs, many more burials than are found in the shaft tombs. It is possible if not probable that each charnel house was associated with an extended family whose use of the structure extended over at least several generations. The presence of multiple burial structures that appear to exist at the same time argues for some cultural mechanism to determine which charnel house was used for a specific interment event, and family relationship is an important linking factor in any society. Another difference is that the burials and artifacts within the charnel house represent multiple funerary events, unlike the typical shaft tomb chamber, although not unlike what appears to be the case with the shaft tomb where each chamber of the tomb may represent a separate burial event.

A potentially important difference that exists between shaft tomb burials and those in the G1 charnel house is the lack of fetal bones and the paucity of infant and young children's skeletal elements. Although we cannot be certain that the disturbance of the contents both in antiquity and more recently has not distorted the age distribution of G1, we doubt that this is a significant factor affecting this distribution. The presence of a few infant bones argues against special taphonomic processes being the cause of this different age distribution. This, of course, raises an important cultural question about how those who died in this early range of ages were disposed of. If they were simply discarded without the society's funerary ceremony, it marks a major change from the careful recovery and interment granted to all ages of diseased society members in the shaft tomb tradition.

The analytical challenges created by secondary burials are even greater in research on the human remains from G1 than they were for the shaft tomb burials. Multiple burial events over many years associated with the charnel house inevitably resulted in the interment process of more recent burials disrupting whatever order was present in earlier burials with damage and scattering of preceding burials and cultural artifacts. Furthermore, the mixture of skeletal elements from different burials was on a much larger scale than occurred in the shaft tomb chambers. All of these complications were enhanced by vandalism that occurred in antiquity and more recently. This made it impossible to associate any skeletal elements from a single burial.

PHYSICAL CHARACTERISTICS OF THE PEOPLE

The primary objective of both the metric and nonmetric analyses of the EB I skeletal samples was to define the basic osteology of the EB I people of Bâb edh-Dhrâ' and identify where the closest biological links were with other human groups in northeastern Africa and the Middle East. Smith (1989) provides a review comparing some human skeletal samples dated to the Early Bronze Age in the Levant. She concludes that the evidence supports the presence of fairly stable human populations with little intrusion from areas outside the Levant. There also is evidence suggesting a decline in health during the transition between the Chalcolithic and Early Bronze Age.

In Chapters 10 and 11, comparisons are attempted with skeletal samples from various sites in these areas. The problems in making these comparisons have been discussed in detail in these chapters. The major point to be made is that published data is regularly flawed. The samples were created using varied criteria, research methods, and protocols so that differences identified in statistical analysis may be the result of differences in sample quality or research methodology; they may not be the result of genetic similarities or differences between the populations represented by the samples.

One of the great frustrations of comparing our data to data sets collected by other researchers is that we know so little about how representative these samples are of the living population from which they came. There are many variables that can affect this aspect of a skeletal sample, including: (1) completeness of the burials, (2) care in excavation and curation of burials to ensure that all interments from a tomb are recovered and available for analysis, (3) variability in the burial tradition of the societies being analyzed, and (4) interobserver variation in research methodology.

The EB IA people of Bâb edh-Dhrâ' were slightly above average in height. Stature based on the average stature estimated from all the major long bones (see Chapter 10 for details) is 166 centimeters (5'5") for males and 156 centimeters (5'2") for females. The primary osteological index of robusticity (muscularity) is a function of the midshaft size (mediolateral diameter + anterio-posterior diameter × 100) of the femur divided by the bicondylar (physiological) length of the femur (Bass 1987:214). For the Bâb edh-Dhrâ' EB IA people, this index is 13.6 for males and 13.0 for females. Not surprisingly, males are more robust than females. These index values indicate that the people of Bâb edh-Dhrâ' were relatively robust.

Both male and female skulls place the Bâb edh-Dhrâ' people within the long-headed (dolichocephalic) category (Bass 1987:69). Measurable male skulls had a cranial index of 71, and female skulls had a cranial index of 73. This general shape of the skull is widespread in the Middle East during the latter part of the fourth millennium BC (Smith 1989) and tells us little about the relationship of the Bâb edh-Dhrâ' sample to skeletal samples from the Levant or Egypt. Multivariate analysis of metric data discussed in Chapter 10 supports but does not confirm an association between the people of Bâb edh-Dhrâ' and those from Egypt dated to a similar time period.

Rast (2003) has suggested at least limited cultural links with Egypt on the basis of mace heads found in the EB I shaft tombs. The association with other skeletal samples is insufficiently demonstrated on the basis of data available to us to reach any sound scientific conclusions about biological relationships. This finding perhaps highlights as much as possible the need for greater attention to not only the degree that human bioarchaeological skeletal samples are representative of the original population but also the degree to which research methods used in data collection are consistent between samples and observers.

GENERAL HEALTH OF THE EB I PEOPLE

The Middle East provides the earliest context for the major transition in human economy from hunting and gathering to agriculture. The transition began at the start of the Holocene; this geological period saw the emergence of sedentism and urbanism with all the profound changes in human ecology that these epochal adjustments in culture meant (Ortner 2001). Ortner et al. (2007) have argued that increased concentrations of people living in an urban environment would certainly increase exposure to various infectious pathogens and one should expect to see increased evidence of infectious disease in the human remains. However, as we will see, the evidence of this trend among the EB I skeletal samples from Bâb edh-Dhrâ' remains elusive.

Making inferences about the health of the EB I people of Bâb edh-Dhrâ' on the basis of osteological and paleopathological evidence needs to be done with a full awareness of the limitations of such inferences. There are several variables that are the basis of such inferences: (1) evidence of skeletal disorders; (2) the relative significance of skeletal evidence of disease; (3) demographics of the skeletal sample, such as life expectancy and subadult mortality; (4) stature and robusticity; and (5) how representative the skeletal sample being studied is of all the people who died in a given society. The most fundamental of these variables is the degree to which a skeletal sample accurately represents the people who died in the society whose biology one is attempting to reconstruct. Despite these limitations, we argue that combinations of data on variables potentially related to the health of the Bâb edh-Dhrâ' people can be studied and some observations made.

In the case of the EB IA Bâb edh-Dhrâ' sample, the age and sex distribution of the sample does support the conclusion that the sample is reasonably representative of those who died during that time period. We recognize that this conclusion is not justified with regard to the skeletal sample from the EB IB G1 Charnel House where very few fetuses, infants, and young children were included in the sample. This obviously affects estimates of subadult mortality and life expectancy. There are a few burials from EB IB shaft tombs, but the number is too small to provide reliable comparisons with the EB IA sample, and we do not know if some of the biases apparent in the charnel house burial sample were present in the EB IB shaft tomb burials.

Most bioarchaeological human skeletons do not show any evidence of abnormality. Although this is true for the EB I burials, this is not the case with the teeth. We have seen in Chapter 13 that there is a very high prevalence of linear enamel hypoplasia in the EB IA dentitions. This is indicative of one or more episodes of significant stress during childhood development. Dental abnormalities, in combination with evidence of disorder in the bony skeleton in several burials, support the probability of a human society in which stress from disease was a fairly common event.

In the EB IA shaft tomb sample, we estimate that, if we include all the burials from that time period, the life expectancy at birth is about 17 years. If we don't include the burials from the fetal phase of development, the estimated life expectancy at birth increases to about 19 years. It is only when we remove the fetal and infant burials from the sample that estimated life expectancy at birth increases to about 26 years. This life expectancy compares with the lowest modern-day life expectancy at birth, which is 28.6 years in Sierra Leone (World Health Organization 2004). Regardless of which estimate of life expectancy one uses, the implication of this low life expectancy and the evidence of widespread stress seen in the abnormalities of the teeth is that significant stress in the EB IA society at Bâb edh-Dhrâ' was fairly common.

This view of health among the EB IA people is further supported by the prevalence of metabolic diseases, including scurvy and rickets. Additional evidence of stress is apparent in the cases of early onset osteoporosis, one cause of which is severe protein/calorie malnutrition. It seems likely that there were at least fairly common periods of famine within the EB IA in which women and young children would be particularly vulnerable.

One needs to be cautious in making broad generalizations about the overall health of the people throughout the entire EB IA period, which lasted at least 100 years. The evidence of significant biological stress in which malnutrition was certainly a factor is not entirely congruent with the cultural evidence of continuous occupation of the site throughout the various Early Bronze Age phases (see Chapter 4). Living in an environment in which serious stress was a common event raises questions about the human ecology associated with the site and how the society survived, which it clearly did.

The evidence of tuberculosis and possibly brucellosis and their association with domestic cattle, sheep, and goats all point to a society that had serious endemic infectious diseases that affected many more people than those for which skeletal evidence of disease is present. Sheep and goats had been an important component of the food supply in the Middle East for several millennia, and both milk and meat from bovids were also an important source of nutrition in the Middle East well before EB I. However, the presence of these sources of high quality nutrition does not mean that all members of Middle Eastern human societies had equal or even adequate access to these foods.

The EB IB skeletal sample from the G1 Charnel House is much smaller (N = 115) and, of even greater importance, the human remains are badly fragmented and disassociated. Furthermore, the almost complete absence of fetuses, infants, and young children indicates an important difference in the burial tradition of the EB IB society. This means that direct comparisons with the EB IA skeletal sample need to be made with caution and that the tentativeness of conclusions compared with those that are possible for the EB IA people should be recognized.

Without doubt the most unexpected finding in comparing the health of the EB IA sample with the sample from EB IB was the absence of evidence of infectious disease in the latter. Given the cultural evidence of increasing sedentism associated with the EB IB people, we hypothesized an increased exposure to pathogens, particularly infectious organisms. Since tuberculosis was identified with reasonable certainty in the EB IA sample, as well as possible presence of brucellosis and other infectious diseases including ulcer and osteomyelitis, we expected evidence for these disorders to increase in prevalence in the hypothesized more sedentary EB IB sample.

The disturbed and fragmentary nature of the G1 sample does limit the confidence we have in interpreting the lack of evidence of disease. Nevertheless, there was a large number of unassociated but relatively intact vertebrae available for analysis (see Chapter 7 for details), and none of these vertebrae showed any evidence of infectious skeletal disorder.

The striking fact is that there is no evidence of either infection or malnutrition in the skeletal elements from the G1 Charnel House. Although it is true that the skeletal remains are broken and incomplete, there are sufficient numbers of vertebrae that some evidence of infection such as tuberculosis or brucellosis should be present if these disorders were common in the EB IB people.

The contrast between evidence of disease prevalence in the EB IA and EB IB skeletal samples is troublesome. With the development of a more sedentary society in EB IB, as the archaeological evidence suggests, there should be increased exposure to infectious diseases. With a heavier dependence on agricultural products and a less varied diet, an increase in the prevalence of metabolic diseases related to a more restricted diet is certainly a possible result. This change in diet is a well-established link associated with the transition to agriculture (Cohen and Armelagos 1984).

Several lines of evidence point to a substantial level of marginal to poor health in the EB IA skeletal sample and minimal evidence of poor health in the EB IB Charnel House sample. In the EB IA remains, the evidence for this stress is apparent in both subadult and adult skeletal remains. In subadult burials, for example, the difference between estimated dental age and estimated age based on long bone length is evidence of factors adversely affecting length growth, while dental development, which is less subject to environmental factors, remains relatively normal. Additional evidence of stress is apparent in the high prevalence of dental enamel hypoplasia, which is present in 87% of the adult dentitions (see Chapter 13). This abnormality in dental enamel occurs when the cells (ameloblasts) that form enamel during tooth formation are affected, usually by systemic disorders. Among the multiple factors that can affect ameloblast function is abnormally high body temperature, so any disorder resulting in high fever could result in defective enamel formed at the time of the illness, if the child survived.

In many of the burials from the shaft tombs, there is evidence of multiple linear defects in enamel formation that are indicative of multiple stress events during childhood dental development. There is no evidence of linear enamel hypoplasia before the age of two. Part of the reason for this is that the incisal edges and occlusal surfaces of teeth form first (Hillson and Bond 1997; Reid and Dean 2006). It is only after these portions of the teeth are formed that surfaces in which linear enamel hypoplasia can occur will develop. The onset of stress events for a young child may have begun at age two, but the lack of linear enamel hypoplasia before that age will also be affected by the fact that enamel of nonocclusal surfaces is not forming at that age. One of the stress events that is associated with early childhood is weaning and the transition to adult foods that typically occurs between two and three years of age.

The change from breast milk to a diet in which cultivated grains were a significant component could easily have led to dietary deficiencies that would have caused health problems related to malnutrition directly and would also have created conditions in which the child was more vulnerable to other diseases, particularly infection. The presence of both scurvy and rickets in subadult burials supports the conclusion that malnutrition was a significant factor for the EB IA children. The one probable case of tuberculosis occurs in a subadult burial (A100E, Burial 73). Before the advent of antibiotics, the peak prevalence of tuberculosis was in children (Ortner 2003), and it is likely that this pattern would have been present in the Bâb edh-Dhrâ' sample. Burial 50 in Tomb A89SE provides additional possible evidence of tuberculosis. The skeleton is that of a very gracile male, a condition that may reflect long and debilitating illness beginning in childhood.

Evidence of marginal diet is also present in at least some adults. There is unmistakable evidence of early onset osteoporosis in both young female and young male adults in the EB IA burials. We were not sensitive to the presence of this disorder until well into the research on evidence for skeletal disease, so it is probable that we missed some cases of early onset osteoporosis. Although there are several disorders that can result in early onset osteoporosis, the evidence of malnutrition in the EB IA society in subadult burials makes protein/calorie malnutrition a plausible diagnostic option.

Endocranial depressions occur in some skulls from the EB IA tombs, such as Burial 1b from Tomb A80W, described in Chapter 6. A specific cause for this abnormality cannot be determined unless there is other evidence of disease in an archaeological skeleton. The pathogenesis of this abnormality in subadults is that the growth of the skull is not in phase with the development of the brain. The indentations apparent in the endocranial surface are from pressure of the brain on the endocranial surface of the skull. We have called attention to the disparity of dental development versus long bone length in subadult skeletons. We hypothesize that this is the result of long bone growth being more affected by various disorders than was dental eruption. In addition to tumor and other disorders that could result in endocranial depressions, it is possible that malnutrition would adversely affect skull development but have less effect on the growth of the brain. This disparity in growth could result in endocranial depressions and should be considered in situations where there is other evidence of malnutrition in the archaeological skeletal sample.

World Health Organization (2004) statistics for developing countries today rank maternal mortality as a major cause of death in women of childbearing age. It seems very likely that the hazards of pregnancy and childbirth were present in human groups throughout the evolutionary process but became a greater problem in the transition from hunter-gatherers to agriculturists. The reason for this is the increased exposure to infectious agents that accompanied the more sedentary culture of the agriculturists. Infectious pathogens would commonly contaminate water supplies. In addition, cultivation of the soil would have increased exposure to soil pathogens. In settled communities, there would be increased potential of human-to-human transmission of infectious pathogens.

The presence of fetal bones in the EB IA burials could be the result of aborted pregnancy, but also could be associated with the death of the mother during pregnancy, which would have resulted in the death of the developing fetus. As revealed in Table 10.6, of the 35 chambers containing fetal skeletal elements, 33 also had adult female skeletal elements, presumably in the childbearing age range. Only two chambers containing fetal skeletons had no evidence of adult female skeletal elements being interred in the chamber. It seems likely that the fetal skeletal elements represent examples of maternal mortality where the fetus died with the mother and was later recovered from the primary burial site for interment in the shaft tomb chambers. However, the two fetuses in chambers that did not contain adult female skeletons raise an intriguing possibility that some were the result of spontaneous abortion. If this is the case in these fetal burials, it highlights what is certainly apparent in burials of infants and young children, that is, fetuses were accorded the same ceremonies as older individuals in the EB IA society.

One of the inconsistencies that is apparent in the data on health is that enthesopathy is common in the Bâb edh-Dhrâ' EB I shaft tomb burials (see descriptions in Chapters 6, 8, and 9). Enthesopathy is a condition in which tendon and ligament attachments to bone become slightly mineralized at the point of attachment with bone. This gives rise to fine spicules of bone at this site. The pathogenesis of this condition is not completely established, and there may be a genetic predisposition to form bone when ligaments and tendons are stressed. Nevertheless, enthesopathy is generally considered to be evidence of biomechanical stress. The high prevalence of this condition is consistent with the substantial robusticity that we have identified in the long bone dimensions (Chapter 10). However, physical stress is somewhat inconsistent with the picture that has emerged of a society that carried a fairly heavy burden of disorders and had a short life expectancy. How hard physical activity could be sustained in a society in which chronic and acute illness was common remains an intriguing question.

We have noted in Chapter 12 that evidence of skeletal fracture is uncommon. The most common site for fracture is the skull vault (Table 12.2). In most cases, the lesions are depressed fractures caused by a blow to the head during interpersonal violence. In every case, the lesions have healed, indicating long-term survival following the injury. We have very few skulls from the EB IB G1 Charnel House, and none were complete with the facial bones or mandible. In the fragmentary skulls we do have, there is a high prevalence of skull injury, including axe wounds. In all cases, there is evidence of healing, indicative of survival following the injury. The one complete adult skull from the EB IB is Burial 1 from the A100N shaft tomb. In this skull, there are two depressed lesions with smooth walls that probably are the result of trauma followed by long-term survival.

Proportionate to the total number of skulls available for study, the prevalence of skull injury is greater in the EB IB period than in EB IA. The data are not sufficiently definitive for the EB IB sample to be certain that this trend reflects changing conditions during the EB I period. However, when combined with the evidence of charnel house destruction and the development of the massive walled city starting in EB II, our data are consistent with a transition from a relatively peaceful period in EB IA to one characterized by increasing violence during EB IB.

CONCLUSIONS

The authors are fully aware of the inherent limitations that Wood et al. (1992) defined regarding making inferences about past living populations on the basis of human bioarchaeological skeletal samples. We are in fundamental agreement with Wood et al. that the issues they identified need to be resolved, despite the uncertainty regarding how significant these limitations are. A human skeletal sample never can, by definition, be the skeletal equivalent of the living population from which the sample is derived. Nevertheless, skeletal samples represent one of the few sources of information available to science as we attempt to reconstruct the biology and evolution of ancestral human populations. An obvious first step in this quest for knowledge of our past biology is to establish, as much as possible, the representativeness of existing bioarchaeological skeletal samples. As we have discovered in the context of comparing our data on the EB I skeletal samples from Bâb edh-Dhrâ' to other skeletal samples, establishing representativeness is a frustrating task on the basis of current evidence about most of these other samples.

Related to this step is the need to be rigorous during field excavations to ensure that burials excavated are as representative of those who died during a specific range of time as is possible in archaeological excavation. This means determining the boundaries of the cemetery and the distribution of burials within the cemetery area. While one is doing this, one also needs to excavate in several areas of the cemetery to ensure that potentially different subsamples of those who died in antiquity are included in the remains recovered for analysis. This care in sampling ancient cemeteries (if the entire cemetery is not excavated) needs to be accompanied by recovering all the contents of tombs excavated, including the very small bones of fetuses and infants that may be present.

In addition, cemetery excavations need to emphasize the recovery and analysis of samples sufficiently large that the effects of within-sample variation are minimal. Our experience with the Bâb edh-Dhrâ' sample indicates that recovery of a minimum of 1,000 burials should be the target for cemetery excavations. Adequate bioarchaeological skeletal samples are particularly crucial when the study of skeletal disease is a particular focus of the research and analysis. Evidence of significant skeletal disorder occurs in about 15% of typical archaeological samples. Even with a sample of 1,000 burials, one can only expect about 150 cases of skeletal disorder, and these will be distributed through a large range of disease categories, making comparisons of specific disease prevalence between samples risky at best. One can, of course, use a more inclusive set of categories, as used in the Global History of Health Project at Ohio State University in the United States (global.sbs.ohio-state.edu), and develop a single index of health based on these categories. This strategy certainly has merit, as demonstrated in the North American Module of the project (Steckel and Rose 2002), but there are important problems in understanding the role of disease in past human populations that require more specific definitions of disease. We will be suggesting a few potential research topics in the final section of this chapter.

In the Bâb edh-Dhrâ' EB IA sample, multiple lines of evidence indicate the presence of several disorders affecting the skeleton, such as a high prevalence of metabolic diseases, including scurvy, rickets, postmenopausal osteoporosis, and early onset osteoporosis. Scurvy and rickets clearly are the result of malnutrition, and a plausible cause of early onset osteoporosis is malnutrition as well, although other disorders are possible. An important question in interpreting these observations is to what extent this malnutrition was pervasive in the Bâb edh-Dhrâ' society throughout the EB IA period. It is possible that the evidence of diseases present in the sample is the result of one or more relatively short episodes scattered over the 100 years associated with the EB IA sample. The presence of skeletal evidence of these disorders in some skeletons does not necessarily indicate that malnutrition was a constant problem throughout the EB IA period. It is at least possible that some of the evidence of metabolic diseases may represent episodes of famine rather than reflecting continuous problems in the food supply throughout the entire EB IA period.

Nevertheless, the relatively high prevalence of skeletal evidence for scurvy and rickets in children does suggest a society that was having difficulty in providing adequate nutrition for its developing members. Malnutrition is particularly troublesome for children who have fewer physiological resources to accommodate temporary periods of inadequate diet and are more vulnerable to secondary disorders such as infection when suffering from malnutrition. The evidence we have for infectious diseases such as tuberculosis implies a much greater prevalence of these diseases in the living population.

FUTURE RESEARCH

The fieldwork to obtain the Bâb edh-Dhrâ' skeletal sample and subsequent research the authors have conducted on the EB I skeletal samples since the first author's initial excavations at the Bâb edh-Dhrâ' cemetery in 1977 have highlighted some theoretical and methodological problems. We have already discussed these problems in the previous sections of this chapter and in earlier chapters. In the final section of this chapter, we suggest some areas of research that need the attention of the bioarchaeological scientific community.

Research on Malnutrition

It seems likely that episodes of malnutrition at least were a significant problem that had to be addressed by our agricultural ancestors. Crop failure is an obvious issue and undoubtedly did result in malnutrition and death in past human societies as it does today. However, a more constant problem is the nutritional deficiencies that developed as the varied diet of our hunter-gatherer ancestors gave way to the less varied diet of the agriculturists. The role of vitamins only began to be understood 200 years ago. Hunter-gatherers did not need to know about vitamins because their diet usually provided all the vitamins needed for human nutrition. This changed with the advent of agriculture, and various types of malnutrition became an important factor in human adaptation.

The use of stable isotopes in reconstructing some aspects of past diets has been a rewarding scientific breakthrough in understanding some of the dietary variability between human groups. However, very little is known about severe malnutrition from famine that undoubtedly affected our agriculturist ancestors. Late in our research on abnormalities in the EB I human remains, we noticed the presence of very light bones in some burials of young adults. Initially we thought this might have been the result of differential taphonomic conditions in the primary and/or secondary burial sites. Toward the end of our research program we prepared plain film radiographs of one or two of these cases and found diminished cortical thickness. Further radiographic analysis using CT technology revealed greatly reduced trabecular bone in the metaphyseal ends of long bones and reduced density of the cortical bone.

Multiple possibilities exist as the cause of this abnormality, including prolonged debilitating illness. However, we argue that the clear evidence of other types of malnutrition make a strong case for this early onset osteoporosis as the result of severe protein/calorie malnutrition. If this is the case, this type of evidence has the potential of providing a very important window on the health of past human societies. Our hypothesis that protein/calorie malnutrition was a significant factor in past human populations needs to be tested using a variety of methods, including careful radiographic analysis of young adult bones from archaeological sites. This will establish the presence of early onset osteoporosis, but additional methods are likely to be needed to distinguish between osteoporosis resulting from starvation and osteoporosis caused by debilitating illness. Biomolecular research may provide useful data in resolving this question.

Research on Methodology
Perhaps the greatest frustration we faced was the difficulty in finding published bioarchaeological skeletal samples for comparison with the Bâb edh-Dhrâ' EB IA burials. We encountered five fundamental problems. The first of these was skeletal samples that were clearly not representative of those who had died in a given society. Major differences between the prevalence of male and female skeletons is evidence of bias in most archaeological skeletal samples. The second problem was inadequate sample sizes in the comparative sample. A third problem was missing data in our own and other samples. The fourth problem was differences in data collection methodology. The final problem was poor knowledge of the archaeological context for comparative samples.

Many of the skeletal collections that have data potentially useful in comparative studies were excavated in the 19th and early 20th centuries. Many of the theoretical and methodological issues that we now recognize had not been identified, and the research objectives associated with these collections were different from those basic to research today. Given the challenges in recovering skeletal collections today, these collections remain a potential source of useful information. However, we do need to understand the biases that are inherent in these skeletal samples. This means a careful reappraisal of field methods used in excavating the samples and a careful review of curatorial decisions that may have affected the representativeness of the sample.

Stimulated by the legislation in the United States mandating repatriation of many skeletal collections of Native Americans, there was a major initiative to standardize the data protocols used in osteological analysis (Buikstra and Ubelaker 1994). However, additional methodological research is needed to define more carefully skeletal abnormalities and genetic features apparent in human remains.

Clarification of Theoretical Issues
The Global History of Health Project has highlighted the value of using multiple variables to derive a single index of health. In attempting to evaluate the relative health of the Bâb edh-Dhrâ' EB IA and EB IB samples, we also conclude that multiple indicators of skeletal disorder are important in defining health factors in archaeological skeletal samples. There are, however, important issues that need clarification before this type of research can achieve its potential.

Different diseases affect the skeleton in different ways. For example, metastatic carcinoma results in skeletal lesions in at least 90% of the individuals with this disorder. By contrast, tuberculosis results in skeletal lesions in only 5–10% of cases. Much remains to be understood about how often different diseases that can affect the skeleton actually do so. This is an important variable in reconstructing the prevalence of disease in the living society represented by the skeletal sample. Related to this observation is the fact that there are differences in the social and biological significance of different diseases. Some diseases, such as brucellosis, usually result in no more than mild morbidity and rarely cause death. Tuberculosis pathogens commonly result in no symptoms in people exposed to the organism but can cause severe disability and death. These two infectious diseases illustrate the fact that a more complete understanding of the effect of disease on human populations in the past benefits from the recognition that different diseases have different potential impacts on the human population in which they occur.

The research we have conducted on the Bâb edh-Dhrâ' EB I skeletal samples reveals much about the biology of the people. With future improvements in research design and methodology, this type of research has the potential to make even greater contributions to our knowledge of human adaptation in past societies.

References

Acsadi, G., and J. Nemeskeri. 1970. *History of Human Lifespan and Mortality.* Budapest, Hungary: Akademai Kiado.

Adams, R. M. 1970. The origin of cities. *Scientific American* 203:153–69.

Adams, R. M., and H. J. Nissen. 1972. *The Uruk Countryside.* Chicago: University of Chicago Press.

Ahlquist, J., and O. Damsten. 1969. A modification of Kerley's method for the microscopical determination of age in human bone. *Journal of Forensic Sciences* 14:205–217.

Albright, W. F. 1924. The archaeological results of an expedition to Moab and the Dead Sea. *Bulletin of the American Schools of Oriental Research* 14:2–12.

Albright, W. F. 1924–1925. The Jordan Valley in the Bronze Age. *Annual of the American School of Oriental Research* 6:13–74.

American Geological Institute. 1985. Archeological geology: A combined science speciality helps scientists reveal Bronze Age burial mounds and map an ancient buried city. *Earth Science* (Winter):13–15.

Amiran, R., and N. Porat. 1984. The basalt vessels of the Chalcolithic Period and Early Bronze Age I. *Tel Aviv* 11:11–19.

Anderson, J. E. 1964. The people of Fairty. *National Museum of Canada Bulletin* 193:28–129.

Angel, J. L. 1966. Porotic hyperostosis, anemias, malarias, and marshes in the prehistoric eastern Mediterranean. *Science* 153:760–63.

Angel, J. L. 1969. The bases of paleodemography. *American Journal of Physical Anthropology* 30:427–38.

Angel, J. L. 1971. *The People of Lerna: Analysis of a Prehistoric Aegean Population.* Princeton, NJ: American School of Classical Studies in Athens.

Aufderheide, A., and C. Rodríguez-Martín. 1998. *The Cambridge Encyclopedia of Human Paleopathology.* Cambridge: Cambridge University Press.

Baby, R. S. 1954. Hopewell cremation practices. *Ohio Historical Society Papers in Archaeology* 1:1–7.

Bar-Yosef, M. D. 1999. *The Role of Shells in the Reconstruction of Socio-Economic Aspects of Neolithic through Early Bronze Age Societies in Sinai.* Ph.D. thesis, Hebrew University, Jerusalem.

Barnes, E. 1994. *Developmental Defects of the Axial Skeleton in Paleopathology.* Niwot: University Press of Colorado.

Barnes, E., and D. J. Ortner. 1997 Multifocal eosinophilic granuloma with a possible trephination in a fourteenth century Greek young skeleton. *International Journal of Osteoarchaeology* 7:542–47.

Bartels, P. 1907. Tuberkulosein der jüngeren Steinzeit. *Archiv für Anthropologie* 6:243–55.

Bass, W. M. 1987. *Human Osteology.* 3rd ed. Columbia: Missouri Archaeological Society.

Bass, W. M. 1995. *Human Osteology. A Laboratory and Field Manual. Special Publication no. 2.* Columbia, MO: Missouri Archaeological Society.

Bentley, G. R. 1987. Kinship and Social Structure at Early Bronze Age Bâb edh-Dhrâ', Jordan: A Bioarchaeological Analysis of the Mortuary and Dental Data. Doctoral dissertation, University of Chicago.

Bentley, G. R. 1991. A bioarchaeological reconstruction of the social and kinship systems at Early Bronze Age Bâb edh-Dhrâ', Jordan. In *Between Bands and States*, ed. S. A. Gregg, 5–34. Occasional Paper No. 9. Carbondale: Center for Archaeological Investigations, Southern Illinois University.

Berry, A. C. 1978. Anthropological and family studies on minor variants of the dental crown. In *Development, Function and Evolution of Teeth*, ed. P. M Butler and K. A. Jowsey, 81–99. New York: Academic.

Berry, A. C., and R. J. Berry. 1967. Epigenetic variation in the human cranium. *Journal of Anatomy* 1001(2):361–79.

Betts, A. V. G., ed. 1992. *Excavations at Tell um-Hammad 1982–1984: The Early Assemblages (BE I-II).* Edinburgh: Edinburgh University Press.

Bevan, B. 1983. Electromagnetics for Mapping Buried Earth Features. *Journal of Field Archaeology* 10:47–54.

Binford, L. R. 1972. *An Archaeological Perspective.* New York: Seminar.

Bloch-Smith, E. 2003. Bronze and Iron Age burials and funerary customs in the southern Levant. In *Near Eastern Archaeology*, ed. S. Richard, 105–115. Winona Lake, IN: Eisenbrauns.

Bocquet-Appel, J. 2002. Paleoanthropological traces of a Neolithic cemographic transition. *Current Anthropology* 43: 637–50.

Bocquet-Appel, J., and S. Naji. 2006. Testing the hypothesis of a worldwide Neolithic demographic transition: Corroboration from American cemeteries. *Current Anthropology* 47: 341–65.

Boldsen, J. L. 2007. Early childhood stress and adult age mortality: A study of dental enamel hypoplasia in the medieval Danish village of Tirup. *American Journal of Physical Anthropology* 132:59–66.

Bondioli, L., R. S. Corruccini, and R. Macchiarelli. 1986. Familial segregation in the Iron Age community of Alfedena, Abruzzo, Italy, based on osteodental trait analysis. *American Journal of Physical Anthropology* 71:393–400.

Bourke, S. J. 2001. The Chalcolithic Period. In *The Archaeology of Jordan*, ed. B. MacDonald, R. Adams, and P. Bienkowski, 107–162. Sheffield: Sheffield Academic.

Bradtmiller, B., J. E. Buikstra, A. M. López-Parra, S. Alvarez, M. S. Mesa, F. Bandrés, and E. Arroyo-Pardo. 1984. Effects of burning on human bone microstructure: A preliminary study. *Journal of Forensic Sciences* 29:535–540.

Braun, E. 1990. Basalt bowls of the EB I Horizon in the southern Levant. *Paleorient* 16:87–96.

Braun, E. 2001. Proto, Early Dynastic Egypt, and Early Bronze I–II of the southern Levant: Some uneasy C14 correlations. *Radiocarbon* 43:1279–96.

Brosch, R., S. V. Gordon, M. Marmiesse, P. Brodin, C. Buchrieseer, K. Eiglmeier, T. Garnier, E. Gutierrez, G. Hewinson, K. Kreemer, L. M. Parsons, A.S. Pym, S. Samper, D. Van Spdingen, and S. T. Cole. 2002. A new evolutionary sequence for the *Mycobacterium tuberculosis* complex. *Proceedings of the National Academy of Science* 99:3684–89.

Brothwell, D. R. 1959. Teeth in earlier human populations. *Proceedings of the Nutrition Society* 18:59–65.

Brothwell, D. R. 1965. *Digging up Bones*. Ithaca, NY: Cornell University Press.

Brothwell, D. R. 1981. *Digging up Bones*. 3rd ed. Ithaca, N.Y.: Cornell University Press.

Buikstra, J. E., and L. W. Konigsberg. 1985. Paleodemography: Critiques and controversies. *American Anthropologist* 87: 316–34.

Buikstra, J. E., and J. Mielke. 1985. Demography, diet and health. In *The Analysis of Prehistoric Diets*, ed. J. Mielke and R. Gilberts, 359–422. New York: Academic.

Buikstra, J. E., and M. Swegle. 1989. Bone modification due to burning: Experimental evidence. In *Bone Modification*, ed. R. Bonnichsen and M. H. Sorg, 247–58. Orono, ME: Center for the Study of the First Americans.

Buikstra, J. E., and D. H. Ubelaker. 1994. *Standards for Data Collection from Human Skeletal Remains*. Research Series, no. 44. Fayetteville: Arkansas Archaeological Survey.

Butler, P. M. 1939. Studies of the mammalian dentition. I: Differentiation of the post-canine dentition. *Proceedings of the Zoological Society of London* 109:1–36.

Canci, A., S. Minozzi, and S. Borgognini Tarli. 1996. New evidence of tuberculosis spondylitis from Neolithic Liguria (Italy). *International Journal of Osteoarchaeology* 6:497–501.

Capasso, L. 1999. Brucellosis at Herculaneum (79 AD). *International Journal of Osteoarchaeology* 9:277–88.

Chesson, M. S. 1999. Libraries of the dead: Early Bronze Age charnel houses and social identity at urban Bâb edh-Dhrâ', Jordan. *Journal of Anthropological Archaeology* 18:137–64.

Chesson, M. S. 2001. Embodied memories of place and people: Death and society in an early urban community. In *Social Memory, Identity, and Death: Anthropological Perspectives on Mortuary Rituals*, ed. M. S. Chesson, 100–113. Archaeological Publications of the American Anthropological Association Publication Series, vol. 10. Arlington, VA: American Anthropological Association.

Clark, V. A. 1979. Investigations in a prehistoric necropolis near Bâb edh-Dhrâ'. *Annual of the Department of Antiquities of Jordan* 23:57–77.

Cohen, M. N., and G. J. Armelagos, eds. 1984. *Paleopathology at the Origins of Agriculture*. New York: Academic.

Cole, T. M., III. 1994. Size and shape of the femur and tibia in northern Plains Indians. In *Skeletal Biology in the Great Plains: Migration, Warfare, Health, and Subsistence*, ed. D. W. Owsley and R. L. Jantz, 219–33. Washington, DC.: Smithsonian Institution Press.

Cook, D. C., and J. E. Buikstra. 1979. Health and differential survival in prehistoric populations: Prenatal dental defects. *American Journal of Physical Anthropology* 51:649–64.

Coppa, A., A. Cucina, D. Mancinelli, R. Vargio, and J. M. Calcagno. 1998. Dental anthropology of central-southern, Iron Age Italy: The evidence of metric versus nonmetric traits. *American Journal of Physical Anthropology* 107:371–86.

Coppa, A., and R. Macchiarelli. 1982. The maxillary dentition of the Iron-Age population of Alfedena (Middle-Adriatic Area, Italy). *Journal of Human Evolution* 11:219–35.

Cox, M., and S. Mays. 2000. *Human Osteology in Archaeology and Forensic Sciences*. London: Greenwich Medical Media.

Dahlberg, A. A. 1956. *Materials for Classification of Tooth Characters, Attributes and Techniques in Morphological Studies of the Dentition*. Department of Anthropology, University of Chicago Mimeo.

DeHaan, J. D., S. Nurbakhsh. 2001. Sustained combustion of an animal carcass and its implications for the consumption of human bodies in fires. *Journal of Forensic Sciences* 46: 1076–81.

Dequeker, J., D. J. Ortner, A. I. Stix, X. G. Cheng, P. Brys, and S. Boonen. 1997. Hip fracture and osteoporosis in a XIIth Dynasty female skeleton from Lisht, Upper Egypt. *Journal of Bone and Mineral Research* 12:881–88.

Dever, W. G. 2003. Chronology of the southern Levant. In *Near Eastern Archaeology*, ed. S. Richard, 82–87. Winona Lake, IN: Eisenbrauns.

Dieppe, P., and K. Lim. 1998. Clinical features and diagnostic problems. In *Rheumatology*, 2nd ed., eds. J. Klippel and P. Dieppe, 3.1–3.16. London: Mosby.

Donahue, J. 1981. Geological investigations at Early Bronze sites. In *The Southeastern Dead Sea Plain Expedition: An Interim Report of the 1977 Season*, ed. W. E. Rast and R. T.

Schaub, 137–54. *Annual of the American Schools of Oriental Research* 46.
Donahue, J. 2003. Geology and Geomorphology. In *Bâb edh-Dhrâ': Excavations at the Town Site (1975–1981). Reports of the Expedition to the Dead Sea Plain, Jordan, Vol. 2.*, eds. W. E Rast and R. T. Schaub, 18–55. Winona Lake, IN: Eisenbrauns.
Ehrich, R. W., ed. 1992. *Chronologies in Old World Archaeology*, 3rd ed. Chicago: University of Chicago Press.
Epstein, C. 1975. Basalt pillar figurines from the Golan. *Israel Exploration Journal* 25:193–201.
Epstein, C. 1988. Basalt pillar figurines from the Golan and the Huleh regions. *Israel Exploration Journal* 38:205–223.
Fawsett, C. D. 1902. A second study of the variation and correlation of the human skull, with special reference to the Naqada crania. *Biometrika* 1:408–467.
Fischer, P. 2000. The Early Bronze Age at Tell Abu al-Kharaz, Jordan Valley. In *Ceramics and Change in the Early Bronze Age of the Southern Levant*, ed. G. Philip and D. Baird, 201–232. Sheffield, England: Sheffield Academic.
Flannery, K. V. 1965. The ecology of early food production in Mesopotamia. *Science* 147:1247–56.
Forfar, J. O., and G. C. Arneil. 1978. *Textbook of Paediatrics*. New York: Churchill Livingstone.
Formicola, V., Q. Milanesi, and C. Scarsisi. 1987. Evidence of spinal tuberculosis at the beginning of the fourth millennium BC from Arene Candide cave (Liguria, Italy). *American Journal of Physical Anthropology* 72:1–6.
Frohlich, B. 1986. The human biological history of the Early Bronze Age population in Bahrain. In *Bahrain through the Ages: The Archaeology*, ed. S. H. A. Al Khalifa and M. Rice, 47–63. London: Routledge & Kegan Paul.
Frohlich, B. 1999. GPS and GLONASS for archaeological GIS: Smithsonian project maps colonial burial grounds with advanced GPS surveying technology. *Professional Surveyor* 19(10):8–14.
Frohlich, B., M. Kervran, V. Caruso, and K. McCormick. 1986. Noncontacting Terrain Conductivity Measurements at Qal'at al-Bahrain. In *Proceedings of the 24th International Archaeometry Symposium*, ed. J. S. Olin and M. J. Blackman, 187–200. Washington, DC: Smithsonian Institution Press.
Frohlich, B., and W. Lancaster. 1986. Electromagnetic surveying in current Middle Eastern archaeology: Application and evaluation. *Geophysics* 51(7):1414–25.
Frohlich, B., and D. J. Ortner. 1982. Excavations of the Early Bronze Age cemetery at Bâb edh-Dhrâ', Jordan, 1981: A preliminary report. *Annual of the Department of Antiquities* 26:249–67, 491–500.
Frohlich, B., and D. J. Ortner. 2000. Social and demographic implications of subadult inhumations in the Ancient Near East. In *The Archaeology of Jordan and Beyond: Essays in Honor of James A. Sauer*, ed. L. E. Stager, J. A. Greene, and M. D. Coogan, 122–31. Winona Lake, IN: Eisenbrauns.

Frohlich, B., D. J. Ortner, and H. al-Khalifa. 1987/88. Human disease in the ancient Middle East. *Dilmun, Journal of the Bahrain Historical and Archaeological Society* 14:61–73.
Frohlich, B., and P. O. Pedersen. 1992. Secular changes within Arctic and Subarctic populations: A study of 632 human mandibles from the Aleutian Islands, Alaska and Greenland. *Arctic Medical Research* 51:173–88.
Garn, S. M., A. B. Lewis, R. S. Kerewsky, and A. A. Dahlberg. 1966. Genetic independence of Carabelli's trait from tooth size or crown morphology. *Archives of Oral Biology* 11:745–47.
Glazier, J. 1984. Economic inequality and land tenure change in Mbeere, Kenya. In *Opportunity, Constraint and Change: Essays in Honor of Elizabeth Colson*, ed. J. Glazier, M. Owy, K. T. Molohon, J. U. Ogbu, and A.P. Royce, 76–83. Berkeley, CA: Kroeber Anthropological Society Papers 63 and 64.
Glueck, N. 1946. *The River Jordan*. Philadelphia: Westminster.
Goodman, A. H., and G. J. Armelagos. 1988. Childhood stress and decreased longevity in a prehistoric population. *American Anthropologist* 90:936–44.
Goodman, A. H., D. L. Martin, G. J. Armelagos, and G. Clark. 1984. Indicators of stress from bone and teeth. In *Paleopathology at the Origins of Agriculture*, eds. M. N. Cohen and G. J. Armelagos, 13–49. New York: Academic, 1984.
Gordon, C. G., and J. E. Buikstra. 1981. Soil pH, bone preservation and sampling biases at mortuary sites. *American Antiquity* 46:566–71.
Gorecki, P. P. 1979. Disposal of human remains in the New Guinea highlands. *Archaeology and Physical Anthropology in Oceania* 14:107–114.
Greene, D. L. 1972. Dental anthropology of Early Egypt and Nubia. *Journal of Human Evolution* 13:315–24.
Gustafson, G., and G. Koch. 1974. Age estimation up to 16 years of age based on dental development. *Odontologisk Revy* 25(1974):297–306.
Hammond, N., K. Pretty, and F. P. Saul. 1975. A Classic Maya family tomb. *World Archaeology* 7:57–78.
Hanihara, K. 1960. *Standard Models for Classification of Crown Characters of the Human Deciduous Dentition*. Mimeo, Department of Anthropology, University of Chicago.
Hanna, B. L. 1962. The biological relationships among Indian groups of the Southwest: Analysis of morphological traits. *American Journal of Physical Anthropology* 20:499–508.
Harlan, J. R. 1981. Natural resources of the southern Ghor. In *The Southeastern Dead Sea Plain Expedition: An Interim Report of the 1977 Season*, ed. W. E. Rast and R. T. Schaub, 155–64. *Annual of the American Schools of Oriental Research* 46.
Harlan, J. R. 1985. The Early Bronze Age environment of the southern Ghor and the Moab Plateau. In *Studies in the History and Archaeology of Jordan II*, ed. A. Hadidi, 125–29. Amman, Jordan: Department of Antiquities.

Harlan, J. R. 2003. Natural resources of the Bâb edh-Dhrâ' region. In *Bâb edh-Dhrâ': Excavations at the Town Site (1975–1981). Reports of the Expedition to the Dead Sea Plain, Jordan, Vol. 2.*, ed. W. E. Rast and R. T. Schaub, 56–61. Winona Lake, IN: Eisenbrauns.

Hassan, F. A. 1983. Earth resources and population: An archaeological perspective. In *How Humans Adapt*, ed. D. J. Ortner, 191–216. Washington, DC.: Smithsonian Institution Press.

Helms, S. W. 1992. Introduction. In *Excavations at Tell um-Hammad 1982–1984: The Early Assemblages (BE I-II)*, ed. A. V. G. Betts, 5–14. Edinburgh: Edinbugh University Press.

Hemphill, B. E. 1999. Foreign elites from Oxos civilization? A craniometric study of anomalous burials from Bronze Age Tepe Hissar. *American Journal of Physical Anthropology* 110(4):421–34.

Hennessy, J. B. 1967. *The Foreign Relations of Palestine during the Early Bronze Age*. London: Bernard Quaritch.

Hesse, B., and P. Wapnish. 1981. Animal remains from the Bâb edh-Dhrâ' cemetery. *Annual of the American Schools of Oriental Research* 46:133–36.

Hillson, S. 1996. *Dental Anthropology*. Cambridge: Cambridge University Press.

Hillson, S. 2001. Recording dental caries in archaeological human remains. *International Journal of Osteoarchaeology* 11:249–89.

Hillson, S., and S. Bond. 1997. Relationship of enamel hypoplasia to the pattern of tooth crown growth: A discussion. *American Journal of Physical Anthropology* 104:89–103.

Hoffman, M. A. 1979. *Egypt Before the Pharoahs*. London: Ark Paperbacks.

Hoppa, R. D., and J. W. Vaupel. 2002. *Paleodemography: Age Distributions from Skeletal Samples*. Cambridge: Cambridge University Press.

Howell, N. 1976. Towards a uniformitarian theory of human paleodemography. In *The Demographic Evolution of Human Populations*, ed. R. H. Ward and K. M. Weiss. 25–40. New York: Academic.

Howells, W. W. 1973. Cranial variation in man: A study by multivariate analysis of patterns of difference among recent human populations. *Papers of the Peabody Museum of Archaeology and Ethnology*. Boston: Harvard University.

Hrdlička, A. 1920. *Anthropometry*. Philadelphia: Wistar Institute of Anatomy and Biology.

Jackson, J. W. 1937. The osteology: Report on the human remains. In *Cemeteries of Armant I*, ed. R. Mond and Oliver Meyers, 144–62. Oxford: Egyptian Exploration Society, Oxford University Press.

Jantz, R. L., and D. W. Owsley. 1994. Growth and dental development in Arikara children. In *Skeletal Biology in the Great Plains: Migration, Warfare, Health, and Subsistence*, ed. D. W. Owsley and R. L. Jantz, 247–58. Washington, DC: Smithsonian Institution Press.

Kaifu, Y. 2000. Tooth wear and compensatory modification of the anterior dentoalveolar complex in humans. *American Journal of Physical Anthropology* 113:369–92.

Kantor, H. J. 1992. The relative chronology of Egypt and its foreign correlations before the First Intermediate Period. In *Chronologies in Old World Archaeology*, 3rd ed., ed. R. W. Ehrich, 3–21. Chicago: University of Chicago Press.

Katzenberg, M. A., D. A. Herring, and S. R. Saunders. 1996. Weaning and infant mortality: Evaluating the skeletal evidence. *Yearbook of Physical Anthropology* 39:177–99.

Keene, H. J. 1965. The relationship between third molar agenesis and the morphological variability of the molar teeth. *Angle Orthodontistry* 35:289–98.

Keene, H. J. 1968. The relationship between Carabelli's trait and the size, number and morphology of the maxillary molars. *Archives of Oral Biology* 13:1023–25.

Kenyon, K. M. 1960. *Excavations at Jericho, Vol. 1: The Tombs Excavated in 1952–54*. London: British School of Archaeology in Jerusalem.

Kerley, E. 1965. The microscopic determination of age in human bone. *American Journal of Physical Anthropology* 23:149–63.

Kerley E., and D. Ubelaker. 1978. Revision in the microscopic method of estimating age at death in human cortical bone. *American Journal of Physical Anthropology* 49:545–46.

Kiesewetter, H. 2006. Analyses of the human remains from the Neolithic cemetery at Al-Buhais 18 (excavations 1996–2000). In *The Archaeology of Jebel Al-Bahais, Sharjah, United Arab Emirates*, ed. H.-P. Uerpmann, M. Uerpmann, and S. A. Ja'sim, 103–380. Sharjah, United Arab Emirates: Department of Culture and Information.

Knodel, J. E. 1988. *Demographic Behavior in the Past: A Study of Fourteen German Village Populations in the Eighteenth and Nineteenth Centuries*. Cambridge: Cambridge University Press.

Kramer, S. N. 1963. *The Sumerians: Their History, Culture, and Character*. Chicago: University of Chicago Press.

Krogman, W. M. 1940a. The peoples of early Iran and their ethnic origins. *American Journal of Physical Anthropology* 26:269–308.

Krogman, W. M. 1940b. Racial types from Tepe Hissar, Iran, from the late fifth to early second millennium BC. *Verhandlingen der koninklijke Nederlandsche Akademie van Wetenschappen.* 39(2):1–87.

Krogman, W. M. 1973. *The Human Skeleton in Forensic Medicine*. Springfield, IL: Thomas.

Krogman, W. M. 1989. Representative Early Bronze crania from Bâb edh-Dhrâ'. In *Bab edh-Dhra', Excavations in the Cemetery Directed by Paul W. Lapp, 1965–67*, ed. R. T. Schaub and W. E. Rast. Winona Lake, IN: Eisenbrauns.

Kyle, M. G., and W. F. Albright. 1924. Results of the Archaeological Survey of the Ghor in Search for the Cities of the Plain. *Bibliotheca Sacra* 81:276–91.

Lapp, P. W. 1966. The cemetery at Bâb edh-Dhrâ', Jordan. *Archaeology* 19:104–111.

Lapp, P. W. 1968. Bâb edh-Dhrâ' tomb A76 and Early Bronze I in Palestine. *Bulletin of the American Schools of Oriental Research* 189:12–41.

Lapp, P. E. 1970. Palestine in the Early Bronze Age. In *Near Eastern Archaeology in the Twentieth Century*, ed. J. A. Sanders, 101–131. New York: Doubleday.

Larsen, C. S. 1997. *Bioarchaeology: Interpreting Behavior from the Human Skeleton*. Cambridge: Cambridge University Press.

Lee, J. R. 2003. Worked Stones. In *Bâb edh-Dhrâ': Excavations at the Town Site (1975–1981): Reports of the Expedition to the Dead Sea Plain, Jordan, Vol. 2.*, ed. W. E. Rast and R. T. Schaub, 622–37. Winona Lake, IN: Eisenbrauns.

Levy, T. E. 1986. The Chalcolithic period in Palestine. *Biblical Archaeologist* 49:81–108.

Levy, T. E. 2003. The Chalcolithic of the Southern Levant. In *Near Eastern Archaeology*, ed. S. Richard, 263–73. Winona Lake, IN: Eisenbrauns.

Littleton, J. 1998. *Skeletons and Social Composition, Bahrain 300 BC–AD 250*. Oxford: British Archaeological Reports.

Littleton, J. 2007. The political ecology of health in Bahrain. In *Ancient Health: Skeletal Indicators of Agricultural and Economic Intensification*, ed. M. N. Cohan and G. M. M. Crane-Kramer. Gainesville: University Press of Florida.

Littleton, J., and B. Frohlich. 1993. Fish-eaters and farmers: Dental pathology in the Arabian Gulf. *American Journal of Physical Anthropology* 92:427–47.

Lovell, N. C., and I. Whyte. 1999. Patterns of dental enamel defects at ancient Mendes, Egypt. *American Journal of Physical Anthropology* 110:69–80.

Maat, G. 1982. Scurvy in Dutch whalers buried at Spitsbergen. In *Proceedings of the Paleopathology Association's Fourth European Meeting. Middleberg, Antwerpen*, ed. G. Haneveld and W. Perizonius, 82–93. Utrecht: Paleopathology Association.

MacDonald, B., R. Adams, and P. Bienkowski, ed. 2001. *The Archaeology of Jordan*. Sheffield, England: Sheffield Academic.

Macumber, P. G. 2001. Evolving landscape and environment in Jordan. In *The Archaeology of Jordan*, ed. B. MacDonald, R. Adams, and P. Bienkowski, 1–30. Sheffield, England: Sheffield Academic.

Manji, F., O. Fejerskov, V. Baelum, and N. Nagelkerke. 1989. Dental calculus and caries experience in 14–65 year olds with no access to dental care. In *Recent Advances in the Study of Dental Calculus*, ed. J. M. Ten Cate, 223–34. Oxford: Oxford University Press, 1989.

Martin, R. 1928. *Lehrbuch der Anthropologie*. 2nd ed. Jena: Gustav Fischer.

Masset, C. 1989. Age estimation on the basis of cranial sutures. In *Age Markers in the Human Skeleton*, ed. M. Y. Iscan, 71–103. Springfield: Thomas.

Matthews, R. 2003. *The Archaeology of Mesopotamia: Theories and Approaches*. London: Routledge.

Mays, S. 1998. *The Archaeology of Human Bones*. London: Routledge.

McCreery, D. W. 1980. The Nature and Cultural Implications of Early Bronze Age Agriculture in the Southern Ghor of Jordan: An Archaeological Reconstruction. Ph.D. dissertation, University of Pittsburgh.

McCreery, D. W. 2003. The Paleoethnobotany of Bâb edh-Dhrâ'. In *Bâb edh-Dhrâ': Excavations at the Town Site (1975–1981). Reports of the Expedition to the Dead Sea Plain, Jordan, Vol. 2.*, ed. W. E. Rast and R. T. Schaub, 449–63. Winona Lake, IN: Eisenbrauns.

McKern, T. W., and T. D. Stewart. 1957. *Skeletal Age Changes in Young American Males*. Natick, MA.: Quartermaster Research and Development Command.

McNeill, J. D. 1979. *EM-312 Operating Manual for EM-31 Non-Contacting Terrain Conductivity Meter*. Mississauga, Ontario: Geonics.

Mellanby, M. 1927. The structure of human teeth. *British Dental Journal* 48:737–51.

Meyers, E. 1970. Secondary burials in Palestine. *Biblical Archaeologist* 33:2–29.

Miroschedji, P. 2002. The socio-political dynamics of Egyptian-Canaanite interaction in the Early Bronze Age. In *Egypt and the Levant*, ed. E. C. M. van den Brink and T. E. Levy, 39–57. London: Leicester University Press.

Molleson, T. 1991. Demographic implication of the age structure of early English cemetery samples. *Actes des journees anthropologiques* 5:113–21.

Molnar, S. 1971. Human tooth wear, tooth function and cultural variability. *American Journal of Physical Anthropology* 34:175–90.

Morant, G. M. 1925. A study of Egyptian craniology from prehistoric to Roman times. *Biometrika* 17:(1–2):1–52.

Morant, G. M. 1935. A study of Predynastic Egyptian skulls from Badari based on measurements taken by Miss. B. N. Stoessiger and Professor D. E. Derry. *Biometrika* 27(3–4):293–309.

Myllarniemi, S., and J. Perheentupa. 1978. Oral findings in the autoimmune polyendocrinopathy candedosis syndrome (APECS) and other forms of hypoparathyroidism. *Oral Surgery* 45:721–29.

Neev, D., and K. O. Emery. 1967. *The Dead Sea*. Bulletin no. 41. Jerusalem: State of Israel, Ministry of Development, Geological Survey.

Nelson, R. 1992. A microscopic comparison of fresh and burned bone. *Journal of Forensic Sciences* 37:1055–60.

Nikiforuk, G., and D. Fraser. 1979. Etiology of enamel hypoplasia and interglobular dentin, the roles of hypocalcaemia and hypophosphatemia. *Metabolic Bone Disease and Related Research* 2:17–23.

Nowell, G. W. 1978. An evaluation of the Miles method of ageing using the Tepe Hissar dental sample. *American Journal of Physical Anthropology* 49(2):271–76.

Oliver, W. J., C. L. Owings, W. E. Brown, and B. A. Shapiro.

1963. Hypoplastic enamel associated with the nephritic syndrome. *Pediatrics* 32:339–406.

Ortner, D. J. 1978. Culture change in the Bronze Age. *Smithsonian* 9:82–87.

Ortner, D. J. 1979. Disease and mortality in the Early Bronze Age people of Bâb edh-Dhrâ', Jordan. *American Journal of Physical Anthropology* 51:589–98.

Ortner, D. J. 1981. A preliminary report on the human remains from the Bâb edh-Dhrâ' cemetery, 1977. *Annual of the American Schools of Oriental Research* 46:119–32.

Ortner, D. J. 1982. The skeletal biology of an Early Bronze IB charnel house at Bâb edh-Dhrâ', Jordan. In *Studies in the History and Archaeology of Jordan I*, ed. A. Hadidi, 93–95. Amman: Department of Antiquities, Hashemite Kingdom of Jordan.

Ortner, D. J. 1998. Male/female immune reactivity and its implications for interpreting evidence in human skeletal paleopathology. In *Sex and Gender in Paleopathological Perspective*, ed. A. L. Grauer and P. Stuart-Macadam, 79–92. New York: Cambridge University Press.

Ortner, D. J. 1999. Palaeopathology: Implications for the history and evolution of tuberculosis. In *Tuberculosis: Past and Present*, ed. G. Pálfi, O. Dutour, J. Deák, and I. Hutás, 255–61. Budapest and Szeged, Hungary: Golden Book, Tuberculosis Foundation.

Ortner, D. J. 2001. Human palaeobiology: Disease ecology. In *Handbook of Archaeological Sciences*, ed. D. R. Brothwell and M. A. Pollard, 225–35. London: John Wiley & Sons.

Ortner, D. J. 2002. Palaeopathology in the twenty-first century. In *Bones and the Man: Studies in Honour of Don Brothwell*, ed. K. Dobney and T. O'Connor, 5–13. Oxford: Oxbow Books.

Ortner, D. J. 2003. *Identification of Pathological Conditions in Human Skeletal Remains*. 2nd ed. Amsterdam: Academic.

Ortner, D. J., W. Butler, J. Cafarella, and L. Milligan. 2001. Evidence of probable scurvy in subadults from archeological sites in North America. *American Journal of Physical Anthropology* 11:343–51.

Ortner, D. J., and M. F. Ericksen. 1997. Bone changes in the human skull probably resulting from scurvy in infancy and childhood. *International Journal of Osteoarchaeology* 7:212–20.

Ortner, D. J., and B. Frohlich. 2007. The EB IA tombs and burials of Bâb edh-Dhrâ', Jordan: A bioarchaeological perspective on the people. In Papers in Honor of Aidan and Eve Cockburn from the Paleopathology Association Meeting, Durham 2004, C. Roberts, M. Lucas Powell, and J. Buikstra, eds. *International Journal of Osteoarchaeology* 17: 358–68.

Ortner, D. J., and B. Frohlich. In Press. The EB IA People of Bâb edh-Dhrâ', Jordan. In *Festschrift in Honor of Drs. Walter E. Rast and R. Thomas Schaub*, ed. M. Chesson.

Ortner, D. J., E. M. Garofalo, and M. K. Zuckerman. 2007. The EB IA burials of Bâb edh-Dhrâ', Jordan: Bioarchaeological evidence of metabolic disease. In *Faces from the Past: Papers in Honor of Patricia Smith*, ed. M. Faerman, L. K. Horwitz, T. Kahana, and U. Zilberman, 181–94. BAR International Series 1603. Oxford: Archaeopress.

Ortner, D. J., E. Kimmerle, and M. Diez. 1999. Skeletal evidence of scurvy in archeological skeletal samples from Peru. *American Journal of Physical Anthropology* 108: 321–31.

Ortner, D. J., and S. Mays. 1998. Dry-bone manifestations of rickets in infancy and early childhood. *International Journal of Osteoarchaeology* 8:45–55.

Ortner, D. J., and C. Ribas. 1997. Bone changes in a human skull from the Early Bronze site of Bâb edh-Dhrâ', Jordan, probably resulting from scalping. *Journal of Paleopathology* 9:137–42.

Ortner, D. J., and G. Theobald. 1993. Diseases in the pre-Roman world. In *The Cambridge World History of Human Disease*, ed. K. F. Kiple, 247–61. New York: Cambridge University Press.

Ortner, D. J., and G. Theobald. 2000. Paleopathological evidence of malnutrition. In *The Cambridge World History of Food*, ed. K. F. Kiple and K. M. Ornelas, 34–44. New York: Cambridge University Press.

Pennington, R. 2001. Hunter-gatherer demography. In *Hunter-gatherers: An Interdisciplinary Perspective*, ed. C. Panter-Brick, R. Layton, and P. Rowley-Conway. Cambridge: Cambridge University Press.

Perry, V. J. 2005. An Analysis of Enamel Hypoplasia and Other Dental Conditions in an Early Bronze Age IA Population from Bâb edh-Dhrâ', Southern Jordan. Master's thesis. Department of Anthropology, University College London.

Philip, G. 2001. The Early Bronze I–III Ages. In *The Archaeology of Jordan*, ed. B. MacDonald, R. Adams, and P. Bienkowski, 163–232. Sheffield, England: Sheffield Academic.

Philip, G., and O. Williams-Thorpe. 1993. A provenance study of Jordanian basalt vessels of the Chalcolithic and Early Bronze I periods. *Paleorient* 19(2):51–63.

Philip, G., and O. Williams-Thorpe. 2000. The production and distribution of ground stone artefacts in the southern Levant during the 5th–4th millennia BC: Some implications of geochemical and petrographic analysis. In *Proceedings of the First International Congress on the Archaeology of the Ancient Near East, Rome, Italy, May 18–23*, ed. P. Matthiae, A. Enea, L. Peyronel, and F. Pinnock, 1379–92. Rome: Universita degli studi de Roma "La Sapienza," Dipartimento di scienze storiche, archeologiche e antropologiche dell'antichita.

Philip, G., and O. Williams-Thorpe. 2001. The production and consumption of basalt artefacts in the southern Levant during the 5th–4th millennia BC: A geochemical and petrographic investigation. In *Archaeological Sciences '97. Proceedings of the Conference Held at the University Durham 2nd–4th September 1997*, ed. A. Millard. BAR International Series 939.

Pollock, S. 1999. *Ancient Mesopotamia*. Cambridge: Cambridge University Press.

Postgate, J. N. 1992. *Early Mesopotamia: Society and Economy at the Dawn of History*. London: Routledge.

Potts, T. 1994. Mesopotamia and the East: An Archaeological and Historical Study of Foreign Relations ca. 3400–3000 BC. Monograph 37. Oxford: Oxford University Committee for Archaeology.

Prowse, T. L., and N. C. Lovell. 1996. Concordance of cranial and dental morphological traits and evidence for endogamy in ancient Egypt. *American Journal of Physical Anthropology* 101:237–46.

Purvis, R. J., W. J. Barrie, G. S. MacKay, E. M. Wilkinson, F. Cockburn, and N. R. Belton. 1973. Enamel hypoplasia of teeth associated with neonatal tetany: A manifestation of maternal vitamin D deficiency. *Lancet* 2(7833): 811–14.

Rashidi, J. S., D. J. Ortner, B. Frohlich, and B. Jonsdottir. 1991. Brucellosis in Early Bronze Age Jordan and Bahrain: An analysis of possible cases of brucella spondylitis. *American Journal of Physical Anthropology* (Suppl.) 32:122–23. Abstract.

Rast, W. E. 1981. Settlement at Numeira. In *The Southeastern Dead Sea Plain Expedition: An Interim Report of the 1977 Season*, ed. W. E. Rast and R. T. Schaub, 35–44. *Annual of the American Schools of Oriental Research* 46.

Rast, W. 1999. Society and mortuary customs at Bâb edh-Dhrâʿ. In *Archaeology, History and Culture in Palestine and the Near East. Essays in Memory of Albert E. Glock*, ed. T. Kapitan, 164–82. Atlanta, GA: Scholars.

Rast, W. E. 2003. Archaeology of the Dead Sea Plain in Jordan. In *Near Eastern Archaeology*, ed. S. Richard, 319–30. Winona Lake, IN: Eisenbrauns.

Rast, W. E., and R. T. Schaub. 1974. Survey of the Southeastern Plain of the Dead Sea, 1973. *Annual of the Department of Antiquities* 19:5–53, 175–85.

Rast, W. E., and R. T. Schaub. 1980. Preliminary report of the 1979 Expedition to the Dead Sea Plain. *Bulletin of the American Schools of Oriental Research* 240:21–61.

Rast, W. E., and R. T. Schaub, eds. 1981. The Southeastern Dead Sea Plain Expedition: An interim report of the 1977 season. *Annual of the American Schools of Oriental Research* 46.

Rast, W. E., and R. T. Schaub. 2003. Bâb edh-Dhrâʿ: Excavations at the Town Site (1975–1981). *Reports of the Expedition to the Dead Sea Plain, Jordan, Vol. 2*. Winona Lake, IN: Eisenbrauns.

Rathbun, T. A. 1982. Morphological affinities and demography of Metal-Age southwest Asian populations. *American Journal of Physical Anthropology* 59(1):47–60.

Reid, D. J., and M. C. Dean. 2000. The timing of linear hypoplasia on human anterior teeth. *American Journal of Physical Anthropology* 113:135–39.

Reid, D. J., and M. C. Dean. 2006. Variation in modern human enamel formation times. *Journal of Human Evolution* 50:329–46.

Richard, S. 2003. The Early Bronze Age in the southern Levant. In *Near Eastern Archaeology*, ed. S. Richard, 286–302. Winona Lake, IN: Eisenbrauns.

Richard, S. ed. 2003. *Near Eastern Archaeology*. Winona Lake, IN: Eisenbrauns.

Roberts, C. A., and J. E. Buikstra. 2003. *The Bioarchaeology of Tuberculosis*. Gainesville: University Press of Florida.

Rosen, A. M. 2003. Paleoenvironments of the Levant. In *Near Eastern Archaeology*, ed. S. Richard, 10–16. Winona Lake, IN: Eisenbrauns.

Rosenzweig, K. A. 1970. Tooth form as a distinguishing trait between sexes and human populations. *Journal of Dental Research* 49:1423–26.

Ruff, C. 1987. Sexual dimorphism in human lower limb bone structure: Relationship to subsistence strategy and sexual division of labor. *Journal of Human Evolution* 16:391–416.

Ruff, C. 1992. Biomechanical analyses of archaeological human skeletal samples. In *Skeletal Biology of Past Peoples: Research Methods*, ed. S. Saunders and M. A. Katzenberg, 37–58. New York: Wiley-Liss.

Ruff, C. 1994. Biomechanical analysis of northern and southern Plains femora: Behavioral implications. In *Skeletal Biology in the Great Plains: Migration, Warfare, Health, and Subsistence*, ed. D. W. Owsley and R. L. Jantz, 235–45. Washington, DC.: Smithsonian Institution Press.

Saller, S. 1964–1965. Bâb edh-Dhrâʿ. *Liber Annuus* 15:137–219.

Saunders, S. 1992. Subadult skeletons and growth related studies. In *Skeletal Biology of Past Peoples: Research Methods*, ed. S. R. Saunders, and A. Katzenberg, 1–20. New York: Wiley-Liss.

Saunders, S. R., and L. Barrans. 1999. What can be done about infant category in skeletal samples? In *Human Growth in the Past. Studies from Bones and Teeth*, ed. R. D. Hoppa and C. M. Fitzgerald, 183–209. Cambridge: Cambridge University Press.

Saunders, S. R., D. A. Herring, and G. Boyce. 1995. Can skeletal samples accurately represent the living population they come from? St Thomas cemetery site, Bellesville. In *Bodies of Evidence: Reconstructing History through Skeletal Analysis*, ed. A. L. Graver, 69–99. New York: Wiley-Liss.

Saunders, S. R., and R. D. Hoppa. 1993. Growth deficit in survivors and non-survivors: Biological mortality bias in subadult skeletal samples. *Yearbook of Physical Anthropology* 36:127–52.

Schaub, R. T. 1981a. Patterns of burial at Bâb edh-Dhrâʿ. In *The Southeastern Dead Sea Plain Expedition: An Interim Report of the 1977 Season*, ed. W. E. Rast and R. T. Schaub, 45–68. *Annual of the American Schools of Oriental Research* 46.

Schaub, R. T. 1981b. Ceramic sequences in the tomb groups at Bâb edh-Dhrâʿ. In *The Southeastern Dead Sea Plain Expedition: An Interim Report of the 1977 Season*, ed. W. E. Rast and R. T. Schaub, 69–118. *Annual of the American Schools of Oriental Research* 46.

Schaub, R. T. 1993. Bâb edh-Dhrâ'. In *New Encyclopaedia of Archaeological Excavations in the Holy Land*, ed. E. Stern, 1:130–36. Carta, Jerusalem: Israel Exploration Society.

Schaub, R. T. 1996. Pots as containers. In *Retrieving the Past: Essays on Archaeological Research and Methodology in Honor of Gus van Beek*, ed. J. D. Seger, 231–44. Winona Lake, IN: Eisenbrauns.

Schaub, R. T. 2000. Terminology and typology of carinated vessels of the Early Bronze I–II of Palestine. In *The Archaeology of Jordan and Beyond: Essays in Honor of James A. Sauer*, ed. L. E. Stager, J. A. Greene, and M. D. Coogan, 444–64. Winona Lake, IN: Eisenbrauns.

Schaub, R. T. In press. Basalt bowls in Early Bronze IA shaft tombs at Bâb edh-Dhrâ': Placement, production and symbol. In *New Approaches to Old Stones: Recent Studies of Ground Stone Artifacts*, ed. Y. Rowan and J. Eberling. Approaches to Anthropological Archaeology Series. London: Equinox.

Schaub, R. T., and W. E. Rast. 1984. Preliminary report of the 1981 Expedition to the Dead Sea Plain, Jordan. *Bulletin of the American Schools of Oriental Research* 254:35–60.

Schaub, R. T., and W. E. Rast. 1989. Bâb edh-Dhrâ': Excavations in the cemetery directed by Paul W. Lapp (1965–1967). *Reports of the Expedition to the Dead Sea Plain, Jordan, Vol 1*. Winona Lake, IN: Eisenbrauns.

Schaub, R. T., and W. E. Rast. 2000. The Early Bronze Age I stratified ceramic sequences from Bâb edh-Dhrâ'. In *Ceramics and Change in the Early Bronze Age of the Southern Levant*, ed. G. Philip and D. Baird, 73–90. Sheffield, England: Sheffield Academic.

Schaub, R. T., and W. E. Rast. 2003. Bâb edh-Dhrâ': Excavations at the Town Site (1975–1981). *Reports of the Expedition to the Dead Sea Plain, Jordan*. vol. 2. Winona Lake, IN: Eisenbrauns.

Schmidt, E. F. 1937. *Excavations at Tepe Hissar, Damghan*. Philadelphia: University of Pennsylvania Press.

Schultz, M. 2003. Light microscopic analysis in skeletal paleopathology. In *Identification of Pathological Conditions in Human Skeletal Remains*, 2nd ed., ed. D. J. Ortner, 73–107. Amsterdam: Academic.

Scott, G. R. 1975. Association between non-metrical tooth crown characteristics. *American Journal of Physical Anthropology* 42:328.

Scott, G. R. 1977. Classification of sex dimorphism, association, and population variation of the canine distal accessory ridge. *Human Biology* 49:453–69.

Scott, G. R. 1980. Population variation of Carabelli's trait. *Human Biology* 52:63–78.

Smith, P. 1989. The skeletal biology and paleopathology of Early Bronze Age populations in the Levant. In *L'urbanisation de la Palestine à l'âge du Bronze Ancien*, ed. Pierre de Miroschedji, 297–313. BAR International Series 527. Oxford: B.A.R.

Smith, P., O. Bar-Yosef, and A. Sillen. 1984. Archaeological and skeletal evidence for dietary change during the Late Pleistocene/Early Holocene in the Levant. In *Paleopathology at the Origins of Agriculture*, ed. M. N. Cohen and G. J. Armelagos, 101–136. New York: Academic.

Smith, P., and L. K. Horwitz. 2007. Ancestors and inheritors: A bio-cultural perspective on the transition to agro-pastoralism in the southern Levant. In *Ancient Health: Skeletal Indicators of Agricultural and Economic Intensification*, ed. M. N. Cohan and G. M. M. Crane-Kramer. Gainesville: University Press of Florida.

Sofaer, J. A., J. D. Niswander, C. J. Maclean, and P. L. Workman. 1972. Population studies on Southwestern Indian tribes. Part V. Tooth morphology as an indicator of biological distance. *American Journal of Physical Anthropology* 37:357–66.

Stager, L. E. 1992. The periodization of Palestine from Neolithic through Early Bronze times. In *Chronologies in Old World Archaeology*, 3rd ed., ed. R. W. Ehrich, 22–41. Chicago: University of Chicago Press.

Steckel, R. H., and J. C. Rose, eds. 2002. *The Backbone of History: Health and Nutrition in the Western Hemisphere*. New York: Cambridge University Press.

Stewart, T. D. 1940. A report on the skeletal remains from the Piscataway Creek ossuary. *American Antiquity* 6:4–18.

Stewart, T. D., ed. 1952. *Hrdlička's Practical Anthropometry*, 4th ed. Philadelphia: Wistar Institute of Anatomy and Biology.

Stewart, T. D. 1968. Identification by the skeletal structure. In *Gradwohl's Legal Medicine*, ed. F. E. Camps, 123–54. Bristol: Williams and Wilkins.

Stewart, T. D. 1979. *Essentials of Forensic Anthropology*. Springfield, IL: Thomas.

Stewart, T. D., and M. Trotter, eds. 1954. *Basic Readings on the Identification of Human Skeletons: Estimation of Age*. New York. Wenner-Gren Foundation for Anthropological Research.

Stoessiger, B. N. 1927. A study of the Badarian crania recently excavated by the British School of Archaeology in Egypt. *Biometrika* 19:110–50.

Suzuki, M., and T. Sakai. 1960. A family study of *torus palatinus* and *torus mandibularis*. *American Journal of Physical Anthropology* 18:263–72.

Swindler, D. R. 1956. *A Study of the Cranial and Skeletal Material Excavated at Nippur*. Part 12. Philadelphia: University of Pennsylvania Monograph.

Tanner, J. M. 1981. *A History of the Study of Human Growth*. Cambridge: Cambridge University Press.

Thomlinson, R. 1965. *Population Dynamics and Consequences of World Demographic Change*. New York: Random House.

Todd, T. W., and D. W. Lyon. 1925. Cranial suture closure, its progress and age relationship. *American Journal of Physical Anthropology* 8(1):23–45.

Tosi, M. 1986. Early maritime cultures of the Arabian Gulf and the Indian Ocean. In *Bahrain through the Ages: The Archaeology*, ed. S. H. A. Al Khalifa and M. Rice, 94–107. London: Routledge and Kegan Paul.

Townsend, G. C., and T. Brown. 1979. Family studies of tooth size factors in the permanent dentition. *American Journal of Physical Anthropology* 50:183–90.

Trotter, M., and G. Glesser. 1952. Estimation of stature from long bones of American whites and negroes. *American Journal of Physical Anthropology* 10:463–514.

Turner, C. G., II. 1967. Dental genetics and microevolution in prehistoric and living Koniag Eskimo. *Journal of Dental Research* 46:911–17.

Turner, C. G., II. 1971. Three-rooted mandibular first permanent molars and the question of American Indian origins. *American Journal of Physical Anthropology* 34:229–41.

Turner, C. G., II., C. R. Nichol, and G. R. Scott. 1991. Scoring procedures for key morphological traits of the permanent dentition: The Arizona State University Dental Anthropology System. In *Advances in Dental Anthropology*, ed. M. A. Kelly and C. S. Larsen, 13–31. New York: Wiley-Liss.

Turner, C. G., II., and G. R. Scott. 1977. Dentition of Easter Islanders. In *Orofacial Growth and Development*, ed. A. A Dahlberg and T. M. Graber, 229–49. The Hague: Mouton.

Ubelaker, D. H. 1974. *Reconstruction of Demographic Profiles from Ossuary Skeletal Samples*. Smithsonian Contributions to Anthropology, No. 18. Washington, DC: Smithsonian Institution Press.

Ubelaker, D. H. 1989. *Human Skeletal Remains: Excavation, Analysis, Interpretation*. 2nd ed. Washington, DC: Taraxacum.

Ubelaker, D. H. 1997. Taphonomic applications in forensic anthropology. In *Forensic Taphonomy*, ed. W. D. Haglund and M. H. Sorg, 77–90. Boca Raton, FL: CRC.

Ucko, P. J. 1969. Ethnographical and archaeological interpretation of funeral remains. *World Archaeology* 1:262–80.

Ullinger J. M., S. G. Sheridan, D. E. Hawkey, C. G. Turner II., and R. Cooley. 2005. Bioarchaeological analysis of cultural transition in the southern Levant using dental nonmetric traits. *American Journal of Physical Anthropology* 128:466–76.

Von Endt, D. W., and D. J. Ortner. 1984. Experimental effects of bone size and temperature on bone diagenesis. *Journal of Archaeological Science* 11:247–53.

Ward, R. H., and K. M. Weiss, eds. 1976. *The Demographic Evolution of Human Populations*. New York: Academic.

Warren, E. 1898. An investigation on the variability of the human skeleton: With special reference to the Naqada race discovered by Professor Flinders Petrie in his exploration of Egypt. *Philosophical Transactions of the Royal Society of London, Series B* 189:135–227.

Webb, S. G. 1990. Prehistoric eye disease (trachoma?) in Australian Aborigines. *American Journal of Physical Anthropology* 81:91–100.

Weinstein, J. 1984. Radiocarbon dating in the southern Levant. *Radiocarbon* 26:297–366.

Weiss, K. M. 1973. *Demographic Models for Anthropology*. Washington DC.: Memoirs of the Society for American Archaeology.

Weiss, K. M., and P. E. Smouse. 1976. The demographic stability of small human populations. In *The Demographic Evolution of Human Populations*, ed.: R. H. Ward and K. M. Weiss, 59–73. London: Academic.

Wood, J. 1994. *Dynamics of Human Reproduction. Biology, Biometry, Demography*. New York: Aldyne de Gruyter.

Wood, J., G. Milner, H. Harpending, and K. Weiss. 1992. The osteological paradox. *Current Anthropology* 33: 343–70.

World Health Organization. 2004. Healthy Life Expectancy 2002. World Health Report. Available at www.who.int.

World Health Organization. 2004. Maternal mortality in 2000. Estimates developed by WHO, UNICEF and UNFPA. Department of Reproductive Health and Research. World Health Organization, Geneva. Available at www.who.int.

Yekutieli, Y. 2000. Early Bronze Age I pottery in southwestern Canaan. In *Ceramics and Change in the Early Bronze Age of the Southern Levant*, ed. G. Philip and D. Baird, 129–52. Sheffield, England: Sheffield Academic.

Zarins, J. 1992. Archaeological and chronological problems within the Greater Southwest Asian arid zone, 8500–1850 B.C. In *Chronologies in Old World Archaeology*, 3rd ed., ed. R.W. Ehrich, 42–62. Chicago: University of Chicago Press.

Index

abnormal bone destruction, 264
abnormal bone formation, 264
abnormal bone shape, 264
abnormal bone size, 264
abortion, 178, 259, 305
Abydos, 247
Afghanistan, 247
age distribution, 231, 234–236
Akkadian Period, 10–11
Al Amrah, 247
Albright, William F., 1
American Center of Oriental Research (ACOR), ix
American dentists' Universal Numbering System, 282
American Schools of Oriental Research (ASOR), ix
amino acids, 48, 256
AMTL. *See* antemortem tooth loss
analytical methods:
    description, 20–23
    Fisher's combined test, 285, 286
    F-matrix, 244, 248
    general linear model (GLM), 285
    Kendall's tau, 285
    Kruskal-Wallis test, 242, 285–286, 289
    metric data, 21–22, 236–238
    missing data, 22, 238, 242–243, 247–249, 285, 307
    multivariate analysis, 22, 243, 246–249, 302
    multivariate difference, 22–23
    nonmetric data, 21, 236–238, 247, 302
    nonparametric analysis, 22, 238, 242
    parametric statistics, 22, 238, 242
    SAS, 22, 237, 243
    Shapiro-Wilkinson test, 238
    SPSS, 237, 284
    Systat, 22, 237, 243
    test for normality, 22, 238, 247
    t-test, 242, 244
    univariate analysis, 22, 242, 244, 285
Anatolia, 10–11
anemia, 71, 73, 108, 116, 119, 150, 159, 173, 177, 180, 181, 265, 274–275
antemortem tooth loss, 67, 68, 75, 79, 101, 118, 152, 155, 159, 160, 168, 172, 188, 196, 202, 205, 209, 210, 213, 221, 226, 282, 292, 294
Arab Potash Company, x
Aratta, 247
Arizona State University Dental Anthropology System, 282

Armant, 239
arthritis, 3, 86, 88, 149, 155, 170, 265, 273
arthropathy, 107, 108, 116, 151, 158, 175, 203, 226, 264, 273
articulated burials, 183
aurochs, 270
axe wound, 80, 142, 146, 226, 267, 305

Badari, 240, 246, 249
Bahrain, 10, 183, 215, 234, 239, 256
Binford, Lewis, 243, 297–298
biological distance, 245, 281, 285, 290
breast feeding, 259
brucellosis, 61, 96, 140, 146, 269, 271–272, 303
burial statistics, 231
burning, effect on bones, 235, 129, 135–136

calcined bones, 135–136
cancer, 59, 63, 88, 151, 157–160, 263, 275
cartilaginous tumor, 101
categories of skeletal disease, 265
cavities. *See* dental caries
cemetery, Bâb edh-Dhrâ':
    A Cemetery, 13, 26
    C Cemetery, 26
    definition of areas, 26
    discovery, 13
    G Cemetery, 26
cemetery areas, 229
chamber statistics, 231
charnel house:
    adult age distribution, 137
    burning, 125, 129–136
    construction, 49–50, 125
    description, 2–3, 123–146
    development, 49
    distribution of skeletal elements, 130–135, 138–139
    excavation, 15–18
    sub-adult mortuary practice, 129
chronic bleeding, 68, 209, 265, 276
chronic hyperextension, 88, 101
chronology, archaeological, 2
classification, 27, 32, 33, 244, 264–265
club foot, 149
Colles' fracture, 197
compression fracture, 67, 121, 146, 173, 176, 205, 267
computed tomography, 86, 229, 238

319

conservation, 13, 18–20
consumption. *See* tuberculosis
corporate group structure, 295
coxa vara, 87
cranial index, 239–241, 302
cranial tumors, 71
cranial variables, 237, 243, 246, 247
craniostenosis, 71
cribra orbitalia, 59, 63, 64, 68, 71, 73, 88, 97, 101, 106, 113, 114, 116, 118, 119, 150, 161, 168, 169, 170, 177, 178, 181, 201, 202, 203, 212, 213, 221, 225, 275
CT scan. *See* computed tomography
Cyprus, 247
cystic lesion, 63, 84, 159, 161, 175, 180, 191, 197, 202, 203, 205, 209, 221, 222, 224, 225, 226, 267

DAM Server. *See* Digital Access Management Server
Dead Sea, 7–8, 15, 40, 49, 231
degenerative joint disease, 86, 191, 203, 221
demography, 129, 156
dental abscess, 67, 77, 95, 119, 161, 170–171, 178, 188, 189, 198, 209, 210, 213, 221, 222, 225, 226, 279, 282, 288, 294
dental attrition, 281–282, 290, 292
dental calculus, 159, 160, 162, 198, 282, 284, 289, 294
dental caries, 68, 75, 77, 87, 93, 95, 119, 159, 188, 196, 197, 198, 209, 210, 212, 222, 279, 282, 286–288, 291, 292, 294, 296
dental hypoplasia, 55, 62–64, 67, 71, 77, 87, 88, 95, 104, 108, 114, 116, 119, 150–152, 157–161, 169, 170, 187, 188, 190, 193, 194, 196–199, 202, 210, 212, 213, 218, 222, 223, 256, 265, 279, 280, 281, 282, 284, 285, 289, 294–296, 302, 304
dental pathologies, 286–289
dental plaque, 93, 281, 282, 294
depressed lesions, 78, 80, 93, 104, 114, 116, 159, 169, 172, 180, 267, 305
dermestid beetle, effect on bone, 79, 299
Dicom, 238
diffuse idiopathic skeletal hyperostosis, 77, 107, 175, 226
Digital Access Management Server, 229, 238
Dilmon civilization, 10
DISH. *See* diffuse idiopathic skeletal hyperostosis
disk degeneration, 68, 189, 193, 210, 213, 226
dolichocephalic, 240, 302
Donahue, Jack, 7

earthquake, 7, 51, 104
Echinococcus, 271
EDSP. *See* Expedition to the Dead Sea Plain
egalitarian societies, 6, 9, 243, 298
Egypt, 7–10, 38, 45, 150, 240, 245–249, 292, 302
EM-31. *See* geophysical surveying methods
enamel hypoplasia. *See* dental hypoplasia
encephalitis, 142

enthesopathies, 67, 71, 79, 84, 95, 112, 118, 150, 176, 189, 193, 194, 197, 198, 203, 205, 206, 209, 210, 222, 223, 225, 226, 305
erosive arthropathy, 107, 108, 116, 151, 175, 264, 273
excavation methods, 13, 15–17, 18
Expedition to the Dead Sea Plain, 1, 25–29, 33, 36, 39, 40
extended family, 46, 236, 298, 300, 301

fertility rate, 259–261
fetal loss, 259–261
fibrous cortical defect, 79
figurine, 2, 25, 27, 40–41, 45, 48, 56, 94, 116, 187, 218, 298
Fisher's combined test, 285, 286. *See also* analytical methods
F-matrix, 244, 248. *See also* analytical methods
fracture, 48, 55, 67, 73, 80, 82, 86, 88, 97, 114, 121, 135, 146, 151, 154, 158, 159, 168, 171, 190, 194, 197, 198, 203, 205, 210, 212, 214, 223, 224, 227, 267, 268, 277, 305

general linear model (GLM), 285. *See also* analytical methods
geophysical surveying methods, 5, 15–17, 183, 184, 200, 214–216, 261
Glueck, Nelson, 1
Gompertz model, 252
gout, 151, 273
grapes, 90
Great Rift Valley, 7, 51, 104
growth rate, 258–262

Hadidi, Adnan, ix
Hellenistic period, 234, 255, 256
hematoma, 205
herniation, 70, 112, 113, 121, 154, 171, 197, 205
hip dysplasia, 94, 160, 181, 268, 278
Holocene period, 263, 269
hormonal changes, 80, 222, 276
hydrocephalus, 61, 71, 101
hyperostosis, 77, 107, 173, 175, 191, 212, 213, 223, 226, 274, 275
hypertrophy, 86
hypocalcemia, 295
hypoparathyroidism, 295
hypoplastic lines. *See* dental hypoplasia

India, 247
infant mortality rate, 235, 261
infection, 269–272
infectious pathogens, 142, 146, 261, 269, 273, 274, 302, 305
Institute for Bioarchaeology, ix
Iran, 10–11, 240, 246, 247, 249
iron deficiency anemia, 265, 274, 275
Israel, 125, 247

Jebel Al-Buhars, 239
Jericho, 35, 36, 42, 145, 247, 271, 273

jewelry, 25, 27, 42, 298
Jordanian Department of Antiquities, ix, x, 23, 183, 224
juxta-articular erosion, 80, 151

Kendall's tau, 285. *See also* analytical methods
Kerak, ix, 7, 40, 183, 224
Kiesewetter, Henrieke, 241, 244, 246
kinship, 8, 231, 281, 289–291, 295
Kish, 247
Koch, Robert, 6
Krogman, Wilton, 229, 247
Kruskal-Wallis test, 242, 285–286, 289. *See also* analytical methods

labor, child birth, 198, 241
Lake Lisan, 7
lapis lazuli, 247
Lapp, Paul, 1, 2, 13, 25–27, 33, 39–40, 183, 229
Lapp herders, 272
LEH. *See* dental hypoplasia
leprosy, 235
Lerna, 240, 246, 247
Levant, 7–10, 25, 32, 39, 45, 49, 245, 249, 296, 297, 301
life expectancy, 142, 252, 255–263, 303
linear enamel hypoplasia (LEH). *See* dental hypoplasia
Lisan marl, 8, 9, 13, 15, 18, 45, 46, 47
Lisan Peninsula, 7, 224
Littleton, Judith, 234, 255–256
lung infection, 151
luxation, 64
Luxor, 246

Maadi-Budo, 246, 247, 249
mace head, 2, 25, 27, 40, 42, 48, 218, 246, 298, 300, 302
Mahalanobis D^2, 248
malignant tumor, 152, 278–279
malnutrition, 71, 78, 80, 95, 113, 114, 142, 146, 151, 168–172, 198, 235, 256, 259, 263, 265, 280, 295, 303–307
mandibular variables, 237–238
marrow hypoplasia, 119, 173, 178, 274, 294
McCreery, David, ix, 26, 291
meningitis, 77, 82, 86, 116, 142, 155
menopause, 80, 95, 222
Mesopotamia, 8–11, 45, 245–249
metabolic disease, 59, 63, 71, 114, 157–161, 193, 263, 265, 274, 280, 303, 306
metric data, 21–22, 236–238. *See also* analytical methods
missing data, 22, 238, 242–243, 247–249, 285, 307. *See also* analytical methods
mobility index, 241
mortality rate, 235, 251, 252, 253, 256, 261
multiple cartilaginous exostosis, 168
multiple interment, 83, 127
multivariate analysis, 22, 243, 246–249, 302. *See also* analytical methods

multivariate difference, 22–23. *See also* analytical methods
Mycobacterium, 270
Mycobacterium bovis, 270
Mycobacterium tuberculosis, 270
myositis ossificans, 197

Naqada, 240, 246–249
National Museum of Natural History, Smithsonian, 23
Native American groups, 137, 145, 264, 307
Natuf, 247
Natufian period, 295, 296
Nippur, 239
nomadic pastoralism, 6, 45, 49, 146, 281, 295, 297, 298, 299
nonmetric data, 21, 236–238, 247, 302. *See also* analytical methods
nonparametric statistics, 22, 238, 242. *See also* analytical methods
normality test, 22, 238, 247. *See also* analytical methods
Norway, 272
Notre Dame, University of, 23
Nubia, 245, 292
nutrition, 142, 233, 235, 256, 259, 260, 261, 274, 294, 303

Omari, 9–10
orthopaedic pathology, 264
orthopaedic radiology, 264
osteoarthritis, 61, 64, 67, 68, 70, 71, 73, 75, 85, 86, 94, 97, 108, 112, 113, 118, 121, 142, 148, 150, 151, 152, 155, 160, 162, 168, 170, 171, 173, 175, 176, 177, 178, 203, 210, 222, 226, 264, 268, 271, 273, 278
osteochondritis dissecans, 77, 162, 187, 190, 194, 226, 269
osteoclastic activity, 266
osteomalacia, 160, 168, 171, 203
osteomyelitis, 94, 146, 269, 272, 303
osteophytosis, 271
osteoporosis, 78, 80, 84, 94, 95, 108, 113, 114, 151, 157, 160, 161, 168, 170, 172, 176, 189, 198, 222, 268, 275, 276, 280, 298, 303, 304, 306, 307
ovulation, 259, 260

Paget's disease, 264
Palestine, 7–10, 35, 38, 45, 246, 249
parametric statistics, 22, 238, 242. *See also* analytical methods
parry fracture, 168, 194
pathogens, 6, 142, 146, 235, 261, 263, 264, 269–274, 295, 302–307
periodontal disease, 73, 160, 162, 168, 169, 221, 279, 282, 294
periostitis, 272
periostosis, 95, 104, 107, 108, 114, 121, 149, 151, 154, 157, 265
Persia, 245
pilasteric index, 239–242
pituitary adenoma, 71

pituitary tumor, 101
platymeric index, 239–242
polyvinyl acetate, 19, 20
population decline, 259
population size, 252, 258, 259, 260–262
porotic hyperostosis, 173, 274, 275
postcranial variables, 238
pottery, chronology, 27–40
Pott's disease, 270
pregnancy, 73, 86, 156, 169, 170, 254, 259, 260, 295, 304, 305
pressure erosion, 61, 80, 116, 198, 199, 206, 267, 278
primary burial, 2, 3, 45, 46–49, 51, 55, 61, 68, 73, 92–94, 97, 112, 113, 127, 136, 139–141, 144–146, 169, 171, 179, 181, 186, 211, 230, 297, 299, 300, 305
protein, 6, 15, 47, 51, 68, 90, 135, 146, 169, 230, 256, 263, 276, 280, 295, 303, 307
pseudarthrosis, 64
psoriatic arthritis, 86
PVA. See polyvinyl acetate
PVA-AYAT. See polyvinyl acetate

Rabadi, Sami, ix, x, 183, 224
Rast, Walter E., x, 2, 13, 183, 229, 262, 281, 302
reed matting, 2, 17, 52, 93, 125, 185, 218, 225
rheumatoid arthritis, 86, 264, 273
rickets, 86, 104, 107, 152, 154, 160, 162, 168, 169, 171, 175, 193, 203, 205, 275–277, 280, 294, 303, 304, 306
robbery, 5, 17, 43, 183, 215, 231
robusticity index, 236, 238–242, 302, 305

Sandved, Kjell, ix
SAS, 22, 237, 243. See also analytical methods
Sauer, James, ix
scalping, 77, 266
scaphocephaly, 177
Schaub, R. Thomas, x, 2, 13, 183, 229, 262, 281
Schmorl's nodes, 190, 199, 212
sclerosis, 181, 182, 226, 270
scoliosis, 160, 170, 194, 197, 199, 203, 206, 221
scurvy, 56, 59, 68, 73, 80, 83, 86, 88, 106, 116, 119, 121, 140, 150, 154–157, 159, 169, 170, 173, 177, 178, 180, 198, 199, 206, 209, 224, 275, 280, 294, 303, 304, 306
sebaceous cyst, 116, 159, 161, 267
secondary burial, 2–5, 45–49, 48, 51, 68, 79, 83, 94, 127, 137, 142, 144, 145, 147, 149, 171, 186, 210, 230, 247, 262, 281, 299–301, 307
sedentary behavior, 146, 242, 256, 261, 281
Sedment, 240, 246, 249
septic arthritis, 108
sex distribution, 236
sexual dimorphism, 242–244
sexual maturity, 234
Shapiro-Wilkinson test, 238. See also analytical methods
Shar-I Sokhta, 11, 246, 247, 249
shroud, textile, 188, 218

sinus infection, 68, 85, 118, 160, 202
skin ulcer, 94, 95, 272
slipped epiphysis, 67, 268, 273
Smith, Patricia, 251, 301
Smithsonian Institution, ix, x, 20, 23, 25, 184, 229, 236, 238, 281
Smithsonian Scholarly Studies Program, ix
social composition, 252, 261–262
spondylolysis, 59, 154, 190, 203, 213, 268, 277
SPSS, 237, 284. See also analytical methods
Stanley, Daniel, 194
staph infection, 272
Staphylococcus aureus, 272
stature, 129, 229, 236, 238–239, 302
stereolithography, 238
stereolithography file format (STL), 238
Stix, Agnes, x
STL. See stereolithography file format
Streptococcus, 272
subluxation, 64, 97, 101, 160, 177, 181, 197, 206, 268, 269, 277
Sumerian Period, 10–11
supratrochlear exostosis, 210, 221
syndesmophyte formation, 173, 205, 214, 222, 223, 225, 226
syphilis, 107
Systat, 22, 237, 243. See also analytical methods

taphonomy, 3, 6, 47, 49, 51, 68, 73, 78, 86, 129, 136, 139, 146, 156, 213, 230, 236, 275, 299, 301, 307
TB. See tuberculosis
Tell Sultan, 247
Tepe Hissar, 239, 243, 244, 246, 247, 249
Tepe Yahya, 247
textile, 2, 27, 188, 218, 222
tomb density, 47, 71, 183, 261, 300
trachoma, 150, 151, 181
trauma, 3, 70, 75, 77, 80, 82, 83, 87, 88, 93, 104, 107, 112–114, 116, 121, 142, 146, 154, 158, 159, 161, 162, 169, 170, 172, 175, 178, 180, 181, 190, 193, 194, 197, 199, 202, 203, 210, 213, 214, 221–223, 226, 233, 235, 265–269, 276, 277, 305
t-test, 242, 244. See also analytical methods
tuberculosis, 78, 86, 87, 93, 94, 96, 140, 146, 151, 154, 170, 198, 210, 235, 269–272, 303, 304, 306, 307
tumor, 71, 101, 107, 116, 152, 158, 172, 278–279, 304
Turkey, 10, 247

United Arab Emirates, 239, 240, 241, 242, 244, 246
univariate analysis, 22, 242, 244, 285. See also analytical methods
urbanism, 6, 8, 263, 302
Uruk Period, 10–11

vandalism, 3, 46, 127, 297, 301
vitamin deficiency, 86, 169, 176, 256, 276

Wadi Kerak, 8, 45, 299
Wadi Mujib, 40
walled town, 1, 8, 27, 45, 262
wood objects, 2, 19, 20, 27, 42, 48, 94, 118, 218

World Health Organization, 303, 304
woven textile. *See* textile

Zuckerman, Molly, x

About the Authors

Donald J. Ortner is a biological anthropologist in the Department of Anthropology, National Museum of Natural History (NMNH), Smithsonian Institution, where he has worked during most of his professional career. In 1988 he was appointed Visiting Professor in the Department of Archaeological Sciences at the University of Bradford in Bradford, England. His major research interest is in human adaptation with a specific focus on the effect of disease on human evolution during the Holocene. The latter interest includes an emphasis on the impact of major developments in human society, such as sedentism, urbanism, and the development of agriculture, on human health. He is the author of more than 125 scientific papers, many of which are on the subject of human disease. He is the coauthor of *Identification of Pathological Conditions in Human Skeletal Remains* (1981), a second edition of which was published in 2003. He organized and edited the proceedings of the Smithsonian Institution's Seventh International Symposium, How Humans Adapt: A Biocultural Odyssey (1983) and coedited *Human Paleopathology* (1991). He is currently conducting research on the antiquity of infectious diseases, such as tuberculosis and brucellosis, for which domestic animals are an intermediate host.

Ortner served as the Chairman of the Department of Anthropology, NMNH, for four years (1988–1992) and as acting director of the NMNH for more than two years (1994–1996). He holds a Ph.D. degree from the University of Kansas and an Honorary D.Sc. degree from the University of Bradford. He has done fieldwork in Jordan and has conducted research projects in the United States, Europe, and Australia. From 1999 to 2001, he was president of the Paleopathology Association, an international scientific society that promotes the study of ancient disease.

Bruno Frohlich is a biological anthropologist in the Department of Anthropology, National Museum of Natural History, Smithsonian Institution. He was educated at the University of Copenhagen and the University of Connecticut. In addition to his academic education he has training from the Royal Danish Army and the Connecticut State Police forensic laboratory. He has done extensive fieldwork in the Middle East (Jordan, Syria, Saudi Arabia, Bahrain, Kuwait), North Africa, Iran, Alaska, Greenland, Europe, and most recently in Mongolia. He has also conducted research projects in Indonesia, India, New Zealand, and Mongolia. His research interest focuses on mortuary practices, forensic medicine, criminology, and geographic information systems as it applies to the field of mortuary archaeology and anthropology.

Frohlich is the director of the Museum's Computed Tomography Laboratory where he is developing new techniques and procedures in nondestructive and noninvasive analytical methods. He is the editor of *To the Aleutians and Beyond,* a volume honoring the late Professor William Laughlin, and has published more than 50 scientific and popular articles. His work has been featured in films and reports by the Discovery Channel, BBC2, the History Channel, National Geographic Society, and news media organizations in the United States, Europe, and the Middle East.

He has served as advisor to several museums, medical examiner's offices, colleges, and universities in Europe, the United States, the Middle East, and Central Asia. He also serves as a trustee at Sterling College, Vermont.